D1743936

# Stochastic Optimal Control of Structures

Yongbo Peng · Jie Li

# Stochastic Optimal Control of Structures

Yongbo Peng
Shanghai Institute of Disaster
Prevention and Relief
Tongji University
Shanghai, China

Jie Li
College of Civil Engineering
Tongji University
Shanghai, China

ISBN 978-981-13-6763-2    ISBN 978-981-13-6764-9   (eBook)
https://doi.org/10.1007/978-981-13-6764-9

Jointly published with Shanghai Scientific and Technical Publishers
The print edition is not for sale in China Mainland. Customers from China Mainland please order the print book from: Shanghai Scientific and Technical Publishers.

Library of Congress Control Number: 2019932670

© Springer Nature Singapore Pte Ltd. and Shanghai Scientific and Technical Publishers 2019
This work is subject to copyright. All rights are reserved by the Publisher, whether the whole or part of the material is concerned, specifically the rights of translation, reprinting, reuse of illustrations, recitation, broadcasting, reproduction on microfilms or in any other physical way, and transmission or information storage and retrieval, electronic adaptation, computer software, or by similar or dissimilar methodology now known or hereafter developed.
The use of general descriptive names, registered names, trademarks, service marks, etc. in this publication does not imply, even in the absence of a specific statement, that such names are exempt from the relevant protective laws and regulations and therefore free for general use.
The publisher, the authors and the editors are safe to assume that the advice and information in this book are believed to be true and accurate at the date of publication. Neither the publisher nor the authors or the editors give a warranty, expressed or implied, with respect to the material contained herein or for any errors or omissions that may have been made. The publisher remains neutral with regard to jurisdictional claims in published maps and institutional affiliations.

This Springer imprint is published by the registered company Springer Nature Singapore Pte Ltd.
The registered company address is: 152 Beach Road, #21-01/04 Gateway East, Singapore 189721, Singapore

# Preface

In recent years, structural control has demonstrated its value for mitigating natural hazards and enhancing the safety and serviceability of structural systems. The associated theory and technologies receive extensive attention and rapid development, and are often categorized into four modalities, i.e., the passive control, the active control, the semiactive control and the hybrid control. Nevertheless, a variety of practical challenges still remain open as to this field, which are derived from the strong uncertainty, high nonlinearity and large dimensionality inherent in civil engineering structures and infrastructural systems. High-performance structural control is thus hardly to be implemented by conventional control policies, since these policies are typically devised based on simple linear models in a deterministic manner or in a stochastic manner merely considering measurement noises.

A typical case refers to the vibration control of seismic structures. In practice, the control law and the control parameters pertaining to control devices are designed using recorded and artificial ground accelerations as the input. Due to the randomness, however, inherent in the occurring time, the occurring space, and the amplitude of earthquakes, the seismic performance of structures arises to be random and nonlinear. It might result in a situation that the structural control system even operates as an adverse manner when the structural system is subjected to a real seismic ground motion that is distinguishedly different from the one referred to the control system design. Stochastic optimal control of structures, therefore, is still a critical challenge that needs to be circumvented in practical engineering.

As an important application of thought of the physical stochastic system, the probability density evolution method (PDEM) has played a critical role in the fields such as the civil engineering, mechanical engineering and ocean engineering for the stochastic response analysis and reliability assessment of structures in the past decades. We believe that the PDEM is also one of the most promising means for exploring stochastic optimal control of structures.

This book is devoted to the systematic development of theory and methods of stochastic optimal control of structures under the engineering excitations as non-stationary and non-Gaussian random processes. In the chapter of introduction, i.e., Chap. 1, the advances of structural control, the history, and status of classical stochastic optimal control are illustrated. The challenges of structural control are then addressed. Following that, the concept of physically based stochastic optimal control is introduced. A brief description as to the scope of this book is included as well. The pertinent theoretical principles are presented in Chap. 2. In this chapter, the classical stochastic optimal control, the random vibration of structures, the dynamic reliability of structures, and the modeling of random dynamic excitations are illustrated. This section provides a solid foundation for the successive development of theory and methods of stochastic optimal control of structures. The following three chapters, i.e., Chaps. 3–5, address the basic principles, probabilistic criteria, and generalized optimal control policy, which are the critical ingredients of the physically based stochastic optimal control. These three chapters constitute the core of the book. The stochastic optimal control of nonlinear structures is discussed in Chap. 6. Applications of the physically based stochastic optimal control are introduced in Chaps. 7 and 8, including the viscous damper control, as a passive modality, for habitability enhancement of high-rise buildings subjected to strong winds, the magnetorheological damper control, as a semiactive modality, for performance improvement of seismic structures. An experimental verification upon the theory and methods of stochastic optimal control of structures is introduced in Chap. 9. Shaking table tests of a steel moment-resisting frame structure deployed with optimally designed viscous dampers and subjected to random seismic ground motions are carried out. Besides, five appendixes are provided to assist the interested reader to better understand the topics involved in this book.

This work was motivated by our interest in developing successful strategies for structural control involving inevitable randomness inherent in dynamic excitations so as to maintain a desired structural performance. We are indebted to many of our colleagues who have provided invaluable assistance in preparation of the book. Special thanks are due to Professor Alfredo H-S. Ang (University of California, Irvine), Professor Pol D. Spanos (Rice University), Professor Georgios Deodatis (Columbia University), Professor Satish Nagarajaiah (Rice University), Professor Biswajit Basu (Trinity College Dublin), Professor Michael Beer (Leibniz Universität Hannover), Professor Jianbing Chen (Tongji University), Professor Soren R.K. Nielsen (Aalborg University), Professor John D. Sorensen (Aalborg University), Professor Michael H. Faber (-Nielsen) (Aalborg University), Professor Roberto Villaverde (University of California, Irvine), Professor Hector Jensen (Santa Maria University), Professor Hanping Hong (University of Western Ontario), and Professor Radoslaw M. Iwankiewicz (Hamburg University of Technology) for their concerns, encouragements, and comments for the book. The first author is particularly indebted to Professor Roger Ghanem for the impeccable supervision and consistent support when he visited University of Southern California from 2007 to 2009.

We wish also to take this opportunity to express our sincere gratitude to the financial support provided by the National Natural Science Foundation of China and the Ministry of Science and Technology of China in the past fifteen years. We are also indebted to Elsevier, Wiley-Blackwell, SAGE, American Society of Civil

Engineers (ASCE), and Techno-Press, respectively, for their permissions to reproduce copyright materials included in our previously published journal papers.

Finally, we would like to thank our families for their long-lasting support, encouragement, and love.

Shanghai, China                                                                          Yongbo Peng
                                                                                                    Jie Li

# Contents

# Chapter 1
# Introduction

## 1.1 Preliminary Remarks

Dynamic loads acting on engineering structures typically arise a significant randomness inherent in their occurring times, spaces, and amplitudes (Housner 1947). The mechanical behaviors of structural materials often feature a remarkable uncertainty as well (Ang and Tang 2006). Therefore, the performance of engineering structure under external excitations always remains a degree of randomness. Traditional deterministic methods for structural design and analysis cannot satisfy with the developing demand of structural engineering. Since the late of 1950s, the random vibration theory (Lin 1967; Zhu 1992) taking into account the randomness inherent in external excitations of engineering structures and the stochastic structure theory (Li 1996) taking into account the randomness inherent in mechanical behaviors of structural materials have received extensive attention. These two theories serve as the milestones of research of stochastic dynamics of structures and provide a solid foundation for the development of new design principle of structural engineering (Li and Chen 2009).

In order to increase the resistance of structures, on the other hand, the base isolation system for seismic mitigation was pioneeringly investigated more than one hundred years ago; while the control technology in concepts of vibration isolation, vibration absorption, and vibration damping was first embraced by structural engineering community in 1960s (Housner et al. 1997). Until to the early of 1970s (Yao 1972), the proposal of structural control concept led to the wide eye-catching and rapid development of the associated theory and technologies with vibration mitigation of structures in practice (Soong 1990; Soong and Dargush 1997; Chu et al. 2005). Structural control thus proved its value and became as one of the promising means that are capable of improving structural behaviors, reinforcing structural safety, and enhancing structural performance. However, owing to the fact that there are definitely uncertainties inherent in external excitations and in structural and control systems, the structural control in terms of the deterministic design principle cannot guaran-

© Springer Nature Singapore Pte Ltd. and Shanghai Scientific and Technical Publishers 2019
Y. Peng and J. Li, *Stochastic Optimal Control of Structures*,
https://doi.org/10.1007/978-981-13-6764-9_1

tee structural safety (Roussis et al. 2003; Cao et al. 2012). It thus arises a practical demand of stochastic optimal control of structures in terms of the probabilistic design principle by quantifying the randomness inherent in structural and control systems in an elegant manner.

It has been more than half a century since the establishment of the stochastic optimal control theory, which was gradually in extension from the information control associated with the measurement noises to the structural control associated with both the random excitations and stochastic structures. A variety of formulations underlying the theory and methods of stochastic optimal control of structures were proposed, e.g., the linear quadratic Gaussian (LQG) control and the covariance control (Yong and Zhou 1999). However, the classical stochastic optimal control theory just relies upon the state equation and the Itô calculus, where the measurement noise and random excitations are often formulated as the white noise or the filtered white noise in mathematics. The cost function in the classical stochastic optimal control theory is thus defined as a function of white Gaussian noise. However, this treatment cannot logically contain the influence of nonstationary random excitations such as seismic ground motions and high winds (Yang 1975). Besides, the classical random vibration theory merely attains the probability density of linear systems or of the extraordinary less-degree-of-freedom nonlinear systems. As to the accurate control of general nonlinear stochastic systems, the probability density solutions are still far from attaining in the framework of classical random vibration theory. Meanwhile, the stochastic optimal control involving the randomness of structural materials and control systems just focuses upon exploring the stochastic stability of systems, which has not yet been well resolved.

For circumventing the dilemma encountered by the classical stochastic dynamics, the probability density evolution method (PDEM) has been proposed in recent years in terms of the probability preservation principle. The probability density evolution method builds the essential connection between the system state evolution and the probability density evolution. From this perspective, the deterministic system and the stochastic system can be solved in a unified framework. On this basis, a generalized probability density evolution equation (GDEE) was derived (Li and Chen 2006a, 2008, 2009). These advances provide a new way to attain the stochastic response and reliability of general multi-degree-of-freedom systems, and also make it possible to carry out the accurate control of linear and nonlinear stochastic structural systems under nonstationary and non-Gaussian engineering excitations. This is a natural extension, in essence, of the probability density evolution method to the optimal control of stochastic systems. In order to differentiate from the classical stochastic optimal control, the structural control advocated in this book from the perspective of probability density evolution is referred to as the physically based stochastic optimal (PSO) control.

## 1.2   Advances of Structural Control

Modern control theory was originally from the Wiener filter theory and the Wiener control theory in mid-twentieth century (Wiener 1948, 1949). It became a fairly complete formulation through a rapid development in the subsequent 20 years. Beginning with the early of 1970s, the deterministic optimal control and stochastic optimal control of structures have been developed into almost independent two branches.

Structural control is devised to reduce or mitigate the responses of structures by utilizing the specified devices or facilities deployed in structures that are capable of shifting, dissipating, absorbing, and supplying energy. A diagram of typical logic of structural control is shown in Fig. 1.1. It is seen that a complete control system involves the sensor for monitoring the external excitation and the structural response, the controller for deducing the feedback law to regulate the structural state, and the actuator for implementing the control gain. According to whether a moderate power supply system is required for the operation of control system, structural control can be categorized into four modalities, i.e., the passive control, the active control, the semiactive control, and the hybrid control (Housner et al. 1997; Ou 2003; Teng 2009).

Passive control refers to a modality that allocates base isolation systems or energy dissipation devices to a structure so as to enhance the damping, stiffness, and strength of the structure. The control gain provided by passive systems just replies upon the structural response to the external excitation, and generally does not need power supplies. In early of 1970s, Kelly and his colleagues proposed and experimentally proved the value of metal yield and energy dissipation component in the seismic mitigation of structures (Kelly et al. 1972; Skinner et al. 1975). The pioneering works played a significant role in the development of ductile design of seismic structures, which later formulated an important branch of structural control using energy dissipation schemes. Thereafter, an amount of energy dissipation components and devices were invented and applied into the vibration control of engineering

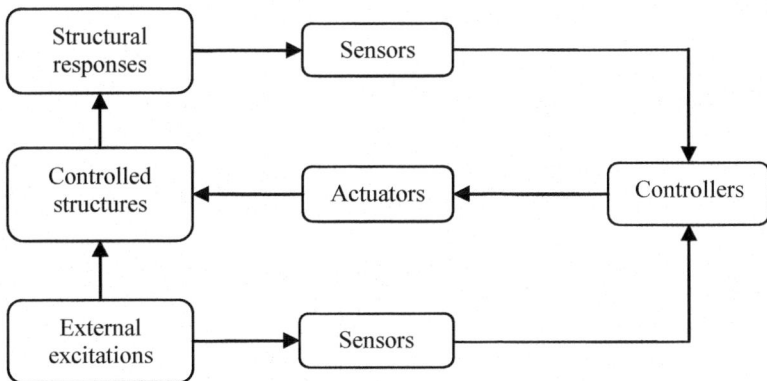

**Fig. 1.1**   Diagram of typical logic of structural control

structures under earthquakes and high winds, such as friction-type energy dissipation devices derived as the principle of automobile brakes (Pall and Marsh 1982; Li and Reinhorn 1995; Qu et al. 2001; Bhaskararao and Jangid 2006), viscoelastic dampers early used in the fatigue control of plane body (Zhang et al. 1989; Zhang and Soong 1992; Shen et al. 1995; Palmeri and Ricciardelli 2006; Xu 2007), viscous dampers or damping walls widely applied in the fields of military and aviation (Constantinou et al. 1993; Reinhorn et al. 1995; Museros and Martinez-Rodrigo 2007), tuned mass dampers (TMDs) and multiple-tuned mass dampers (MTMDs) (Villaverde 1994; Setareh 1994; Li 2000; Guo and Chen 2007; Wong and Johnson 2009), and tuned fluid dampers (Wakahara et al. 1992; Tamura et al. 1995; Tait et al. 2008). Until now, the research and development of energy dissipation devices with excellent performance are still one of the hotspot issues in the field of structural control (Chan and Albermani 2008; Jung et al. 2010; Zhang et al. 2013; Berardengo et al. 2015; Amjadian and Agrawal 2018).

Base isolation system is the most widely accepted passive control system, which is often used in important buildings and bridges in high-intensity seismic zones. Modern isolation technologies originated in the 1970s as well. Kelly proposed an earthquake isolation system with natural rubber bearings (laminated rubber bearing) (Kelly 1978), which was later refined to the so-called high-damping rubber bearing (HDRB) (Kelly et al. 1987) and the lead rubber bearing (LRB) (Aiken et al. 1989). In addition to the laminated rubber bearing, a variety of base isolation technologies, such as the friction sliding bearing (FSB) (Dolce et al. 2007), the curved surface slider (CSS) (Quaglini et al. 2017), and the smart base isolation system (Casciati et al. 2007), are widely used in practice. Benefiting from the knowledge of friction-controllable principle (Feng et al. 1993) and eddy-current damping mechanism (Kriezis et al. 1992), a sliding hydromagnetic bearing (HMB) was proposed recently (Villaverde 2017), which has been proved to be effectiveness of enhancing the capacity of the sliding isolation system in aspects of energy dissipation and deformation constraint (Peng et al. 2019a, b).

The associated control algorithms with the passive control modality focus upon the optimal deployment and parameter design of dampers and isolators so as to attain the best trade-off between cost and effect. For instance, Chen et al. carried out the optimal deployment of dampers in a truss using the simulated annealing algorithm where the maximum cumulative energy dissipation in a finite period was considered as the optimization criterion (Chen et al. 1991). Zhang and Soong proposed a sequential method for the optimal deployment of viscoelastic dampers based on the concept of controllability and the mean-square interstory drift (Zhang and Soong 1992). Using the sequential method, Wu et al. addressed the optimal parameters and the optimal placement of energy dissipation devices in a three-dimensional asymmetric structure involving the coupling effect between translations and retortions (Wu et al. 1997). Shukla and Datta analyzed the influence of the viscoelastic damper models and the input seismic ground motions on the device deployment by integrating the root-mean-square interstory drift and the concept of controllability (Shukla and Datta 1999). Takewaki et al. proposed a gradient method for the optimal damper deployment through minimizing the amplitude of transfer function of the interstory drift to better the structural performance (Takewaki et al. 1999). Using the gradient method as well, Singh and Moreschi optimized the damping coefficients of viscous and viscoelastic dampers so as to gain a maximum reduction of structural responses (Singh and Moreschi 2001). Kim et al. adopted the performance-based design principle and the capacity spectrum method to carry out the parameter design of viscous dampers (Kim et al. 2003). In definition of the active optimal control gain as the design objective, Ou optimized the placement and parameters of passive control devices (Ou 2003). Since the gradient method might result in a local convergence problem, Singh and Moreschi optimized the placement of viscous and viscoelastic dampers using the genetic algorithm (Singh and Moreschi 2002). Park et al. employed the minimum cost of structural systems in their life cycles as the design criterion, and optimized the parameters, number, and placement of viscoelastic dampers by virtue of the genetic algorithm (Park et al. 2004). Silvestri and Trombetti addressed the efficiency of mass-proportional damping scheme through a comparative study against the stiffness-proportional damping scheme using the genetic algorithm, and also stressed the importance of the objective function in the procedure (Silvestri and Trombetti 2007). The genetic algorithm was also employed by Rama Raju et al., in addressing the efficiency of various damper deployments (Rama Raju et al. 2014). In order to attain a better trade-off between energy dissipation and excessive deformation prevention of friction pendulum systems, Bucher performed probability-based optimal design of friction coefficient and radius of curvature pertaining to the devices by applying a Pareto-type optimization approach (Bucher 2009).

The previous works show that the parameter design and placement optimization of energy dissipation devices and base isolation systems following the performance-based principle are the general thought of passive control modality. Owing to the simple logic, ready realization, and meeting with the system stability requirements, the passive control modality has been widely used in recent years, while the practical challenges show that this control modality lacks a sufficient efficiency for strengthening performance of high-rise buildings subjected to seismic ground motions and high winds (Housner et al. 1997).

Active control provides a more efficient means to gain acquirement by exerting a certain compensative force to reduce the structural response. It often involves an online measurement of structural state and a real-time estimation of some physical quantities of concern. Active control force is calculated according to the designed control law, whereby the actuator is driven to implement the gain on the structure (Soong 1990). During the past 40 more years, a variety of active control devices have been researched and developed, where the active tendon system (ATS) (Roorda 1975; Soong et al. 1988; Chung et al. 1988, 1989; Soong 1990) and the active mass damper (AMD) (Chang and Soong 1980; Spencer and Nagarajaiah 2003) are most applied in practice. According to the specified control law, the active modality can theoretically implement the adaptive control (Dewey and Jury 1963), intelligent control (Fu 1971), sliding mode control (Utkin 1977), stochastic control (Stengle 1986), optimal control (Anderson and Moore 1990), and robust control (Suhardjo 1990).

As to the optimal control, the classical linear quadratic regulator (LQR) associated with the linear structural system has been still the control algorithm widely used in practice (Chang and Soong 1980; Soong 1990; Stavroulakis et al. 2006). In this context, Yang et al. proposed a simultaneous optimal control algorithm involving the influence of external excitations upon the feedback control gain (Yang et al. 1987). The optimal control of nonlinear structural systems was concerned as well. Maris et al. proposed the optimal pulse control method for reducing the vibration of nonlinear flexible systems under general dynamic loads (Masri et al. 1981). Abdel-Rohman and Rayfeh addressed the feasibility of the perturbation method for devising the active control devices so as to mitigate the nonlinear vibration of a hinged single-span bridge (Abdel-Rohman and Nayfeh 1987). Kamat employed a variable metric algorithm to carry out the active optimal control of nonlinear structural systems (Kamat 1988). Yang et al. extended the simultaneous optimal control algorithm to the hybrid control of nonlinear and hysteretic structural systems under seismic ground motions (Yang et al. 1992a, b). Based on the sequential expansion of performance function and the functional framework of Hamilton–Jacobi optimal control, Suhardjo et al. developed a family of nonlinear optimal control schemes (Suhardjo et al. 1992). Utkin proposed a sliding mode control method suitable for the uncertainty conditions (Utkin 1992). Yang et al. further proposed a generalized optimal method for linear and nonlinear controls, where the dynamic effect of actuators was treated logically during the optimization process so as to reduce the adverse effect of a system time delay (Yang et al. 1994). Krishnan et al. performed the modeling, simulation, and analysis of active control of nonlinear structural systems using the neural network algorithm (Krishman et al. 1995). It is thus revealed that the active optimal control algorithms have almost been settled down in the mid-1990s. While the challenges from the time delay and from the stability associated with the active control systems still remain open. Since then, the issues pertaining to the optimal deployment of active control devices received extensive attention. For instance, Xu et al. proposed an optimal method for the gain design and actuator placement in a feedback control system (Xu et al. 1994). Furuya and Haftka optimized the actuator placement in a spatial structure using a hybrid algorithm by integrating the generic and simulate anneal schemes (Furuya and Haftka 1996). Using the algebraic Riccati equation solution, Hiramoto et al. carried out the optimal deployment of actuators and sensors in a controlled flexible structural system (Hiramoto et al. 2000). Tan et al. proposed an integrated strategy on the actuator placement design and the gain parameter optimization of control systems by virtue of the generic algorithm and acceleration feedback method (Tan et al. 2005). Yan et al. developed a control method to analyze and optimize an adaptive truss structure where the generic algorithm was employed to design the active system (Yan et al. 2005).

Though the active control can attain a satisfactory structural performance, it often requires high-level power supplies which might be up to thousands of watts. However, the power supply system probably suffers from a serious damage under the hazardous actions such as earthquakes (Patten et al. 1998). Moreover, the entire structural system tends to dynamic instability due to the time delay inherent in actuator dynamics and numerical calculations, the measurement noise, and the system modeling errors (Soong 1990). Naturally, a combination modality by virtue of the active and the passive control systems was proposed. This modality has the advantage of reducing the risk derived from power requirement and dynamics instability, in the case of meeting with the expected control gain. This concept prompted the development of the semiactive and the hybrid control modalities since the 1990s (Chu et al. 2005).

The semiactive control aims at accommodating the stiffness and the damping of the controlled structural system, and mitigating the vibration through the real-time adjustment of the control device. In comparison with the active control, the semiactive control requires less external power supply but is able to implement an almost same control effectiveness as the active control (Jansen and Dyke 2000). After more than 20 years' development, the semiactive control devices have been formed into a variety of types, such as the active variable stiffness system (Kobori et al. 1990; Yang et al. 2007), the active variable damping system (Kawashima et al. 1992; Mizuno et al. 1992; Shinozuka and Ghanem 1992; Patten 1997; Kobori 2003), the electrorheological damper (Ehrgott and Masri 1992; Gavin et al. 1993), the magnetorheological damper (Spencer et al. 1996, 1997; Tu et al. 2011), the piezoelectric actuator (Kamada et al. 1997), the piezoelectric variable friction damper (Zhao and Li 2010), and the shape memory alloy (Aiken et al. 1993).

Due to the essential nonlinearities inherent in the semiactive control devices, the control gain of structural systems highly relies upon the algorithm employed in the semiactive control. Karnopp et al. pioneeringly proposed a force generator for implementing the Skyhook variable damping feedback control (Karnopp et al. 1974). Thereafter, Hrovat et al. addressed the concept of the semiactive control, and proposed the so-called Hrovat algorithm thereafter for the variable damping control referring to the Bang–Bang control of relay action (Hrovat et al. 1983). They first provided the idea of designing a semiactive control system so as to trace the performance of an active control system. Brogan proposed a control strategy suitable for adjusting the electrorheological damper according to the Lyapunov stability theory, whereby the structural response can be reduced by minimizing the changing rate of the Lyapunov function (Brogan 1991). Sack and Patten developed a clipped-optimal control algorithm suitable for the semiactive hydraulic damping devices (Sack and Patten 1994). McClamroch and Gavin proposed a decentralized Bang–Band control algorithm so as to minimize the total energy of structures (McClamroch and Gavin 1995). Dyke et al. suggested an LQG clipped-optimal control strategy for strengthening the seismic performance of structural systems attached with the magnetorheological damper according to an acceleration feedback control algorithm (Dyke et al. 1996a, b). Inaudi developed a modulated homogenous friction algorithm suitable for the variable friction damper (Inaudi 1997). Jansen and Dyke investigated various semiactive control strategies such as the Lyapunov stability theory, the LQG clipped-optimal control, the decentralized Band-Bang control, the modulated homogenous friction, and the maximum energy dissipation. They proposed the maximum-energy-cost algorithm (Jansen and Dyke 2000). Following the idea of optimizing the parameters of the semiactive control to trace the optimal active control, Ou and Li performed the parameter design of the magnetorheological damper employed in vibration mitigation of structural systems (Ou and Li 2010). In their work, the magnetorheological damping force was designed to approach the optimal active control force, and meanwhile, the magnetorheological damping control effectiveness was assumed to be the same as the active optimal control. Xu and Li developed a multiple-step prediction strategy for the magnetorheological damping system, so as to ensure the structural stability and to reduce the influence of system time delay in a certain extent (Xu and Li 2008). Xu and Guo proposed a fuzzy-neuro control strategy so as to implement the accurate and rapid definition of the current input in the magnetorheological damper. Their work attempted to bypass the challenges associated with the accuracy and time delay brought about by the traditional two-mode control strategy, i.e., Passive-off and Passive-on (Xu and Guo 2008).

Hybrid control is often referred to as a combined modality between the passive and active or semiactive control systems. This modality aims at enhancing the effectiveness of the passive modality in conjunction with the active or semiactive modalities, using as few as possible power supplies in the operation of control systems. This treatment can strengthen the stability of the controlled structures (Chu et al. 2005). The hybrid control can enlarge the benefits of active, passive, and semiactive controls, bypassing the constraints and limitations related to the onefold control modality, while the hybrid control needs more space in practice, which is thus often used in the structural control for multiple-level defense systems. At present, the application of active structural control is implemented mostly utilizing the hybrid modality. The hybrid tuned mass damper (HTMD) and the hybrid base isolation device (HBID) are the two main control systems in practice. For instance, a hybrid arch tuned mass damper was developed and applied in the wind-induced vibration control of long-span bridges and building structures, and in the swing control of steamer navigations (Tanida et al. 1991). Cheng et al. proposed a hybrid tuned mass damper system combining a control actuator and a passive tuned liquid damper for wind-induced vibration control of Nanjing television tower (Cheng et al. 1994). Watakabe et al. developed a hybrid tuned mass damper with switch mode between active and passive control modalities, which was applied in the vibration control of high-rise building subjected to seismic ground motions and high winds (Watakabe et al. 2001). Feng et al. proposed a semiactive friction-controllable fluid bearing, and constructed a hybrid base isolation system for seismic mitigation of structures (Feng et al. 1993). Reinhorn and Riley addressed the validity of sliding base-isolated structural system on a small-scale bridge model through theoretical analysis and experimental investigations (Reinhorn and Riley 1994). Lin et al. applied the magnetorheological damper to the large-scale base-isolated structures and proceeded with a series of shaking-table tests (Lin et al. 2007). In recent years, Asai et al. investigated the seismic mitigation performance of a smart base isolation system using the real-time hybrid simulation technique by integrating the physical substructure of the magnetorheological damper and the numerical substructure of the controlled structure (Asai et al. 2015).

Similar to the semiactive control, the hybrid control algorithm is related to the application of control devices. In regard to the hybrid tuned mass damper, a series of control algorithms have been proposed considering the limitations of device stroke and output, such as the gain scheduling technique (Tamura et al. 1994) and the ad hoc control algorithm (Niiya et al. 1994). In regard to the hybrid base isolation system, a variety of nonlinear control strategies have been developed considering the essential nonlinear behaviors inherent in the base isolation systems, including the acceleration feedback control employing the instantaneous optimal control algorithm (Nagarajaiah et al. 1993), fuzzy control (Nagarajaiah 1994; Lin et al. 2007), neural network-based control (Venini and Wen 1994), the adaptive nonlinear control (Rodellar et al. 1994), and the LQG/H2 control (Asai et al. 2015).

## 1.3 Stochastic Optimal Control of Structures

Since randomness inherent in the external excitations, the structural systems, and the control devices, the traditional deterministic control cannot guarantee the safety of structures. Stochastic optimal control of structures has thus become a challenging issue in the area of structural control (Housner et al. 1997). This topic can be traced back to the earlier researchers on the relevant fundamental theory. In 1940s, the stochastic differential theory had been established on the basis of the Itô calculus (Sobczyk 1991). Until to the late of 1950s and the early of 1960s, however, the development of the state-space method (Kalman 1960a), the maximum principle (Pontryagin 1962), the dynamic programming (Bellman 1957), and the Kalman–Bucy filtering theory (Kalman 1960b; Bucy and Kalman 1961) were eventually proposed and developed, which underlined the establishment of the modern optimal control theory. Stochastic optimal control then formed as a new branch of optimal control discipline.

### *1.3.1 Classical Stochastic Optimal Control*

The motivation of stochastic optimal control is to define the optimal control so that the transition probability density function (Spencer and Bergman 1993) or the statistical moments (Wojtkiewicz et al. 1996) are limited in a specified range of errors. In the theoretical framework of classical LQG control, there have formed the stochastic maximum principle (Yong and Zhou 1999) and the stochastic dynamic programming (Stengel 1994). Meanwhile, a series of design criteria for defining the stochastic optimal control law in accordance with the optimal control gain have been developed. The minimum variance criterion, for example, seeks for a control force with the objective of minimizing the variance of performance functions subjected to the mean constraint (Sain 1966). The optimal neighborhood feedback criterion involves an interaction scheme based on an initial costate estimation in view of the control equation and the performance objective (Stengel 1994). The statistical moment assessment criterion seeks for an optimal control law in the sense of trade-off, between the statistics of quantities of concern (Zhang and Xu 2001). The reliability-based design criterion is to attain a control law leading to a minimum failure probability of structural systems, through iteratively solving the limit state equation pertaining to the performance objective (Spencer et al. 1994a; May and Beck 1998). The probability density tracking criterion employs the Markov process theory to attain the analytical solution of probability density of a family of specified systems, so as to define the control law approaching the desired probability density (Elbeyli and Sun 2002; Elbeyli et al. 2005; Sun 2006).

In aspect of stochastic optimal control of linear systems, the present moment reliability-based structural optimal control has a wide application. For example, Spencer et al. addressed the design and optimization method of a single-degree-of-freedom system attached with active tendon systems. In their work, a performance function directly relevant to the failure probability was defined, and the first-crossing principle and a stationary white noise model of seismic ground motions were employed. Utilizing the first-order reliability method/the second-order reliability method (FORM/SORM), the reliability-based control of structures was then implemented so as to minimize the failure probability of structural displacement (Spencer et al. 1994a). Battaini et al. further employed the FORM/SORM to carry out the reliability assessment and experimental analysis of a controlled multi-degree-of-freedom system (Battaini et al. 2000). It was indicated in their work that the computational efficiency of moment-based reliability method was better than the Monte Carlo simulation, while the former merely attained an approximate solution in the case that the performance function of structures featured a high nonlinearity in the neighborhood of checking point, or that the distribution of basic random variables was far from the Gaussian distribution. The non-Gaussian properties of structural response might enlarge the error between the approximate solution and the accurate solution due to the high-order nonlinear mapping from the basic random variables to the structural responses. Therefore, the moment-based reliability method cannot be extended into the large-scale and complex structural systems. In 1998, May and Beck explored the first generation of Benchmark model pertaining to the probabilistic control problem, so as to maximize the reliability of the uncertain structures and control systems subjected to seismic ground motions. In this work, an acceleration feedback control scheme was employed aiming at the design of optimal active mass damper; the performance measure on the design objective includes the interstory drift, story acceleration of structures, and the deployment positions of sensors. An approximate expansion estimating the failure probability of structures was then proposed based on the level-crossing process theory (May and Beck 1998). However, their work involved the search iterations of the optimal value with respect to the approximate expansion and of the stochastic optimal control law, which will result in an increased computational cost.

In aspect of nonlinear stochastic optimal control, Zhu and his colleagues have been devoted to the extension in the framework of Hamilton systems since the late of 1990s. They proposed the optimal control strategy of nonlinear stochastic dynamical systems in the reference of the stochastic averaging method and the stochastic dynamic programming (Zhu et al. 2000, 2001, 2006). However, the nonlinear stochastic optimal control of large-size engineering structures in this theoretical framework still remains a challenge.

Beginning with the middle of 1980s, Skelton and his colleagues systematically developed a covariance control theory (Skelton 1988). The basic idea is to seek for a certain control law sets so as to maintain a good consistency between the covariance and the objective of stationary processes of linear systems (Collins and Skelton 1985; Hotz and Skelton 1987; Skelton and Ikeda 1989). Thereafter, the covariance control theory was extended into the control of bilinear systems (Yasuda et al. 1990). On the basis of this framework, Field and Bergman developed a linear covariance control method with the constraint of reliability, whereby the covariance structure of systems can be readily derived in assumption of the Poisson-process level-crossing of stationary responses (Field and Bergman 1998). Elbeyli and Sun further proposed a covariance control method for the nonlinear stochastic dynamical system, of which the covariance control gain can be constructed through minimizing the performance index relevant to the stationary probability density of structural responses (Elbeyli and Sun 2004).

In general cases, the state estimation problem resulted from the measurement noise is a critical issue that needs to be settled down in the stochastic optimal control of structures. On the form of mathematics, the state estimation and optimal control of linear systems are dual that both involve solving independently the Riccati equation with initial values. This family of stochastic optimal control problems thus can be circumvented by virtue of the separation principle (Astrom 1970). However, in regard to the stochastic optimal control of nonlinear systems involving the state estimation, the associated issue becomes to be more complicated since the separation principle is not more available (Stengel 1986; Housner et al. 1997).

In aspect of robustness assessment of stochastic systems, the reliability of system stability is an argument of interest which is usually derived from the distribution of real eigenvalues of the system over the positive and negative semi-planes. Due to the fact that the distribution of eigenvalues cannot be attained through a ready solving procedure, the random simulation method is a preferable means, whereby the distribution of eigenvalues of sample structures over the complex plane is attained, and the instability probability of controlled linear systems can be then estimated (Stengel and Ray 1991). Another alternative means is to redefine the stability criterion of controlled systems into the formulation of ultimate state equation, and then to solve the issue of stochastic stability of structural systems with random parameters in the context of classical reliability theory (Spencer et al. 1992; Spencer et al. 1994b). Using similar methods, Field, Breitung, and Taflaidis et al. individually carried out the analysis of stochastic stability of controlled systems (Field et al. 1995, 1996; Breitung et al. 1998; Taflanidis et al. 2008).

## 1.3.2  Challenges of Structural Control in Civil Engineering

The control theory and methods were thrived in the fields of electronics and information engineering, mechanical engineering, aerospace engineering, etc. They focus on the state regulation of systems under distributions such as random excitations and measurement noise, while new challenging issues have to be encountered when these achievements are applied into the field of civil engineering. Different from the practical demands, however, as emerged in the mechanical engineering and aerospace engineering, there are more uncertainty and higher complexity inherent in civil engineering. The structural control in civil engineering involves a variety of practical challenges such as the structural safety, system durability, structural comfortability, etc. Moreover, large output and high performance are claimed as to the control devices. The challenging issues of structural control in civil engineering that are distinguished from the classical control theory and methods lie in dynamic excitations, structural parameters, nonlinear effects, control law formulas, and control modalities.

(i)   Challenges related to dynamic excitations

In the period of service, the civil engineering structures usually suffer from the dynamic excitations, especially from the risk of hazardous dynamic actions. The hazardous dynamic actions such as strong earthquakes, high winds, and huge waves exhibit significant randomness inherent in their occurring time, occurring space, and occurring intensity. The influences of random excitations upon the accurate quantification of structural state and the logical design of control systems are thus prominent. The classical stochastic optimal control theory, derived from the Itô stochastic differential equation, is restricted to the assumption of white Gaussian noise excitations and measurement noise. It still lacks the sufficient exploration into the case under the general random excitations. This limitation owes to the fact the classical stochastic optimal control theory has been mostly applied in the nonmechanical problems such as those raised from the mechanical engineering and aerospace engineering. While the challenges related to the random excitation become predominant, the stochastic optimal control theory is used to deal with the mechanical problems that occur in the civil engineering. In fact, the seismic ground motion exhibits significant nonstationarities, and the high wind even just the stable airflow exhibits certain nonstationarities. However, the random excitations in the classical stochastic optimal control are almost assumed to be stationary white Gaussian noise, which is obviously far from the hazardous dynamic actions upon the civil engineering structures.

(ii)   Challenges related to structural parameters

Due to the uncertainties inherent in the structural materials and manufactures, the basic parameters of civil engineering structures usually exhibit randomness. This brings about a series of new issues to the structural control. The influences of random parameters upon the stochastic optimal control of structures give rise to two aspects. One is the state estimation. The Kalman filter theory is a celebrated method for dealing with the measurement noise and the incomplete measurement in the classical system control. How this method is applied to state estimation of structures with random parameters constitutes a new challenge. The other is the stability of control system. The presence of random parameters leads to the issue of stochastic eigenvalues, which also brings about a new challenge to the Lyapunov stability theory based stability analysis of classical control systems.

(iii)  Challenges related to nonlinear effects

It is well knowledged that the nonlinearities inherent in the control system of civil engineering structures are derived from two causes, i.e., the nonlinearities of structural components and the nonlinearities of control devices, which are typically categorized into material nonlinearity and geometric nonlinearity. The structural systems with large deformation belong to a family of systems with geometric nonlinearity, which are usually modeled as harden and soften oscillators. The structural and control device systems with hysteretic behaviors belong to a family of systems with material nonlinearity, which are usually represented by bilinear elastic–plastic models and Bouc–Wen differential models. The former is widely used in the modeling of mechanical and aerospace structures, while the latter refers to as the main formulation of nonlinear modeling of civil engineering structures. Nonlinear effects especially the coupling influences between the hysteresis and the randomness result in an extreme difficulty of stochastic optimal control of engineering structures.

(iv)  Challenges related to control law formulas

In the classical stochastic optimal control, the elementary variables of control law in modality of state feedback are usually defined as the empirical manner, which lacks a rational criterion. Therefore, the classical stochastic optimal control often involves a forward control law. Although the proposal of backward stochastic differential equations provides a basis for the design of backward control law, it is still within the theoretical framework of Itô stochastic differential equations. In fact, how to design a logical backward control law in conjunction with the nonlinear effects of civil engineering structures constitutes another challenging issue.

(v)  Challenges related to control modality

Due to a small size of controlled objects and the controllability of working conditions, the classical stochastic optimal control mostly focuses on the centralized control modality. However, the civil engineering structure exhibits a large size, a complex architecture, and a significant influence from the hazardous dynamic actions. It is usually protected by a multiple-level defense system as a distributed control modality. Therefore, the centralized control modality based traditional control theory always falls short. The research on the distributed control modality or on the distributed–centralized control modality prompts the internal demand of structural control in civil engineering. Besides, the classical stochastic optimal control always is concerned with a family of control systems exhibiting input and output feedbacks, while the structural control in civil engineering has a more wide range, including a non-feedback control system, i.e., passive modality, and an input or output feedback control system, i.e., active, semiactive, and hybrid modalities.

In summary, the classical stochastic optimal control theory which was derived from the Itô stochastic differential equation hinges upon a hypothesis that the external disturbance is viewed as a white Gaussian noise or a filtered white Gaussian noise. It thus cannot rationally assess the influence of random excitations as the nonstationary and non-Gaussian noise upon the optimization and design of system parameters of structural control. This situation results in the limitation of the classical stochastic optimal control theory in engineering applications, especially for the optimal control of complicated dynamic systems of nonlinear structures under strongly nonstationary random excitations, where the desired performance is hardly to be attained. Therefore, it is necessary to explore a new theory and associated methods suitable for the stochastic optimal control of civil engineering structures.

## *1.3.3  Physically Based Stochastic Optimal Control*

It is readily recognized that the relevant theory and methods of classical stochastic optimal control are all developed on the basis of Itô calculus, which underlies the state equation of systems. This treatment allows an exclusive assumption that the external excitation is viewed as a white Gaussian noise or a filtered white Gaussian noise, which is far from the real engineering excitations. This assumption thus limits the engineering application of the classical stochastic optimal control in practice. In fact, the assumption hinders the development of the modern theory of stochastic dynamical system as well. Just in view of this situation, the probability density evolution method was developed based on the probability preservation principle. The PDEM bridges the essential relation between the probability density evolution and the physical state evolution of systems, that is, the physical state evolution of systems drives the probability density evolution. The deterministic system and the stochastic system can thus be summarized into a unified framework (Li and Chen 2009). Moreover, this progress profoundly reveals that the physical evolution mechanism of systems is still the critical content of stochastic system researches, which underlies the theory of physical stochastic system. In this framework, a novel theory and the associated methods for the stochastic optimal control of structures are expected to develop.

In the end nineteenth century, the research of practical systems with random initial state formed the basis of the Gibbs-Liouville theory, and proved the celebrated Liouville equation (Syski 1967). It is Einstein who addressed the special cases of diffusion processes and established the diffusion equation for Brownian motion in 1905 (Einstein 1905). Then it was extended by Fokker and Planck who derived the classical Fokker–Planck equation (Fokker 1914; Planck 1917). In 1931, Kolmogorov independently deduced a same formulation as the Fokker–Planck equation, and a backward Kolmogorov equation was then derived (Kolmogorov 1931). Owing to the rigorous mathematical basis, the Kolmogorov equation is so-called the Fokker—Planck–Kolmogorov equation (FPK equation). Thereafter, the FPK equation and its solutions formed the primary topics of random vibration theory. In 1957, Dostupov and Pugachev attempted to quantify the randomness inherent in the system input through introducing the Karhunen–Loeve decomposition (Dostupov and Pugachev 1957). It is the so-called Dostupov–Pugachev equation. It is regret; however, the equations mentioned above are all high-dimensional and strong-coupling partial differential equations, of which the analytical solutions are hardly derived. Li and Chen explored the probability preservation principle in an elegant manner, and secured the essential relation between probability density evolution and physical state evolution of systems. A family of decoupling probability density evolution equations, i.e., the so-called generalized probability density evolution equation (GDEE), was then proposed in the past 15 years (Li and Chen 2006a, b, c, 2008, 2009). It is recognized that the GDEE accommodates the randomness both inherent in external excitations and in structural systems, which provides a new way to carry out the response analysis and reliability assessment of stochastic systems subjected to general random excitations, and also allows a potential for stochastic optimal control of linear and nonlinear multi-degree-of-freedom systems.

It should be noted that the probability density evolution method reveals the essential relation between the deterministic evolution of sample orbits and the probability density evolution of statistical assembles, and fulfills the decoupling and the dimension reduction of probability density evolution equation of general stochastic systems. A ready solving procedure for the stochastic systems can be thus proposed where the solution of physical equation and probability density evolution equation are readily attained in a separate manner. As to the controlled stochastic dynamical system, the physical equation includes the term of control gain, while the control gain just relies upon the physical state of systems. Thus, the probability density evolution equation of controlled system is a coupled equation set with respect to the system state and the control gain, which still remains a dependence on the physical equation of controlled systems. This finding breaks through the limitation of the Itô calculus and provides a new way toward the stochastic optimal control of engineering structures subjected to general random excitations such as seismic ground motions, high winds, and huge waves.

Design and optimization of control gain is the critical task of the physically based stochastic optimal control, in which three steps are often involved, i.e., the design of control law, the optimization of controller parameters, and the optimization of control device placement. The design of control law hinges upon the associated control modalities with feedback logic such as the active, semiactive, or hybrid controls. The formulation of control law is generally derived from Pontryagin's maximum principle or from Bellman's optimality principle, while the optimization of the controller parameters and of the control device placement serve as the two important aspects of structural control, which are both referred to the probabilistic criteria in function of probability density or reliability of system quantities of concern.

## 1.4 Scope of the Book

In the civil engineering community, the objective of structural control is often definite, while the loads acting on the engineering structures cannot be predicted accurately, especially for the dynamic excitations. Therefore, the stochastic optimal control of structures considering the randomness inherent in engineering excitations ought to be paid sufficient attention. For this reason, the present book focuses on the hazardous dynamic actions, specifically on the random seismic ground motion and the fluctuating wind-velocity field, and devotes to developing a novel theory and the pertinent successful strategies for stochastic optimal control of engineering structures, in conjunction with the probability density evolution method. The outline is sketched as follows: the performance evolution of controlled systems is first investigated, and a family of probabilistic criteria in terms of structural responses is then established; the generalized optimal control policy and the associated control law involving the simultaneous optimization of the controller parameters and the control device placement are then proposed; in order to verify the proposed methodology, a series of engineering applications and experimental studies of controlled structures are then introduced.

The scope of the book is illustrated as follows:

In Chap. 2, the associated theoretical principles pertaining to the physically based stochastic optimal control are addressed, including the classical stochastic optimal control in the framework of the stochastic maximum principle and the stochastic dynamic programming, the random vibration of linear and nonlinear structures, the dynamic reliability of structures, and the modeling of random dynamic excitations. The kernel equation of the PDEM, i.e., the generalized probability density evolution equation, is introduced as well. This chapter devotes to providing a solid foundation for the successive developments of theory and methods of stochastic optimal control of structures.

In Chap. 3, the probability density evolution method of stochastic optimal control is detailed. Performance evolution of controlled structural systems is first investigated. The solution of the physically based stochastic optimal control is deduced according to Pontryagin's maximum principle. Active stochastic optimal control based on the probabilistic criterion on system second-order statistics evaluation is discussed. For validating purposes, comparative studies against the classical LQG control are carried out.

In Chap. 4, a family of probabilistic criteria for the physically based stochastic optimal control is proposed, including the single-objective optimization criteria with respect to the second-order moments such as the mean and the variance, and with respect to the tail of probability density, i.e., the exceedance probability, of equivalent extreme-value responses; and the multiple-objective optimization criteria with respect to the mean and the exceedance probability of equivalent extreme-value responses in performance trade-off and in energy trade-off, respectively. Numerical examples are studied to prove the applicability of the proposed probabilistic criteria.

In Chap. 5, the concept of generalized optimal control policy is proposed. This concept indicates a unified formula of the optimal control law with optimized controller parameters pertaining to passive, active, semiactive, and hybrid controls, and with optimized control device placement. In order to attain the optimal placement of control devices at each sequential step, a probabilistic controllability index in argument of exceedance probability is defined. Comparative studies between control device deployment strategies using the minimum controllability index gradient criterion and the maximum controllability index criterion are then carried out.

In Chap. 6, the theory and methods of physically based stochastic optimal control are extended to the nonlinear structures. Stochastic optimal polynomial control of a family of hardening Duffing oscillators and two classes of nonlinear structural systems with Clough hysteretic components and with Bouc–Wen hysteretic components are investigated. The exceedance probability-based probabilistic criterion is employed as well. The control effectiveness of the optimal polynomial controller on different types of nonlinear stochastic dynamical systems is addressed. For validating purposes, comparative studies against the statistical linearization-based LQG control are carried out.

In Chap. 7, practical application of the proposed methodology in the wind-resistant structures with viscous dampers is addressed. Equivalent linearization techniques for smoothing the viscously damped structures are introduced. Design criteria for optimal deployment of viscous dampers are proposed. For illustrative purposes, the wind-induced comfortability control of a high-rise building in the typhoon-prone area is investigated. The details on the parameter design and placement optimization of viscous dampers are provided. The effectiveness of mitigating wind-induced roof acceleration and enhancing structural habitability is proved.

In Chap. 8, the stochastic optimal control of seismic structures using the magnetorheological damper is investigated. The bound Hrovat algorithm and the parametrized model for the magnetorheological damper are addressed, and thereby the control law for variable damping on structures is defined. Using the technique of molecular dynamics simulations, the bulk behaviors of microscale magnetorheological suspensions are investigated, and their relevance to the control current is discussed as well. On this basis, the gain design, parameter optimization, and control effectiveness analysis of a controlled structure attached with the magnetorheological damper are carried out.

In Chap. 9, experimental studies of stochastic optimal control are introduced. A six-story steel moment-resisting frame structure deployed with viscous dampers is investigated. By virtue of the versatile facilities held by the State Key Laboratory of Disaster Reduction in Civil Engineering at Tongji University, the shaking-table test of a controlled structural model subjected to the random seismic ground motion is carried out. Experimental investigations show the applicability and effectiveness of the proposed theory and methods of stochastic optimal control of structures.

In order to facilitate the readers' understanding, five appendixes related to the topic included in this book are provided to the end.

# References

Åström KJ (1970) Introduction to stochastic control theory. Academic Press, New York

Abdel-Rohman M, Nayfeh AH (1987) Active control of nonlinear oscillations in bridges. ASCE J Eng Mech 113:335–348

Aiken ID, Kelly JM, Tajirian FF (1989) Mechanics of low shape factor elastomeric seismic isolation bearings. Report No. UCB/EERC-89/13, Earthquake Engineering Research Center, University of California, Berkeley, CA

Aiken ID, Nims DK, Whittaker AS, Kelly JM (1993) Testing of passive energy dissipation systems. Earthq Spectra 9(3):335–368

Amjadian M, Agrawal AK (2018) Modeling, design, and testing of a proof-of-concept prototype damper with friction and eddy current damping effects. J Sound Vib 413:225–249

Anderson Brian DO, Moore J (1990) Optimal control: linear quadratic methods. Prentice-Hall, Englewood Cliffs

Ang AH-S, Tang WH (2006) Probability concepts in engineering: emphasis on applications to civil and environmental engineering. Wiley

Asai T, Chang CM, Spencer BF Jr (2015) Real-time hybrid simulation of a smart base-isolated building. J Eng Mech 141(3):04014128-1-10

Battaini M, Casciati F, Faravelli L (2000) Some reliability aspects in structural control. Probab Eng Mech 15:101–107

Bellman R (1957) Dynamic programming. Princeton University Press, Princeton

Berardengo M, Cigada A, Guanziroli F, Manzoni S (2015) Modelling and control of an adaptive tuned mass damper based on shape memory alloys and eddy currents. J Sound Vib 349:18–38

Bhaskararao AV, Jangid RS (2006) Seismic analysis of structures connected with friction dampers. Eng Struct 28:690–703

Breitung K, Casciati F, Faravelli (1998) Reliability based stability analysis for actively controlled structures. Eng Struct 20(3):211–215

Brogan WL (1991) Modern control theory. Prentice-Hall, Englewood Cliffs

Bucher C (2009) Probability-based optimal design of friction-based seismic isolation devices. Struct Saf 31(6):500–507

Bucy RS, Kalman RE (1961) New results in linear filtering and prediction theory. ASME Trans J Basic Eng 83:95–108

Cao M, Tang H, Funaki N, et al (2012) Study on a real 8F steel building with oil damper damaged during the 2011 Great East Japan Earthquake. In: Proceeding of world conference on earthquake engineering, Lisbon, Portugal

Casciati F, Faravelli L, Hamdaoui K (2007) Performance of a base isolator with shape memory alloy bars. Earthq Eng Eng Vib 6(4):401–408

Chang James CH, Soong TT (1980) Structural control using active tuned mass dampers. ASCE J Eng Mech 106(6):1091–1098

Chan RWK, Albermani F (2008) Experimental study of steel slit damper for passive energy dissipation. Eng Struct 30(4):1058–1066

Chen GS, Robin JB, Salama M (1991) Optimal placement of active/passive members in truss structures using simulated annealing. AIAA J 29(8):1327–1334

Cheng W, Qu W, Li A (1994) Hybrid vibration control of Nanjing TV tower under wind excitation. In: Proceedings of 1st world conference on structural control, Los Angeles, USA, vol 1, pp WP2-32–WP2-41

Chu SY, Soong TT, Reinhorn AM (2005) Active, hybrid and semi-active structural control. Wiley, New York

Chung LL, Lin RC, Soong TT, Reinhorn AM (1989) Experimental study of active control for MDOF seismic structures. J Eng Mech 115(8):1609–1627

Chung LL, Reinhorn AM, Soong TT (1988) Experiments on active control of seismic structures. ASCE J Eng Mech 114(2):241–256

Collins EG Jr, Skelton RE (1985) Covariance control of discrete systems. In: Proceedings of 24th IEEE conference on decision and control, Ft. Lauderdale, FL, pp 542–547

Constantinou MC, Symans MD, Tsopelas P, Taylor DP (1993) Fluid viscous dampers in applications of seismic energy dissipation and seismic isolation. In: Proceedings of seminar on seismic isolation, passive energy dissipation, and active control, Applied Technology Council, Palo Alto, USA No. ATC-17-1, vol 2, pp 581–592

Dewey A, Jury E (1963) A note on Aizerman's conjecture. IEEE Trans Autom Control AC-19:482–483

Dolce M, Cardone D, Palermo G (2007) Seismic isolation of bridges using isolation systems based on flat sliding bearings. Bull Earthq Eng 5:491–509

Dostupov BG, Pugachev VS (1957) The equation for the integral of a system of ordinary differential equations containing random parameters. Autom i Telemekhanika 18:620–630

Dyke SJ, Spencer BF Jr, Quast P et al (1996a) Implementation of an active mass driver using acceleration feedback control. Microcomput Civil Eng 11(5):305–323

Dyke SJ, Spencer BF Jr, Sain MK, Carlson JD (1996b) Modeling and control of magnetorheological dampers for seismic response reduction. Smart Mater Struct 5:565–575

Ehrgott RC, Masri SF (1992) Use of electrorheological materials in intelligent systems. In: Proceedings of US-Italy-Japan workshop/symposium on structural control and intelligent systems, Sorrento, Italy, pp 87–100

Einstein A (1905) In: Furth R (ed) Investigations on the theory of the Brownian movement (1956). Dover Publications, New York

Elbeyli O, Hong L, Sun JQ (2005) On the feedback control of stochastic systems tracking pre-specified probability density functions. Trans Inst Meas Control 27(5):319–329

Elbeyli O, Sun JQ (2002) A stochastic averaging approach for feedback control design of nonlinear systems under random excitations. J Vib Acoust 124:561–565

Elbeyli O, Sun JQ (2004) Covariance control of nonlinear dynamic systems via exact stationary probability density function. J Vib Acoust 126:71–76

Feng Q, Shinozuka M, Fujii S (1993) Friction-controllable sliding isolated systems. ASCE J Eng Mech 119(9):1845–1864

Field RV Jr, Bergman LA (1998) Reliability-based approach to linear covariance control design. ASCE J Eng Mech 124(2):193–199

Field RV Jr, Bergman LA, Hall WB (1995) Computation of probabilistic stability measures for a controlled distributed parameter system. Probab Eng Mech 10:181–192

Field RV Jr, Voulgaris PG, Bergman LA (1996) Probabilistic stability robustness of structural systems. ASCE J Eng Mech 122(10):1012–1021

Fokker AD (1914) Die mittlere Energie rotierender elektrischer Dipole im Strahlungsfeld. Ann Phys (Leipz) 43:810–820 (in German)

Fu KS (1971) Learning control systems and intelligent control systems: an intersection of artificial intelligence and automatic control. IEEE Trans Autom Control 16:70–72

Furuya H, Haftka RT (1996) Combining genetic and deterministic algorithms for locating actuators on space structures. J Spacecr Rockets 33(3):422–427

Gavin HP, Ortiz DS, Hanson RD (1993) Testing and Modeling of a proto-type ER damper for seismic structural response control. In: Proceedings of international workshop on structural control, Honolulu, USA, pp 166–180

Guo YQ, Chen WQ (2007) Dynamic analysis of space structures with multiple tuned mass dampers. Eng Struct 29:3390–3403

Hiramoto K, Doki H, Obinata G (2000) Optimal sensor actuator placement for active vibration control using explicit solution of algebraic Riccati equation. J Sound Vib 229(5):1057–1075

Housner GW (1947) Characteristics of strong-motion earthquakes. Bull Seismol Soc Am 37(1):19–31

Housner GW, Bergman LA, Caughey TK, Chassiakos AG, Claus RO, Masri SF, Skelton RE, Soong TT, Spencer BF Jr, Yao James TP (1997) Structural control: past, present, and future. ASCE J Eng Mech 123(9):897–971

Hotz A, Skelton RE (1987) Covariance control theory. Int J Control 46(1):13–32

Hrovat D, Barak P, Rabins M (1983) Semi-active versus passive or active tuned mass dampers for structural control. ASCE J Eng Mech 109(3):691–705

Inaudi JA (1997) Modulated homogeneous friction: a semi-active damping strategy. Earthq Eng Struct Dyn 26(3):361–376

Jansen LM, Dyke SJ (2000) Semi-active control strategies for MR dampers comparative study. ASCE J Eng Mech 126(8):795–803

Jung HJ, Jang DD, Lee HJ, Lee IW, Cho SW (2010) Feasibility test of adaptive passive control system using MR fluid damper with electromagnetic induction part. J Eng Mech 136(2):254–259

Kalman RE (1960a) On the general theory of control systems. In: Proceedings of 1st IFAC Moscow congress. Butterworth Scientific Publications

Kalman RE (1960b) A new approach to linear filtering and prediction problems. ASME Trans J Basic Eng 82:35–45

Kamada T, Fujita T, Hatayama T, Arikabe T, Murai N, Aizawa S, Tohyama K (1997) Active vibration control of frame structures with smart structures using piezoelectric actuators (Vibration control by control of bending moments of columns). Smart Mater Struct 6:448–456

Kamat MP (1988) Active control of structures in nonlinear response. J Aerosp Eng 1:52–62

Karnopp DC, Crosby MJ, Harwood RA (1974) Vibration control using semi-active force generators. J Eng Ind 96(2):619–626

Kawashima K, Unjoh S, Iida H, Mukai H (1992) Effectiveness of the variable damper for reducing seismic response of highway bridge. In: Proceedings of second US-Japan workshop on earthquake protective systems for bridges, PWRI, Tsukuba Science City, Japan, pp 479–493

Kelly JM (1978) Experimental results of an earthquake isolation system using natural rubber bearing. Report No. UCB/EERC-78/03, Earthquake Engineering Research Center, University of California, Berkeley, CA

Kelly JM, Buckle IG, Koh CG (1987) Mechanical characteristics of base isolation bearings for a bridge deck model test. Report No. UCB/EERC-86/11, Earthquake Engineering Research Center, University of California, Berkeley, CA

Kelly JM, Skinner RI, Heine AJ (1972) Mechanisms of energy absorption in special devices for use in earthquake-resistant structures. Bull N Z Natl Soc Earthq Eng 5(3):63–88

Kim J, Choi H, Min KW (2003) Performance-based design of added viscous dampers using capacity spectrum method. J Earthq Eng 7(1):1–24

Kobori T (2003) Past, present and future in seismic response control in civil engineering structures. In: Proceedings of 3rd world conference on structures control. Wiley, New York, pp 9–14

Kobori T, Takahashi M, Nasu T (1990) Experimental study on active variable stiffness system-active seismic response controlled structure. In: Proceedings of 4th world congress council on tall buildings and urban habitat, Hong Kong, pp 561–572

Kolmogorov AN (1931) Über die analytischen Methoden in der Wahrscheinlichkeitsrechnung. Math Ann 104:415–458 (in German)

Kriezis EE, Tsiboukis TD, Panas SM, Tegopoulos JA (1992) Eddy currents: theory and applications. Proc. IEEE 80(10):1559–1589

Krishnan R, Nerves AC, Singh MP (1995) Modeling, simulation and analysis of active control of structures with nonlinearity using neural networks. In: Proceedings of 10th engineering mechanics specialty conference, vol 2. ASCE, Boulder, Colorado, pp 1054–1057

Li C (2000) Performance of multiple tuned mass dampers for attenuating undesirable oscillations of structures under the ground acceleration. Earthq Eng Struct Dyn 29:1405–1421

Li C, Reinhorn AM (1995) Experimental and analytical investigation of seismic retrofit of structures with supplemental damping: part 2-friction devices. NCEER Report 95-0009, State University of New York at Buffalo, Buffalo, New York

Li J (1996) Stochastic structural system: analysis and modelling. Science Press, Beijing (in Chinese)

Li J, Chen JB (2006a) The probability density evolution method for dynamic response analysis of non-linear stochastic structures. Int J Numer Meth Eng 65:882–903

Li J, Chen JB (2006b) The dimension-reduction strategy via mapping for probability density evolution analysis of nonlinear stochastic systems. Probab Eng Mech 21(4):442–453

Li J, Chen JB (2006c) Probability density evolution method for stochastic dynamical systems. Adv Natl Sci 16(6):712–719 (in Chinese)

Li J, Chen JB (2008) The principle of preservation of probability and the generalized density evolution equation. Struct Saf 30:65–77

Li J, Chen JB (2009) Stochastic dynamics of structures. Wiley, Singapore

Lin PY, Roschke PN, Loh CH (2007) Hybrid base-isolation with magnetorheological damper and fuzzy control. Struct Control Health Monit 14:384–405

Lin YK (1967) Probabilistic theory of structural dynamics. McGraw-Hill, New York

Masri SF, Bekey GA, Caughey TK (1981) On-linear control of nonlinear flexible structures. J Appl Mech 49:871–884

May BS, Beck JL (1998) Probabilistic control for the active mass driver benchmark structural model. Earthq Eng Struct Dyn 27:1331–1346

McClamroch NH, Gavin HP (1995) Closed loop structural control using electrorheological dampers. In: Proceedings of the American control conference, American Automatic Control Council, Washington, D.C., pp 4173–4177

Mizuno T, Kobori T, Hirai J, Yoshinori M, Niwa N (1992) Development of adjustable hydraulic damper for seismic response control of large structures. In: DOE facilities programs, systems interaction, and active/inactive damping, ASME, New Orleans, LA, vol 229, pp 163–170

Museros P, Martinez-Rodrigo MD (2007) Vibration control of simply supported beams under moving loads using fluid viscous dampers. J Sound Vib 300:292–315

Nagarajaiah S (1994) Fuzzy controller for structures with hybrid isolation system. In: Proceedings of 1st world conference on structural control, vol TA2, pp 67–76

Nagarajaiah S, Riley MA, Reinhorn AM (1993) Hybrid control of sliding isolated bridge. ASCE J Eng Mech 119(11):2317–2332

Niiya T, Ishimaru S, Koizumi T, Takai S (1994) A hybrid system controlling large amplitude vibrations of high-rise buildings. In: Proceedings of 1st world conference on structural control, vol FA2, pp 43–52

Ou JP (2003) Structural vibration control: active, semi-active and intelligent modalities. Science Press, Beijing (in Chinese)

Ou JP, Li H (2010) Analysis of capability for semi-active or passive damping systems to achieve the performance of active control systems. Struct Control Health Monit 17(7):778–794

Pall AS, Marsh C (1982) Response of friction damped braced frames. ASCE J Struct Div 108(6):1313–1323

Palmeri A, Ricciardelli F (2006) Fatigue analyses of buildings with viscoelastic dampers. J Wind Eng Ind Aerodyn 94:377–395

Park KS, Koh HM, Hahm D (2004) Integrated optimum design of viscoelastically damped structural systems. Eng Struct 26:581–591

Patten WN (1997) New life for the Walnut Creek Bridge via semi-active vibration control. Newsl Int Assoc Struct Control 2(1):4–5

Patten WN, Mo C, Kuehn J, Lee J (1998) A primer on design of semi-active vibration absorbers (SAVA). ASCE J Eng Mech 124(1):61–68

Peng YB, Ding LC, Chen JB (2019a) Performance evaluation of base-isolated structures with sliding hydromagnetic bearings. Struct Control Health Monit 26:e2278

Peng YB, Ding LC, Chen JB, Villaverde R (2019b) Experimental study of sliding hydromagnetic isolators for seismic protection. J Struct Eng 145(5):04019021

Planck M (1917) Uber einen Satz der statistichen Dynamik und eine Erweiterung in der Quantumtheorie. Sitzungberichte der Preussischen Akadademie der Wissenschaften 324–341 (in German)

Pontryagin LS (1962) The mathematical theory of optimal processes (trans: Trirogoff KN). Interscience, New York

Qu WL, Chen ZH, Xu YL (2001) Dynamic analysis of wind-excited truss tower with friction dampers. Comput Struct 79:2817–2831

Quaglini V, Gandelli E, Dubini P (2017) Experimental investigation of the re-centring capability of curved surface sliders. Struct Control Health Monit 24(2):e1870

Rama Raju K, Ansu M, Iyer NR (2014) A methodology of design for seismic performance enhancement of buildings using viscous fluid dampers. Struct Control Health Monit 21(3):342–355.

Reinhorn AM, Li C, Constantinou MC (1995) Experimental and analytical investigation of seismic retrofit of structures with supplemental damping: part 1-fluid viscous damping devices. NCEER Report 95-0001, State University of New York at Buffalo, Buffalo, New York

Reinhorn AM, Riley MA (1994) Control of bridge vibrations with hybrid devices. In: Proceedings of 1st world conference on structural control, vol TA2, pp 50–59

22 1 Introduction

Rodellar J, Barbat AH, Molinares N (1994) Response analysis of buildings with a new nonlinear base isolation system. In: Proceedings of 1st world conference on structural control, vol TP1, pp 31–40

Roorda J (1975) Tendon control in tall structures. J Struct Div 101(3):505–521

Roussis PC, Constantinou MC, Erdik M, Durukal E, Dicleli M (2003) Assessment of performance of seismic isolation system of Bolu Viaduct. J Bridge Eng 8(4):182–190

Sack RL, Patten W (1994) Semi-active hydraulic structural control. In: Proceedings of international workshop on structural control, University of Southern California, Los Angeles, pp 417–431

Sain MK (1966) Control of linear systems according to the minimal variance criterion-a new approach to the disturbance problem. Trans Autom Control 118–122 (IEEE)

Setareh M (1994) Use of the doubly-tuned mass dampers for passive vibration control. In: Proceedings of 1st world conference on structural control, vol 1, pp WP4-12–WP4-21

Shen KL, Soong TT, Chang KC, Lai ML (1995) Seismic behaviour of reinforced concrete frame with added viscoelastic dampers. Eng Struct 17(5):372–380

Shinozuka M, Ghanem R (1992) Use of variable dampers for earthquake protection of bridges. In: Proceedings of Second US-Japan workshop on earthquake protective systems for bridges, PWRI, Tsukuba Science City, Japan, pp 507–516

Shukla AK, Datta TK (1999) Optimal use of viscoelastic dampers in building frames for seismic force. ASCE J Struct Eng 4:401–409

Silvestri S, Trombetti T (2007) Physical and numerical approaches for the optimal insertion of seismic viscous dampers in shear-type structures. J Earthq Eng 11(5):787–828

Singh MP, Moreschi LM (2001) Optimal seismic response control with dampers. Earthq Eng Struct Dyn 30:553–572

Singh MP, Moreschi LM (2002) Optimal placement of dampers for passive response control. Earthq Eng Struct Dyn 31(4):955–976

Skelton RE (1988) Dynamic systems control: linear systems analysis and synthesis. Wiley, New York

Skelton RE, Ikeda M (1989) Covariance controllers for linear continuous time systems. Int J Control 49:1773–1785

Skinner RI, Kelly JM, Heine AJ (1975) Hysteresis dampers for earthquake-resistant structures. Earthq Eng Struct Dyn 3:287–296

Sobczyk K (1991) Stochastic differential equations: with applications to physics and engineering. Kluwer Academic Publishers, Dordrecht

Soong TT (1990) Active structural control: theory and practice. Longman Scientific & Technical, New York

Soong TT, Dargush GF (1997) Passive energy dissipation systems in structural engineering. Wiley, New York

Soong TT, Reinhorn AM, Yang JN (1988) Active response control of building structures under seismic excitation. In: Proceedings of 9th world conference on earthquake engineering, Tokyo/Kyoto, Japan, vol 8, pp 453–458

Spencer BF Jr, Bergman LA (1993) On the numerical solution of the Fokker-Planck equation for nonlinear stochastic systems. Nonlinear Dyn 4:357–372 (in German)

Spencer BF Jr, Kaspari DC Jr, Sain MK (1994a) Structural control design: a reliability-based approach. In: Proceedings of the American control conference, Baltimore, Maryland, pp 1062–1066

Spencer BF Jr, Sain MK, Won CH, Kaspari D Jr, Sain PM (1994b) Reliability-based measures of structural control robustness. Struct Saf 15:111–129

Spencer BF Jr, Nagarajaiah S (2003) State of the art of structural control. J Struct Eng 845–856 (Forum)

Spencer BF Jr, Sain MK, Carlson JD (1996) Dynamical model of a magnetorheological damper. In: Proceedings of Structures Congress XIV, ASCE, Chicago, IL, USA, pp 361–370

Spencer BF Jr, Sain MK, Carlson JD (1997) Phenomenological model of magnetorheological damper. ASCE J Eng Mech 123(3):230–238

Spencer BF Jr, Sain MK, Kantor JC, Montemagno C (1992) Probabilistic stability measures for controlled structures subject to real parameter uncertainties. Smart Mater Struct 1:294–305

Stavroulakis GE, Marinova DG, Hadjigeorgiou E, Foutsitzi G, Baniotopoulos CC (2006) Robust active control against wind-induced structural vibrations. J Wind Eng Ind Aerodyn 94:895–907

Stengel RF (1986) Stochastic optimal control: theory and application. Wiley, New York

Stengel RF (1994) Optimal control and estimation. Dover Publications, New York

Stengel RF, Ray LR (1991) Stochastic robustness of linear time-invariant control systems. IEEE Trans Autom Control 36(1):82–87

Suhardjo J (1990) Frequency domain techniques for control of civil engineering structures with some robustness considerations. PhD Dissertation, Department of Civil Engineering, University of Notre Dame, Notre Dame

Suhardjo J, Spencer BF Jr, Sain MK (1992) Nonlinear optimal control of a Duffing system. Int J Non-Linear Mech 27(2):157–172

Sun JQ (2006) Stochastic dynamics and control. Elsevier, Amsterdam

Syski R (1967) Stochastic differential equations. In: Saaty TL (ed) Modern nonlinear equations (Chapter 8). McGraw-Hill, New York

Taflanidis AA, Scruggs JT, Beck JL (2008) Reliability-based performance objectives and probabilistic robustness in structural control applications. J Eng Mech 134(4):291–301

Tait MJ, Isyumov N, EI Damatty AA (2008) Performance of tuned liquid dampers. ASCE J Eng Mech 134(5):417–427

Takewaki I, Yoshitani S, Uetani K, Tsuji M (1999) Non-monotonic optimal damper placement via steepest direction search. Earthq Eng Struct Dyn 28:655–670

Tamura Y, Fujii K, Ohtsuki T, Wakahara T, Koshaka R (1995) Effectiveness of tuned liquid dampers under wind excitations. Eng Struct 17:609–621

Tamura K, Shiba K, Inada Y, Wada A (1994) Control gain scheduling of a hybrid mass damper system against wind response of tall buildings. In: Proceedings of 1st world conference on structural control, vol FA2, pp 13–22

Tan P, Dyke SJ, Richardson A, Abdullah M (2005) Integrated device placement and control design in civil structures using genetic algorithms. ASCE J Struct Eng 131(10):1489–1496

Tanida K, Koike Y, Mutaguchi K, Uno N (1991) Development of hybrid active-passive damper. ASME Act Passive Damping, PVP-Vol. 211:21–26

Teng J (2009) Structural vibration control: theories, technologies and methods. Science Press, Beijing (in Chinese)

Tu JW, Liu J, Qu WL, Zhou Q, Cheng HB, Cheng XD (2011) Design and fabrication of 500-kN large-scale MR damper. J Intell Mater Syst Struct 22(5):475–487

Utkin VI (1977) Variable structure systems with sliding modes. IEEE Trans Autom Control AC-22:212–222

Utkin VI (1992) Sliding modes in control optimization. Springer, New York

Venini P, Wen YK (1994) Hybrid vibration control of MDOF hysteretic structures with neural networks. In: Proceedings of 1st world conference on structural control, vol TA3, pp 53–56

Villaverde R (1994) Seismic control of structures with damped resonant appendages. In: Proceedings of 1st world conference on structural control, vol 1, pp WP4-113–WP4-122

Villaverde R (2017) Base isolation with sliding hydromagnetic bearings: concept and feasibility study. Struct Infrastruct Eng 13(6):709–721

Wakahara T, Ohyama T, Fujii K (1992) Suppression of wind-induced vibration of a tall building using tuned liquid damper. J Wind Eng Ind Aerodyn 41–44:1895–1906

Watakabe M, Tohdo M, Chiba O, Izumi N, Ebisawa H, Fujita T (2001) Response control performance of a hybrid mass damper applied to a tall building. Earthq Eng Struct Dyn 30:1655–1676

Wiener N (1948) Cybernetics: or control and communication in the animal and the machine. The MIT Press, Paris

Wiener N (1949) Extrapolation, interpolation and smoothing of stationary time series, with engineering applications. The MIT Press, Cambridge

Wojtkiewicz SF, Spencer BF Jr, Bergman LA (1996) On the cumulant-neglect closure method in stochastic dynamics. Int J Non-Linear Mech 31(5):657–684

Wong Kevin KF, Johnson J (2009) Seismic energy dissipation of inelastic structures with multiple tuned mass dampers. ASCE J Eng Mech 135(4):265–275

Wu B, Ou JP, Soong TT (1997) Optimal placement of energy dissipation devices for three-dimensional structures. Eng Struct 19(2):113–125

Xu K, Warnitchai P, Igusa T (1994) Optimal placement and gains of sensors and actuators for feedback-control. J Guid Control Dyn 17(5):929–934

Xu LH, Li ZX (2008) Semi-active multi-step predictive control of structures using MR dampers. Earthq Eng Struct Dyn 37:1435–1448

Xu ZD (2007) Earthquake mitigation study on viscoelastic dampers for reinforced concrete structures. J Vib Control 13(1):29–43

Xu ZD, Guo YQ (2008) Neuro-fuzzy control strategy for earthquake-excited nonlinear magnetorheological structures. Soil Dyn Earthq Eng 28(9):717–727

Yan S, Zheng K, Zhao Q, Zhang L (2005) Optimal placement of active members for truss structure using genetic algorithm. Lecture notes in computer science, No 3645. Springer, Berlin, Heidelberg, pp. 386–395

Yang JN (1975) Application of optimal control theory to civil engineering structures. ASCE J Eng Mech Div 101(EM6):819–838

Yang JN, Akbarpour A, Ghaemmaghami P (1987) New optimal control algorithms for structural control. ASCE J Eng Mech 113(9):1369–1386

Yang JN, Bobrow J, Jabbari F, Leavitt J, Cheng CP, Lin PY (2007) Full-scale experimental verification of resetable semi-active stiffness dampers. Earthq Eng Struct Dyn 36(9):1255–1273

Yang JN, Li Z, Danielians A, Liu SC (1992a) Hybrid control of nonlinear and hysteretic systems I. ASCE J Eng Mech 118(7):1423–1440

Yang JN, Li Z, Danielians A, Liu SC (1992b) Hybrid control of nonlinear and hysteretic systems II. ASCE J Eng Mech 118(7):1441–1456

Yang JN, Li Z, Vongchavalitkul S (1994) Generalization of optimal control theory: linear and nonlinear control. ASCE J Eng Mech 120(2):266–283

Yao James TP (1972) Concept of structural control. ASCE J Struct Div 98(ST7):1567–1574

Yasuda K, Kherat S, Skelton RE, Yaz E (1990) Covariance control and robustness of bilinear systems. In: Proceedings of 29th IEEE conference on decision and control, pp 1421–1425

Yong JM, Zhou XY (1999) Stochastic controls: Hamiltonian systems and HJB equations. Springer, New York

Zhang P, Song GB, Li HN, Lin YX (2013) Seismic control of power transmission tower using pounding TMD. J Eng Mech 139(10):1395–1406

Zhang RH, Soong TT (1992) Seismic design of viscoelastic dampers for structural applications. ASCE J Struct Eng 118(5):1375–1391

Zhang RH, Soong TT, Mahmoodi P (1989) Seismic response of steel frame structures with added viscoelastic dampers. Earthq Eng Struct Dyn 18:389–396

Zhang WS, Xu YL (2001) Closed form solution for along-wind response of actively controlled tall buildings with LQG controllers. J Wind Eng Ind Aerodyn 89:785–807

Zhao DH, Li HN (2010) Shaking table tests and analyses of semi-active fuzzy control for structural seismic reduction with a piezoelectric variable-friction damper. Smart Mater Struct 19(10):105031

Zhu WQ (1992) Random vibration. Science Press, Beijing (in Chinese)

Zhu WQ (2006) Nonlinear stochastic dynamics and control in Hamiltonian formulation. ASME Trans 59:230–248

Zhu WQ, Ying ZG, Ni YQ, Ko JM (2000) Optimal nonlinear stochastic control of hysteretic systems. ASCE J Eng Mech 126(10):1027–1032

Zhu WQ, Ying ZG, Soong TT (2001) An optimal nonlinear feedback control strategy for randomly excited structural systems. Nonlinear Dyn 24:31–51

# Chapter 2
# Theoretical Principles

## 2.1 Preliminary Remarks

Stochastic optimal control is a subfield of control theory, which focuses upon the stochastic systems and develops into a cross-discipline between the stochastic process theory and the optimal control theory. The associated theories and technologies with the electronics and information engineering, mechanical engineering, and aerospace engineering, were flourished since 1960s, and just concerned the state adjustment of systems under random disturbances such as random excitations and measurement noise. The development in the field of civil engineering began after the seventies of twentieth century. Different from the requirements of the fields of mechanical engineering and aerospace engineering, the civil engineering structures exhibit a large size and experience a complicated external excitation. They have to encounter a series of challenging issues in regard to the safety, the durability, and the comfortability. These issues become more serious in the case of hazardous actions with uncertainties inherent in the occurring time, occurring space, and occurring intensity. The conventional stochastic optimal control theory, however, originated from the random process theory assumes the white Gaussian noise as the random disturbance, which is obviously far from the hazardous actions of engineering structures. Therefore, it is necessary to explore a logical theory and pertinent methods for the stochastic optimal control of civil engineering structures which circumvents the dilemma encountered by the conventional stochastic optimal control theory.

This chapter aims at addressing the theoretical principles relevant to the succeeding chapters in this book. The remaining sections included in this chapter include the classical stochastic optimal control, the random vibration of structures, and its advances that underlies the solution methods for controlled stochastic dynamical systems, the dynamic reliability of structures that underlies the design basis for probabilistic criteria of stochastic optimal control of structures, and the modeling of random dynamic excitations that underlies the uncertainty quantification and simulation of hazardous actions of engineering structures. Through integrating the involved

© Springer Nature Singapore Pte Ltd. and Shanghai Scientific and Technical Publishers 2019
Y. Peng and J. Li, *Stochastic Optimal Control of Structures*,
https://doi.org/10.1007/978-981-13-6764-9_2

sections, the principle for the theory and methods of stochastic optimal control of structures are provided.


## 2.2   Classical Stochastic Optimal Control

The stochastic optimal control aims at attaining the optimal control law that promotes the stochastic system to an expected state through minimizing a certain cost function by the celebrated optimal control schemes. It is well recognized that the pioneering work on the optimal control theory is the proposal of calculus of variations. In history, Pierre and Fermat introduced firstly the so-called Fermat's least action principle to explore the minimum path of ray propagating through the optical media in 1662. In 1755, Lagrange introduced the delta calculus, and then Euler proposed the elementary definition of variation method. In 1930s, the Hamilton–Jacobi equation was derived in the framework of the variation method owing to Hamilton and Jacobi's contributions. Till the mid-twentieth century, the classical variation theory was completely established. The research of modern optimal control theory began from the late period of World War II. Its theoretical milestones consist of the maximum principle proposed by Pontryagin in 1956, the dynamic programming proposed by Bellman in 1957, the state-space method, and linear filtering theory developed by Kalman in 1960 (Yong and Zhou 1999). In early of 1960s, owing to the developments of the stochastic maximum principle (Kushner 1962) and the stochastic dynamic programming (Florentin 1961), the research of stochastic optimal control theory was marked as the beginning.

In state space, the equation of motion of a controlled stochastic dynamical system can be written as

$$\dot{\mathbf{Z}}(t) = \mathbf{g}[\mathbf{Z}(t), \mathbf{U}(t), \mathbf{w}(t), t], \ \mathbf{Z}(t_0) = \mathbf{z}_0 \qquad (2.2.1)$$

The output equation of the system is given by

$$\hat{\mathbf{Z}}(t) = \mathbf{h}[\mathbf{Z}(t), \mathbf{U}(t), \mathbf{w}(t), t] \qquad (2.2.2)$$

The measure equation of the system is then given by

$$\mathbf{Y}(t) = \mathbf{j}[\hat{\mathbf{Z}}(t), \mathbf{n}(t), t] \qquad (2.2.3)$$

where $\mathbf{Z}(t)$ is the $2n$-dimensional column vector denoting system state; $\hat{\mathbf{Z}}(t)$ is the $m$-dimensional vector denoting system output; $\mathbf{U}(t)$ is the $r$-dimensional vector denoting control force; $\mathbf{w}(t)$ is the $s$-dimensional vector denoting random excitations; $\mathbf{n}(t)$ is the $m$-dimensional vector denoting measurement noise; $\mathbf{Y}(t)$ is the $m$-dimensional measured vector denoting system state; $\mathbf{g}(\cdot)$ is the $2n$-dimensional functional vector denoting system state evolution; $\mathbf{h}(\cdot), \mathbf{j}(\cdot)$ are the $m$-dimensional functional vectors denoting the output and measurement of systems, respectively, which both rely upon the number of sensors.

For the ready solution, the external excitation and measurement noise termed in the classical stochastic optimal control theory are usually assumed to be an additive delta-correlated process, i.e., white Gaussian noise, which exhibits the properties as follows:

$$E[\mathbf{w}(t)] = \mathbf{0}, \ E[\mathbf{w}(t)\mathbf{w}^{\mathrm{T}}(\tau)] = \mathbf{W}(t)\delta(t - \tau) \tag{2.2.4}$$

$$E[\mathbf{n}(t)] = \mathbf{0}, \ E[\mathbf{n}(t)\mathbf{n}^{\mathrm{T}}(\tau)] = \mathbf{N}(t)\delta(t - \tau) \tag{2.2.5}$$

where $\mathbf{W}(t)$, $\mathbf{N}(t)$ are $s \times s$, $m \times m$ symmetric and semi-positive spectral density matrices, respectively; $\delta(\cdot)$ denotes the Dirac delta function which exhibits the behaviors as follows:

$$\delta(t - \tau) = \begin{cases} +\infty, \ t = \tau \\ 0, \quad t \neq \tau \end{cases} \tag{2.2.6}$$

Since the external excitation and measurement noise are random processes, the system state and system output are random processes as well. The cost function of the stochastic optimal control is often defined as the mathematical expectation of Bolza formulation (Housner et al. 1997), i.e.,

$$J = E\left[ \phi[\mathbf{Z}(t_f), t_f] + \int_{t_0}^{t_f} L[\mathbf{Z}(t), \mathbf{U}(t), t]\mathrm{d}t \right] \tag{2.2.7}$$

where $E[\cdot]$ denotes the operator of mathematical expectation; $\phi[\cdot]$ is the terminal cost function; $L[\cdot]$ is the running cost function; $t_0$ is the initial time; $t_f$ is the terminal time; $\mathbf{Z}(t_f)$ is the $2n$-dimensional vector denoting the system state at the terminal moment.

It is seen from Eq. (2.2.1) that the state vector $\mathbf{Z}(t)$ can be determined uniquely in the case of the known control force vector $\mathbf{U}(t)$. The cost function Eq. (2.2.7) merely relies upon the control force vector $\mathbf{U}(t)$. Therefore, the problem of stochastic optimal control can be viewed as the minimization of the cost function $J$ by seeking for a certain control force vector $\mathbf{U}(t)$ in the available domain. In fact, the minimization of the cost function $J$ can be implemented using the variation method involving the dynamical constraint Eq. (2.2.1) (Lanczos 1970; Naidu 2003).

According to the Lagrange multiplier method, the introduction of costate vector $\boldsymbol{\lambda}(t) \in \mathbb{R}^n$ ($\mathbb{R}^n$ denotes the $n$-dimensional Euclidean space) can transfer the problem of functional extreme value with the abovementioned constraint to a problem of functional extreme value without constraints. This transfer is completely equivalent if the state-control equation Eq. (2.2.1) is satisfied in a rigorous condition, i.e.,

$$J = E[\phi(\mathbf{Z}(t_f), t_f)] + \int_{t_0}^{t_f} \left\{ H[\mathbf{Z}(t), \mathbf{U}(t), \boldsymbol{\lambda}(t), t] - E[\boldsymbol{\lambda}^{\mathrm{T}}(t)\dot{\mathbf{Z}}(t)] \right\} dt \qquad (2.2.8)$$

where $H[\cdot]$ denotes the Hamilton function:

$$H[\mathbf{Z}(t), \mathbf{U}(t), \boldsymbol{\lambda}(t), t] = E[L(\mathbf{Z}(t), \mathbf{U}(t), t)] + E[\boldsymbol{\lambda}^{\mathrm{T}}(t)\dot{\mathbf{Z}}(t)] \qquad (2.2.9)$$

Performing integration by parts with respect to the system state $\dot{\mathbf{Z}}(t)$ involved in Eq. (2.2.8), there is

$$J = E[\phi(\mathbf{Z}(t_f), t_f) + \boldsymbol{\lambda}^{\mathrm{T}}(t_0)\mathbf{Z}(t_0) - \boldsymbol{\lambda}^{\mathrm{T}}(t_f)\mathbf{Z}(t_f)]$$
$$+ \int_{t_0}^{t_f} \left\{ H[\mathbf{Z}(t), \mathbf{U}(t), \boldsymbol{\lambda}(t), t] + E[\dot{\boldsymbol{\lambda}}^{\mathrm{T}}(t)\mathbf{Z}(t)] \right\} dt$$
$$(2.2.10)$$

Minimizing the cost function, i.e.,

$$\min\{J\} \rightarrow \delta J = 0, \qquad (2.2.11a)$$

one could have the variation formula of the former three terms in Eq. (2.2.10):

$$\delta\left\{ E[\phi(\mathbf{Z}(t_f), t_f) + \boldsymbol{\lambda}^{\mathrm{T}}(t_0)\mathbf{Z}(t_0) - \boldsymbol{\lambda}^{\mathrm{T}}(t_f)\mathbf{Z}(t_f)] \right\}$$
$$= E\left[ \left. \frac{\partial \phi}{\partial \mathbf{Z}} \right|_{t=t_f} - \boldsymbol{\lambda}^{\mathrm{T}}(t_f) \right] \delta\mathbf{Z}(t_f) \qquad (2.2.11b)$$

In fact, the derivation of Eq. (2.2.11b) utilizes the condition $\delta\mathbf{Z}(t_0) = 0$, i.e., the differential of state vector is zero in the case of the given initial conditions. The variation formulation of the fourth term is given by

$$\delta\left\{ \int_{t_0}^{t_f} \left\{ H[\mathbf{Z}(t), \mathbf{U}(t), \boldsymbol{\lambda}(t), t] + E[\dot{\boldsymbol{\lambda}}^{\mathrm{T}}(t)\mathbf{Z}(t)] \right\} dt \right\}$$
$$= \int_{t_0}^{t_f} \left\{ \frac{\partial H}{\partial \mathbf{Z}}\delta\mathbf{Z}(t) + \frac{\partial H}{\partial \mathbf{U}}\delta\mathbf{U}(t) + E[\dot{\boldsymbol{\lambda}}^{\mathrm{T}}(t)]\delta\mathbf{Z}(t) \right\} dt \qquad (2.2.11c)$$

Then it yields

$$\delta J = E\left[ \left\{ \left. \frac{\partial \phi}{\partial \mathbf{Z}} \right|_{t=t_f} - \boldsymbol{\lambda}^{\mathrm{T}}(t_f) \right\} \right] \delta\mathbf{Z}(t_f) + \int_{t_0}^{t_f} \left\{ \frac{\partial H}{\partial \mathbf{Z}}\delta\mathbf{Z}(t) + \frac{\partial H}{\partial \mathbf{U}}\delta\mathbf{U}(t) + E[\dot{\boldsymbol{\lambda}}^{\mathrm{T}}(t)]\delta\mathbf{Z}(t) \right\} dt$$

$$= E\left[\left\{\frac{\partial\phi}{\partial\mathbf{Z}}\bigg|_{t=t_f} - \boldsymbol{\lambda}^{\mathrm{T}}(t_f)\right\}\delta\mathbf{Z}(t_f) + \int_{t_0}^{t_f}\left\{E\left[\left\{\frac{\partial H}{\partial\mathbf{Z}} + \dot{\boldsymbol{\lambda}}^{\mathrm{T}}(t)\right\}\right]\delta\mathbf{Z}(t) + \frac{\partial H}{\partial\mathbf{U}}\delta\mathbf{U}(t)\right\}dt\right]$$

$$(2.2.12)$$

Therefore, the necessary condition for the minimization of the cost function $J$ used in the variation method is the Pontryagin's maximum principle (Sperb 1981; Liberzon 2012). Its expression is that if the control force $\mathbf{U}^*(t)$ is referred to as an optimal control, and the system state $\mathbf{Z}^*(t)$ corresponds to the optimal trajectory of the optimal control, there must be a costate $\boldsymbol{\lambda}^*(t)$ leading to that under the random excitation $\mathbf{w}(t)$, the control force $\mathbf{U}^*(t)$, the system state $\mathbf{Z}^*(t)$ and the costate $\boldsymbol{\lambda}^*(t)$ can simultaneously satisfy the conditions as follows:

$$\boldsymbol{\lambda}(t_f) = \left(\frac{\partial\phi[\mathbf{Z}(t_f), t_f]}{\partial\mathbf{Z}}\right)^{\mathrm{T}} \tag{2.2.13}$$

$$\dot{\boldsymbol{\lambda}}(t) = -\left(\frac{\partial H[\mathbf{Z}^*(t), \mathbf{U}^*(t), \boldsymbol{\lambda}^*(t), t]}{\partial\mathbf{Z}}\right)^{\mathrm{T}} \tag{2.2.14}$$

$$\frac{\partial H[\mathbf{Z}^*(t), \mathbf{U}^*(t), \boldsymbol{\lambda}^*(t), t]}{\partial\mathbf{U}} = \mathbf{0} \tag{2.2.15}$$

Equations (2.2.13), (2.2.14) and (2.2.15) constitute the classical Euler–Lagrange equation for stochastic optimal control.

In view of Eq. (2.2.9), there is

$$\dot{\mathbf{Z}}(t) = \frac{\partial H[\mathbf{Z}^*(t), \mathbf{U}^*(t), \boldsymbol{\lambda}^*(t), t]}{\partial\boldsymbol{\lambda}^{\mathrm{T}}} \tag{2.2.16}$$

It is noted that Eq. (2.2.14) is so-called costate equation, and Eq. (2.2.16) is so-called state equation; both form into the Hamilton canonical equations. Integrating Eqs. (2.2.14), (2.2.15) and (2.2.16), one can derive the functional relation between the optimal control law and the system state.

Pontryagin's maximum principle is just the necessary condition other than the sufficient condition with respect to the existence of the optimal control. The two-point boundary value problem, consisting of the solution of canonical equations and the known boundary conditions, is more difficult to be dealt with than the usual initial value problem. The challenge lies in that some boundary values are given at the initial moment, while some boundary values are given at the terminal moment, resulting in neither solving forward nor solving backward. The two-point boundary value problem thus exhibits the analytical solution merely in very few situations. Often, a numerical solution by virtue of iteration methods is preferred.

In fact, the functional based conditional extreme-value problem shown in Eq. (2.2.8) usually involves two kinds of solving procedures (Athans and Falb 1966). One is to construct the stochastic Euler–Lagrange equation, i.e., Eqs. (2.2.13), (2.2.14) and (2.2.15), according to the Pontryagin's maximum principle. The other is to derive the stochastic Hamilton–Jacobi–Bellman (HJB) equation according to the

Bellman's optimality principle. The necessary condition of solving the optimal control from the HJB equation is consistent with the Pontryagin's maximum principle (Yong and Zhou 1999).

The essence of the dynamic programming method on solving the HJB equation is the specification of the idea that one part of the optimal path is optimal as well. As far as the cost function Eq. (2.2.7) is concerned, the substitution of $t_0$ with $t, t \in [t_0, t_f]$ will lead to

$$
J = E\left[ \phi[\mathbf{Z}(t_f), t_f] + \int_t^{t_f} L[\mathbf{Z}(\tau), \mathbf{U}(\tau), \tau]d\tau \right] \tag{2.2.17}
$$

An optimal value function is defined as follows:

$$
V[\mathbf{Z}^*(t), t] = \min\{J\} = J[\mathbf{Z}^*(t), \mathbf{U}^*(t), t]
$$

$$
= E\left[ \phi[\mathbf{Z}^*(t_f), t_f] - \int_{t_f}^t L[\mathbf{Z}^*(\tau), \mathbf{U}^*(\tau), \tau]d\tau \right] \tag{2.2.18}
$$

of which the differential formulation is

$$
dV[\mathbf{Z}^*(t), t] = -E\big[L[\mathbf{Z}^*(t), \mathbf{U}^*(t), t]\big]dt \tag{2.2.19}
$$

If the measurement noise is out of concern, then

$$
dV[\mathbf{Z}^*(t), t] = -L[\mathbf{Z}^*(t), \mathbf{U}^*(t), t]dt \tag{2.2.20}
$$

On the other hand, the total differential formulation of the optimal value function is written as

$$
dV[\mathbf{Z}^*(t), t] = E\left[ \frac{\partial V[\mathbf{Z}^*(t), t]}{\partial t}dt + \frac{\partial V[\mathbf{Z}^*(t), t]}{\partial \mathbf{Z}}d\mathbf{Z}(t) + \frac{1}{2}d\mathbf{Z}^{\mathrm{T}}(t)\frac{\partial^2 V[\mathbf{Z}^*(t), t]}{\partial \mathbf{Z}^2}d\mathbf{Z}(t) \right] \tag{2.2.21}
$$

Integrating Eq. (2.2.20) and Eq. (2.2.21), one has

$$
-L[\mathbf{Z}^*(t), \mathbf{U}^*(t), t]dt = \frac{\partial V[\mathbf{Z}^*(t), t]}{\partial t}dt
$$
$$
+ E\left[ \frac{\partial V[\mathbf{Z}^*(t), t]}{\partial \mathbf{Z}}d\mathbf{Z}(t) + \frac{1}{2}d\mathbf{Z}^{\mathrm{T}}(t)\frac{\partial^2 V[\mathbf{Z}^*(t), t]}{\partial \mathbf{Z}^2}d\mathbf{Z}(t) \right] \tag{2.2.22}
$$

Therefore

$$
-\frac{\partial V[\mathbf{Z}^*(t), t]}{\partial t} = L[\mathbf{Z}^*(t), \mathbf{U}^*(t), t] + E\left[ \frac{\partial V[\mathbf{Z}^*(t), t]}{\partial \mathbf{Z}}\dot{\mathbf{Z}}(t) + \frac{1}{2}\dot{\mathbf{Z}}^{\mathrm{T}}(t)\frac{\partial^2 V[\mathbf{Z}^*(t), t]}{\partial \mathbf{Z}^2}\dot{\mathbf{Z}}(t) \right] \tag{2.2.23}
$$

and

$$\frac{\partial V[\mathbf{Z}^*(t), t]}{\partial t} = - \min_{\mathbf{U}} \left\{ L[\mathbf{Z}^*(t), \mathbf{U}(t), t] + E\left[ \frac{\partial V[\mathbf{Z}^*(t), t]}{\partial \mathbf{Z}} \dot{\mathbf{Z}}(t) + \frac{1}{2} \dot{\mathbf{Z}}^{\mathrm{T}}(t) \frac{\partial^2 V[\mathbf{Z}^*(t), t]}{\partial \mathbf{Z}^2} \dot{\mathbf{Z}}(t) \right] \right\}$$

(2.2.24)

Define a Hamilton function

$$H[\mathbf{Z}^*(t), \mathbf{U}(t), t] = L[\mathbf{Z}^*(t), \mathbf{U}(t), t] + E\left[ \frac{\partial V[\mathbf{Z}^*(t), t]}{\partial \mathbf{Z}} \dot{\mathbf{Z}}(t) + \frac{1}{2} \dot{\mathbf{Z}}^{\mathrm{T}}(t) \frac{\partial^2 V[\mathbf{Z}^*(t), t]}{\partial \mathbf{Z}^2} \dot{\mathbf{Z}}(t) \right]$$

(2.2.25)

then

$$\frac{\partial V[\mathbf{Z}^*(t), t]}{\partial t} = - \min_{\mathbf{U}} \left\{ H[\mathbf{Z}^*(t), \mathbf{U}(t), t] \right\} \tag{2.2.26}$$

Equation (2.2.26) is the so-called Hamilton–Jacobi–Bellman equation (HJB equation) in the context of randomness.

Solving the control problem by virtue of HJB equation often involves two steps: first minimizing the right term of Eq. (2.2.26) to obtain the optimal control law; then in conjunction with the state equation, i.e., Eq. (2.2.1), attaining the control gain and the associated responses of the controlled system.

For illustrative purposes, a linear controlled stochastic dynamical system in formulation of the Itô-type stochastic calculus is addressed herein:

$$\dot{\mathbf{Z}}(t) = \mathbf{A}\mathbf{Z}(t) + \mathbf{B}\mathbf{U}(t) + \mathbf{L}w(t) \tag{2.2.27}$$

where $\mathbf{A}, \mathbf{B}, \mathbf{L}$ are the $n \times n$ system matrix, the $n \times r$ matrix denoting the location of control force, and the $n \times s$ matrix denoting the location of white Gaussian noise excitation, respectively.

In the cost function, the terminal function and the Lagrange multiplier are both defined as a quadratic form

$$\phi[\mathbf{Z}(t_f), t_f] = \frac{1}{2}\mathbf{Z}^{\mathrm{T}}(t_f)\mathbf{S}(t_f)\mathbf{Z}(t_f) \tag{2.2.28}$$

$$L[\mathbf{Z}(t), \mathbf{U}(t), t] = \frac{1}{2}[\mathbf{Z}^{\mathrm{T}}(t)\mathbf{Q}\mathbf{Z}(t) + \mathbf{U}^{\mathrm{T}}(t)\mathbf{R}\mathbf{U}(t)] \tag{2.2.29}$$

where $\mathbf{S}(t_f), \mathbf{Q}$ are the $2n \times 2n$ semi-positive and symmetric state weighting matrices; $\mathbf{R}$ is the $r \times r$ positive and symmetric control force weighting matrix.

Substituting Eqs. (2.2.28) and (2.2.29) into Eq. (2.2.25), and utilizing the feature of Itô stochastic differential equation, one can derive the Hamilton function as follows:

$$H[\mathbf{Z}^*(t), \mathbf{U}(t), t] = \frac{1}{2}\left( \mathbf{Z}^{*\mathrm{T}}\mathbf{Q}\mathbf{Z}^* + \mathbf{U}^{\mathrm{T}}\mathbf{R}\mathbf{U} \right)$$

$$+ E\left[\frac{\partial V}{\partial \mathbf{Z}}(\mathbf{AZ}^* + \mathbf{BU} + \mathbf{Lw}) + \frac{1}{2}(\mathbf{AZ}^* + \mathbf{BU} + \mathbf{Lw})^{\mathrm{T}}\frac{\partial^2 V}{\partial \mathbf{Z}^2}(\mathbf{AZ}^* + \mathbf{BU} + \mathbf{Lw})\right]$$

$$= \frac{1}{2}\left(\mathbf{Z}^{*\mathrm{T}}\mathbf{QZ}^* + \mathbf{U}^{\mathrm{T}}\mathbf{RU}\right) + \frac{\partial V}{\partial \mathbf{Z}}(\mathbf{AZ}^* + \mathbf{BU}) + \frac{1}{2}\mathrm{Tr}\left(\frac{\partial^2 V}{\partial \mathbf{Z}^2}\mathbf{LWL}^{\mathrm{T}}\right) \qquad (2.2.30)$$

where $\mathrm{Tr}(\cdot)$ denotes the trace of matrix, i.e., $\mathbf{X}^{\mathrm{T}}\mathbf{AX} = \mathrm{Tr}(\mathbf{AXX}^{\mathrm{T}})$.

The optimal value function is assumed as follows (Li and Chen 2009):

$$V[\mathbf{Z}(t), t] = \frac{1}{2}\mathbf{Z}^{\mathrm{T}}(t)\mathbf{S}(t)\mathbf{Z}(t) + v(t) \qquad (2.2.31)$$

where $v(t)$ is the correction term of stochastic optimal control against the deterministic optimal control.

It is readily seen from Eq. (2.2.31) that the terminal condition of the optimal value function is

$$V[\mathbf{Z}(t_f), t_f] = \frac{1}{2}\mathbf{Z}^{\mathrm{T}}(t_f)\mathbf{S}(t_f)\mathbf{Z}(t_f) \qquad (2.2.32)$$

and

$$\frac{\partial V}{\partial \mathbf{Z}} = \mathbf{Z}^{\mathrm{T}}(t)\mathbf{S}(t), \quad \frac{\partial^2 V}{\partial \mathbf{Z}^2} = \mathbf{S}(t) \qquad (2.2.33)$$

Equation (2.2.26) is then rewritten as

$$\frac{\partial V}{\partial t} = -\min_{\mathbf{U}}\frac{1}{2}\{[\mathbf{Z}^{*\mathrm{T}}\mathbf{QZ}^* + \mathbf{U}^{\mathrm{T}}\mathbf{RU}] + 2\mathbf{Z}^{\mathrm{T}}\mathbf{S}[\mathbf{AZ}^* + \mathbf{BU}] + \mathrm{Tr}(\mathbf{SLWL}^{\mathrm{T}})\} \qquad (2.2.34)$$

Minimization of the right term of Eq. (2.2.34) needs to satisfy the condition $\partial H/\partial \mathbf{U} = \mathbf{0}$, then

$$\mathbf{U}(t) = -\mathbf{R}^{-1}\mathbf{B}^{\mathrm{T}}\mathbf{S}(t)\mathbf{Z}(t) \qquad (2.2.35)$$

Substituting Eqs. (2.2.35) and (2.2.31) into Eq. (2.2.34) yields

$$\frac{\partial V}{\partial t} = \frac{1}{2}\mathbf{Z}^{\mathrm{T}}(t)\dot{\mathbf{S}}(t)\mathbf{Z}(t) + \dot{v}(t)$$

$$= -\frac{1}{2}\{[\mathbf{Z}^{\mathrm{T}}\mathbf{QZ} + \mathbf{Z}^{\mathrm{T}}\mathbf{SBR}^{-1}\mathbf{B}^{\mathrm{T}}\mathbf{SZ}] + 2\mathbf{Z}^{\mathrm{T}}\mathbf{S}[(\mathbf{A} - \mathbf{BR}^{-1}\mathbf{B}^{\mathrm{T}}\mathbf{S})\mathbf{Z}] + \mathrm{Tr}(\mathbf{SLWL}^{\mathrm{T}})\}$$

$$= -\frac{1}{2}\mathbf{Z}^{\mathrm{T}}[\mathbf{Q} + 2\mathbf{SA} - \mathbf{SBR}^{-1}\mathbf{B}^{\mathrm{T}}\mathbf{S}]\mathbf{Z} - \frac{1}{2}\mathrm{Tr}(\mathbf{SLWL}^{\mathrm{T}}) \qquad (2.2.36)$$

In comparison with the coefficient terms in Eq. (2.2.36), there are

$$\dot{\mathbf{S}}(t) = -\mathbf{Q} - 2\mathbf{SA} + \mathbf{SBR}^{-1}\mathbf{B}^{\mathrm{T}}\mathbf{S} \qquad (2.2.37)$$

$$\dot{v}(t) = -\frac{1}{2}\mathrm{Tr}\big(\mathbf{SLWL}^{\mathrm{T}}\big) \tag{2.2.38}$$

It is recognized from Eq. (2.2.37) that owing to the symmetry of $\mathbf{S}(t)$, the matrix product $\mathbf{SA}$ is also a symmetric matrix, and there is $\mathbf{SA} = \mathbf{A}^{\mathrm{T}}\mathbf{S}^{\mathrm{T}} = \mathbf{A}^{\mathrm{T}}\mathbf{S}$, then

$$\dot{\mathbf{S}}(t) = -\mathbf{S}(t)\mathbf{A} - \mathbf{A}^{\mathrm{T}}\mathbf{S}(t) + \mathbf{S}(t)\mathbf{BR}^{-1}\mathbf{B}^{\mathrm{T}}\mathbf{S}(t) - \mathbf{Q} \tag{2.2.39}$$

It is just the differential Riccati equation in the classical optimal control theory.

Therefore, the optimal value function has the formulation as follows:

$$V[\mathbf{Z}(t), t] = \frac{1}{2}\mathbf{Z}^{\mathrm{T}}(t)\mathbf{S}(t)\mathbf{Z}(t) + \frac{1}{2}\int\limits_{t}^{t_f} \mathrm{Tr}\big(\mathbf{S}(\tau)\mathbf{LWL}^{\mathrm{T}}\big)\mathrm{d}\tau \tag{2.2.40}$$

In fact, the Euler–Lagrange equation, i.e., Eqs. (2.2.13), (2.2.14) and (2.2.15), deducted from Pontryagin's maximum principle also exhibits the solution of optimal control law.

Substituting the Hamilton function Eq. (2.2.9) into Eq. (2.2.15), and utilizing the nature of the Itô stochastic differential equation, we have

$$\frac{\partial H}{\partial \mathbf{U}} = \mathbf{U}^{\mathrm{T}}(t)\mathbf{R} + \boldsymbol{\lambda}^{\mathrm{T}}(t)\mathbf{B} = \mathbf{0} \tag{2.2.41}$$

then the control law has a formulation as follows:

$$\mathbf{U}(t) = -\mathbf{R}^{-1}\mathbf{B}^{\mathrm{T}}\boldsymbol{\lambda}(t) \tag{2.2.42}$$

In conjunction with the costate equation, i.e., Eq. (2.2.14), there is

$$\dot{\boldsymbol{\lambda}}(t) = -\left(\frac{\partial H}{\partial \mathbf{Z}}\right)^{\mathrm{T}} = -\mathbf{QZ}(t) - \mathbf{A}^{\mathrm{T}}\boldsymbol{\lambda}(t) \tag{2.2.43}$$

The costate and state vectors are assumed to have the relation as follows:

$$\boldsymbol{\lambda}(t) = \mathbf{P}(t)\mathbf{Z}(t) \tag{2.2.44}$$

then the control law is rewritten as

$$\mathbf{U}(t) = -\mathbf{R}^{-1}\mathbf{B}^{\mathrm{T}}\mathbf{P}(t)\mathbf{Z}(t) \tag{2.2.45}$$

It is seen that the optimal control law derived from the Pontryagin's maximum principle has the same formulation as the optimal control law derived from the

Bellman's optimality principle. Further, substituting Eq. (2.2.44) into Eq. (2.2.43), one has

$$\dot{\boldsymbol{\lambda}}(t) = \dot{\mathbf{P}}(t)\mathbf{Z}(t) + \mathbf{P}(t)\dot{\mathbf{Z}}(t) = -\left[\mathbf{Q} - \mathbf{A}^{\mathrm{T}}\mathbf{P}(t)\right]\mathbf{Z}(t) \qquad (2.2.46)$$

In view of the state-control equation; see Eq. (2.2.27), we have

$$\dot{\mathbf{P}}(t) = -\mathbf{P}(t)\mathbf{A} - \mathbf{A}^{\mathrm{T}}\mathbf{P}(t) + \mathbf{P}(t)\mathbf{B}\mathbf{R}^{-1}\mathbf{B}^{\mathrm{T}}\mathbf{P}(t) - \mathbf{Q} \qquad (2.2.47)$$

where $\mathbf{P}(t)$ is the Riccati matrix. It is noted that Eq. (2.2.47) is so-called the matrix differential Riccati equation, which is consistent with Eq. (2.2.39).

The abovementioned stochastic optimal control in formulation of the Itô stochastic differential equation for white Gaussian noise excited systems just refers to as the classical linear quadratic Gaussian (LQG) control.

## 2.3  Random Vibration of Structures

### 2.3.1  Linear Random Vibration

#### 2.3.1.1  Spectral Transfer Matrix Method

A linear stochastic dynamical system is considered as follows:

$$\mathbf{M}\ddot{\mathbf{X}}(t) + \mathbf{C}\dot{\mathbf{X}}(t) + \mathbf{K}\mathbf{X}(t) = \mathbf{F}(\boldsymbol{\Theta}, t) \qquad (2.3.1)$$

where $\mathbf{M}$, $\mathbf{C}$, and $\mathbf{K}$ are the $n \times n$ mass, damping, and stiffness matrices, respectively; $\ddot{\mathbf{X}}(t)$, $\dot{\mathbf{X}}(t)$, $\mathbf{X}(t)$ are the $n$-dimensional column vectors denoting system acceleration, velocity, and displacement, respectively; $\mathbf{F}(\boldsymbol{\Theta}, t)$ is the $n$-dimensional column vector denoting random excitations, and $\boldsymbol{\Theta}$ is an $n_{\boldsymbol{\Theta}}$-dimensional vector denoting random parameters of system which exhibits the joint probability density function $p_{\boldsymbol{\Theta}}(\boldsymbol{\theta})$.

Defining the $n \times n$ unit impulse response function matrices $\mathbf{h}(t)$, where the component $h_{ij}(t)$ denotes the response of the $i$th degree in the case that the unit impulse acts on the $j$th degree of the system, one can attain the system response $\mathbf{h}_j(t)$ from the equation of motion as follows:

$$\mathbf{M}\ddot{\mathbf{h}}_j(t) + \mathbf{C}\dot{\mathbf{h}}_j(t) + \mathbf{K}\mathbf{h}_j(t) = \mathbf{I}_j\delta(\boldsymbol{\Theta}, t) \qquad (2.3.2)$$

where $\mathbf{I}_j = (0, 0, \ldots, 0, \underset{j}{1}, 0, \ldots, 0)^{\mathrm{T}}$ is $n$-dimensional column vectors denoting the location of the unit impulse $\delta(\boldsymbol{\Theta}, t)$ acting on the $j$th degree of the system.

According to the Duhamel integral, the solution of system displacement can be derived as follows (Lutes and Sarkani 2004):

$$\mathbf{X}(t) = \int_0^t \mathbf{h}(t - \tau)\mathbf{F}(\mathbf{\Theta}, \tau)d\tau \qquad (2.3.3)$$

Therefore, the mean and the correlation function of the system displacement are given by, respectively,

$$\mathbf{\mu_X}(t) = E[\mathbf{X}(t)] = \int_0^t h(t - \tau)\mathbf{\mu_F}(\tau)d\tau \qquad (2.3.4)$$

$$\mathbf{R_X}(t_1, t_2) = E[\mathbf{X}(t_1)\mathbf{X}^{\mathrm{T}}(t_2)]$$

$$= E\left\{ \left[ \int_0^{t_1} \mathbf{h}(t_1 - \tau_1)\mathbf{F}(\mathbf{\Theta}, \tau_1)d\tau_1 \right] \left[ \int_0^{t_2} \mathbf{h}(t_2 - \tau_2)\mathbf{F}(\mathbf{\Theta}, \tau_2)d\tau_2 \right]^{\mathrm{T}} \right\}$$

$$= \int_0^{t_1} \int_0^{t_2} \mathbf{h}(t_1 - \tau_1)\mathbf{R_F}(\tau_1, \tau_2)\mathbf{h}^{\mathrm{T}}(t_2 - \tau_2)d\tau_1 d\tau_2 \qquad (2.3.5)$$

where $\mathbf{\mu_X}(t)$ and $\mathbf{R_X}(t_1, t_2)$ are the mean and the correlation function of the random excitation, respectively.

If the random excitation can be viewed as a stationary process, the steady solution of the system displacement is a stationary process as well. The associated correlation function is rewritten as

$$\mathbf{R_X}(\tau) = \int_{-\infty}^{\infty} \int_{-\infty}^{\infty} \mathbf{h}(\tau_1)\mathbf{R_F}(\tau - \tau_1 - \tau_2)\mathbf{h}^{\mathrm{T}}(\tau_2)d\tau_1 d\tau_2 \qquad (2.3.6)$$

The steady solution shown in Eq. (2.3.6) can be derived from the perspective of frequency-domain analysis. In fact, the frequency response transfer function of the stochastic dynamical system represented by Eq. (2.3.1) is denoted by

$$\mathbf{H}(\omega) = (\mathbf{K} - \omega^2\mathbf{M} + i\omega\mathbf{C})^{-1} \qquad (2.3.7)$$

where i is the imaginary unit, which is defined by its property $i = \sqrt{-1}$.

The Fourier transfer of system displacement $\mathbf{X}(\omega)$ can thus be defined by virtue of the transitive relation in frequency-domain analysis:

$$\mathbf{X}(\omega) = \mathbf{H}(\omega)\mathbf{F}(\mathbf{\Theta}, \omega) \qquad (2.3.8)$$

where $\mathbf{F}(\mathbf{\Theta}, \omega)$ is the Fourier transfer of the random excitation.

Assigning the complex conjugate on both sides of Eq. (2.3.8), there is

$$\mathbf{X}^*(\omega) = \mathbf{F}^*(\mathbf{\Theta}, \omega)\mathbf{H}^*(\omega) \tag{2.3.9}$$

By multiplying Eqs. (2.3.8) and (2.3.9), setting mathematical expectation and taking limitation with respect to the duration, there is

$$\lim_{T \to \infty} \frac{1}{2T} E[\mathbf{X}(\omega)\mathbf{X}^*(\omega)] = \lim_{T \to \infty} \frac{1}{2T} E[\mathbf{H}(\omega)\mathbf{F}(\mathbf{\Theta}, \omega)\mathbf{F}^*(\mathbf{\Theta}, \omega)\mathbf{H}^*(\omega)] \tag{2.3.10}$$

This formulation is just the definition of the power spectral density of system displacement (Li and Chen 2009), i.e.,

$$\mathbf{S}_{\mathbf{X}}(\omega) = \lim_{T \to \infty} \frac{1}{2T} E[\mathbf{X}(\omega)\mathbf{X}^*(\omega)] \tag{2.3.11}$$

Further, we have

$$\mathbf{S}_{\mathbf{X}}(\omega) = \mathbf{H}(\omega)\mathbf{S}_{\mathbf{F}}(\omega)\mathbf{H}^*(\omega) \tag{2.3.12}$$

where $\mathbf{S}_{\mathbf{F}}(\omega)$ is the power spectral density of random excitation.

In view of the Wiener–Khintchine theorem (Wiener 1964; Chatfield 1989), the autocorrelation function of a wide-sense-stationary random process has a spectral decomposition given by the power spectrum of that process, that is, the mean-square solution of system displacement can be denoted by

$$E[\mathbf{X}(t)\mathbf{X}^{\mathrm{T}}(t)] = \mathbf{R}_{\mathbf{X}}(\tau) = \frac{1}{2\pi} \int\limits_{-\infty}^{\infty} \mathbf{S}_{\mathbf{X}}(\omega) \, d\omega \tag{2.3.13}$$

It is noted that the moment transitive relation of stochastic system is clearly revealed by Eqs. (2.3.6) and (2.3.12).

### 2.3.1.2 Modal Superposition Method

In practical applications, the complexity of attaining the analytical solution of the unit impulse response function $\mathbf{h}(t)$ is far more than that of the frequency response transfer function $\mathbf{H}(\omega)$ for a multiple-degree-of-freedom system. Moreover, the solution procedure of mean-square responses often involves high-dimensional integrals on the unit impulse response function and on the frequency response transfer function. The computational cost is unacceptable in most cases. In fact, for the linear stochastic dynamical system, a workload-reduced way refers to the so-called modal superposition method. The basic idea is that the original multiple-degree-of-freedom stochastic system is decoupled into a series of single-degree-of-freedom stochastic systems, so as to significantly reduce the computational cost.

According to the principle of the modal superposition method, the equation of motion of the stochastic dynamical system shown in Eq. (2.3.1) can be rewritten as

$$\bar{\mathbf{M}}\ddot{\mathbf{U}}(t) + \bar{\mathbf{C}}\dot{\mathbf{U}}(t) + \bar{\mathbf{K}}\mathbf{U}(t) = \mathbf{\Phi}^{\mathrm{T}}\mathbf{F}(\mathbf{\Theta}, t) \tag{2.3.14}$$

where $\bar{\mathbf{M}} = \mathbf{\Phi}^{\mathrm{T}}\mathbf{M}\mathbf{\Phi}$, $\bar{\mathbf{C}} = \mathbf{\Phi}^{\mathrm{T}}\mathbf{C}\mathbf{\Phi}$, $\bar{\mathbf{K}} = \mathbf{\Phi}^{\mathrm{T}}\mathbf{K}\mathbf{\Phi}$ are the $n \times n$ modal mass, modal damping, and modal stiffness matrices, respectively; $\mathbf{U} = \mathbf{\Phi}^{\mathrm{T}}\mathbf{X}$ is the $n$-dimensional column vector denoting modal displacement; $\mathbf{\Phi} = [\phi_1, \phi_2, \ldots, \phi_q] = [\phi_{ij}]_{n \times q}$ ($q \leq n$) is the modal matrix.

Assuming that the damping matrix $\mathbf{C}$ is a proportional damping matrix, Eq. (2.3.14) can be then decomposed into $q$ mutually independent single-degree-of-freedom systems, of which the equation of motion of the $j$th-order mode is shown as follows:

$$\ddot{u}_j(t) + 2\zeta_j\omega_j\dot{u}_j(t) + \omega_j^2 u_j(t) = \frac{1}{\bar{m}_j}\phi_j^{\mathrm{T}}\mathbf{F}(\mathbf{\Theta}, t) = \frac{1}{\bar{m}_j}\sum_{k=1}^{n}\phi_{jk}F_k(\mathbf{\Theta}, t), \; j = 1, 2, \ldots, q \tag{2.3.15}$$

where $\bar{m}_j$ is the $j$th-order modal mass; $\omega_j$ is the $j$th-order modal frequency; $\zeta_j$ is the $j$th-order modal damping ratio.

By means of the Duhamel integral, the componental formulation of the displacement of the linear system in the modal space can be derived as

$$u_j(t) = \frac{1}{\bar{m}_j} \int_0^t h_j(t - \tau)\phi_j^{\mathrm{T}}\mathbf{F}(\mathbf{\Theta}, \tau) \, \mathrm{d}\tau \tag{2.3.16}$$

where $u_j(t)$ is referred to as the $j$th-order modal displacement.

The displacement solution of the linear system in the original state space is then given by

$$\mathbf{X}(t) = \sum_{j=1}^{q} \frac{1}{\bar{m}_j} \int_0^t h_j(t - \tau)\phi_j\phi_j^{\mathrm{T}}\mathbf{F}(\mathbf{\Theta}, \tau) \, \mathrm{d}\tau \tag{2.3.17}$$

Further, the mean and correlation function of the displacement can be deduced as follows:

$$\mu_{\mathbf{X}}(t) = E[\mathbf{X}(t)] = \sum_{j=1}^{q} \frac{1}{\bar{m}_j} \int_0^t h_j(t - \tau)\phi_j\phi_j^{\mathrm{T}}\mu_{\mathbf{F}}(\tau) \, \mathrm{d}\tau \tag{2.3.18}$$

$\mathbf{R_X}(t_1, t_2) = E[\mathbf{X}(t_1)\mathbf{X}^T(t_2)]$

$$= E\left\{\left[\sum_{j=1}^{q}\frac{1}{\bar{m}_j}\int_0^{t_1}h_j(t_1 - \tau_1)\phi_j\phi_j^T\mathbf{F}(\Theta, \tau_1)d\tau_1\right]\left[\sum_{k=1}^{q}\frac{1}{\bar{m}_k}\int_0^{t_2}h_k(t_2 - \tau_2)\phi_k\phi_k^T\mathbf{F}(\Theta, \tau_2)d\tau_2\right]^T\right\}$$

$$= \sum_{j=1}^{q}\sum_{k=1}^{q}\frac{1}{\bar{m}_j\bar{m}_k}\int_0^{t_1}\int_0^{t_2}h_j(t_1 - \tau_1)h_k(t_2 - \tau_2)\phi_j\phi_j^T\mathbf{R_F}(\tau_1, \tau_2)\phi_k\phi_k^Td\tau_1 d\tau_2 \qquad (2.3.19)$$

When the steady response of the system can be viewed as a second-order stationary process, the correlation function can be simplified by

$$\mathbf{R_X}(\tau) = \sum_{j=1}^{q}\sum_{k=1}^{q}\frac{1}{\bar{m}_j\bar{m}_k}\int_{-\infty}^{\infty}\int_{-\infty}^{\infty}h_j(\tau_1)h_k(\tau_2)\phi_j\phi_j^T\mathbf{R_F}(\tau - \tau_1 - \tau_2)\phi_k\phi_k^Td\tau_1 d\tau_2$$

$$= \int_{-\infty}^{\infty}\int_{-\infty}^{\infty}\left(\sum_{j=1}^{q}\frac{1}{\bar{m}_j}h_j(\tau_1)\phi_j\phi_j^T\right)\mathbf{R_F}(\tau - \tau_1 - \tau_2)\left(\sum_{k=1}^{q}\frac{1}{\bar{m}_k}h_k(\tau_2)\phi_k\phi_k^T\right)d\tau_1 d\tau_2$$

$$\qquad (2.3.20)$$

In comparison with Eq. (2.3.6), it is readily recognized:

$$\mathbf{h}(t) = \sum_{j=1}^{q}\frac{1}{\bar{m}_j}h_j(t)\phi_j\phi_j^T \qquad (2.3.21)$$

Likewisely, the power spectral density of modal displacement of the linear system can be derived straightforwardly from the power spectral density of the random excitation according to the transfer relation in frequency-domain analysis. The element of the power spectral density of modal displacement is given by

$$S_{U_j U_k}(\omega, t) = \frac{1}{\bar{m}_j\bar{m}_k}\bar{H}_j^*(\omega, t)\bar{H}_k(\omega, t)\phi_j^T\mathbf{S_F}(\omega)\phi_k \qquad (2.3.22)$$

where $\mathbf{S_F}(\omega)$ denotes the power spectral density of the random excitation, which is assumed to be a second-order stationary process herein; $\bar{H}_j(\omega, t)$ denotes the frequency response transfer function of the $j$th-order modal system, which is time-dependent and usually has the formulation as follows:

$$\bar{H}_j(\omega, t) = H_j(\omega)\left\{1 - \left(\cos\omega_d^j t + \frac{\zeta_j\omega_j + i\omega}{\omega_d^j}\sin\omega_d^j t\right)e^{-(\zeta_j\omega_j + i\omega)t}\right\} = H_j(\omega)C_j(\omega, t)$$

$$\qquad (2.3.23)$$

where $H_j(\omega) = 1/(\omega_j^2 - \omega^2 + 2i\zeta_j\omega_j\omega)$; $\omega_d^j = \omega_j\sqrt{1 - \zeta_j^2}$; $C_j(\omega, t)$ denotes the modulation function.

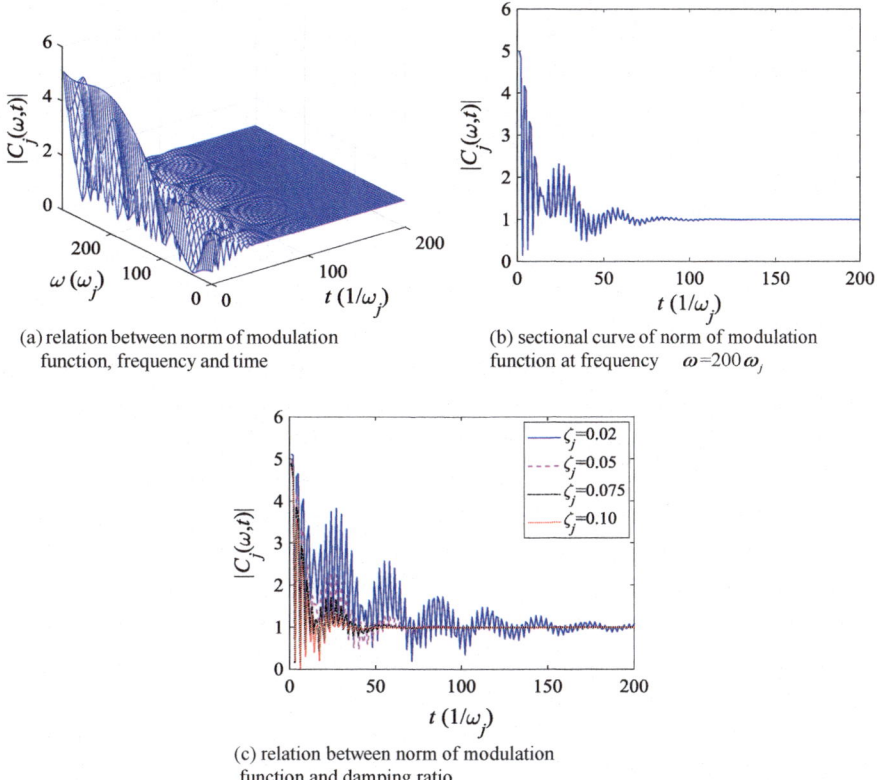

(a) relation between norm of modulation function, frequency and time

(b) sectional curve of norm of modulation function at frequency $\omega = 200\omega_j$

(c) relation between norm of modulation function and damping ratio

**Fig. 2.1** Norm of the modulation function

In order to illustrate the time-variant behaviors of the frequency response transfer function, the relation between the norm of the modulation function $C_j(\omega, t)$, the frequency $\omega$, and the time $t$ is addressed. A schematic on their relation is shown in Fig. 2.1a, where the modal frequency is set as $\omega_j = 2\pi$ rad/s, the modal damping ratio is set as $\zeta_j = 0.05$.

Figure 2.1b shows the sectional curve of the norm of the modulation function at the frequency $\omega = 200\omega_j$. It is indicated that in the first phase, the frequency response transfer function exposes a significant time-dependence, which is the so-called unsteady initial effect. Therefore, the system response remains as a nonstationary process even in the case that the system is subjected to a stationary excitation. As the time $t \to \infty$, there exists $|C_j(\omega, t)| \to 1$, $\bar{H}_j(\omega, t) \to H_j(\omega)$, and the system response approaches to a stationary process. Besides, the damping ratio has an influence on the modulation function as well. It is seen from Fig. 2.1c that along with the increase of the damping ratio, the fluctuation of the modulation function declines and approaches 1 in a shorter time interval. For this reason, whether taking into account the unsteady initial effect of system responses or not depends upon the system properties and the system quantities of interest.

In the original state space, the power spectral density of the displacement can be expressed as a formulation of complete quadratic combination (CQC) as follows (Der Kiureghian 1981):

$$\mathbf{S_X}(\omega, t) = \mathbf{\Phi} \mathbf{S_U}(\omega, t) \mathbf{\Phi}^\mathrm{T} \tag{2.3.24}$$

of which the component is

$$
\begin{aligned}
S_{X_j X_k}(\omega, t) &= \sum_{s=1}^{q} \sum_{r=1}^{q} \phi_{js} S_{U_s U_r}(\omega, t) \phi_{rk} \\
&= \sum_{s=1}^{q} \sum_{r=1}^{q} \phi_{js} \frac{1}{\bar{m}_s \bar{m}_r} \bar{H}_s^*(\omega, t) \bar{H}_r(\omega, t) \phi_s^\mathrm{T} \mathbf{S_F}(\omega) \phi_r \phi_{rk}
\end{aligned} \tag{2.3.25}
$$

As to a second-order stationary process, there is a time-independent solution of the power spectral density when $t \rightarrow \infty$; say

$$S_{X_j X_k}(\omega) = \sum_{s=1}^{q} \sum_{r=1}^{q} \phi_{js} \frac{1}{\bar{m}_s \bar{m}_r} H_s^*(\omega) H_r(\omega) \phi_s^\mathrm{T} \mathbf{S_F}(\omega) \phi_r \phi_{rk} \tag{2.3.26}$$

Further, the mean-square solution of the displacement can be derived according to the Wiener–Khintchine theorem:

$$E[\mathbf{X}(t)\mathbf{X}^\mathrm{T}(t)] = \frac{1}{2\pi} \int_{-\infty}^{\infty} \mathbf{S_X}(\omega) \, d\omega \tag{2.3.27}$$

It is indicated that the mean-square solution of stationary response of linear systems is time-independent under the hypothesis of response duration $t \rightarrow \infty$.

When the damping matrix $\mathbf{C}$ is a non-proportional damping matrix, the equation of motion of the system transfers to a state equation, and the eigenvector still serves as the basis for the state space, while the eigenvalue and eigenvector are often complex in this case. A complex modal analysis is thus required to derive a stationary solution of the system response (Fang et al. 1991; Zhou et al. 2004).

### 2.3.1.3  Pseudo-Excitation Method

When the linear system exhibits a high dimension, solving the power spectral density (PSD) of the system response shown in Eq. (2.3.27) involves a complicated procedure. The pseudo-excitation method (PEM) could be employed to obtain the PSD solution in an elegant manner (Lin et al. 2001). This method decomposes the solving procedure into a series of deterministic harmonic analysis through constructing a pseudo-harmonic excitation. This treatment can enhance the efficiency of numerical schemes significantly.

Denoting the PSD of the random excitation $\mathbf{F}(\mathbf{\Theta}, t)$ as $\mathbf{S_F}(\omega)$, a pseudo-harmonic excitation $\mathbf{F} = \widetilde{\mathbf{F}}_{\sqrt{\mathbf{S}}} e^{\mathrm{i}\omega t}$ can be readily constructed, where $\widetilde{\mathbf{F}}_{\sqrt{\mathbf{S}}}$ satisfies $\widetilde{\mathbf{F}}_{\sqrt{\mathbf{S}}} \cdot \widetilde{\mathbf{F}}^*_{\sqrt{\mathbf{S}}} = \mathbf{S_F}(\omega)$, i denotes the imaginary unit. Replacing the excitation in Eq. (2.3.14) by the pseudo-excitation yields

$$\bar{\mathbf{M}}\ddot{\widetilde{\mathbf{U}}}(t) + \bar{\mathbf{C}}\dot{\widetilde{\mathbf{U}}}(t) + \bar{\mathbf{K}}\widetilde{\mathbf{U}}(t) = \mathbf{F}^{\mathrm{T}}\widetilde{\mathbf{F}}_{\sqrt{\mathbf{S}}} e^{\mathrm{i}\omega t} \qquad (2.3.28)$$

where $\ddot{\widetilde{\mathbf{U}}}(t), \dot{\widetilde{\mathbf{U}}}(t), \widetilde{\mathbf{U}}(t)$ are the $n$-dimensional column vectors denoting the corresponding acceleration, velocity, and displacement to the system subjected to the pseudo-excitation, respectively.

According to the classical random vibration theory, the stationary solution of Eq. (2.3.28) can be deduced as

$$\widetilde{U}_j(\omega, t) = \frac{1}{\bar{m}_j} H_j(\omega)\phi_j^{\mathrm{T}}\widetilde{\mathbf{F}}_{\sqrt{\mathbf{S}}} e^{\mathrm{i}\omega t} \qquad (2.3.29)$$

The auto-power spectral density of system response is then derived as follows:

$$S_{\widetilde{U}_j \widetilde{U}_k}(\omega) = \widetilde{U}_j(\omega, t)\widetilde{U}_k^*(\omega, t) = \frac{1}{\bar{m}_j \bar{m}_k} H_j(\omega)H_k(\omega)\phi_j^{\mathrm{T}}\widetilde{\mathbf{F}}_{\sqrt{\mathbf{S}}}\widetilde{\mathbf{F}}^*_{\sqrt{\mathbf{S}}}\phi_k$$
$$= \frac{1}{\bar{m}_j \bar{m}_k} H_j(\omega)H_k(\omega)\phi_j^{\mathrm{T}}\mathbf{S_F}(\omega)\phi_k = S_{U_j U_k}(\omega) \qquad (2.3.30)$$

It is shown that in the calculation of spectral density function, the factor of pseudo-harmonic excitation $e^{\mathrm{i}\omega t}$ is always paired with its complex conjugate $e^{-\mathrm{i}\omega t}$ which are eventually counteracted by multiplication, revealing the time-independent behaviors of auto- and cross-power spectral densities of stationary processes.

Further, one can attain the mean-square solution of system responses:

$$E[\mathbf{U}(t)\mathbf{U}^{\mathrm{T}}(t)] = \frac{1}{2\pi} \int_{-\infty}^{\infty} \mathbf{S_U}(\omega)\,\mathrm{d}\omega \qquad (2.3.31)$$

Projecting the generalized coordinate space onto the original coordinate space, then

$$E[\mathbf{X}(t)\mathbf{X}^{\mathrm{T}}(t)] = \mathbf{\Phi} E[\mathbf{U}(t)\mathbf{U}^{\mathrm{T}}(t)]\mathbf{\Phi}^{\mathrm{T}} = \frac{1}{2\pi} \int_{-\infty}^{\infty} \mathbf{\Phi}\mathbf{S_U}(\omega)\mathbf{\Phi}^{\mathrm{T}}\,\mathrm{d}\omega \qquad (2.3.32)$$

In the procedure of time-domain analysis using the pseudo-excitation method, the frequency-domain partition of the power spectral density of random excitation $S_F(\omega)$ is a critical step. The number of partition points, however, often arises up to hundreds and even thousands, which results in a same-scale workload to the Fourier transfer. Therefore, it cannot be accepted if all the partition points are included in the calculation.

Although the uniform partition strategy can reduce the frequency point number to hundreds or tens through increasing the frequency distance (spectral interval), the equivalence between the original power spectral density and the sample-based power spectral density would seriously lose. In order to attain a good trade-off between accuracy and efficiency of the pseudo-excitation method, a weighted frequency-domain partition method is used, i.e., defining a spectral window function as follows:

$$W(\omega_n) = \begin{cases} 1, \ \omega_n \in [\omega_1, \omega_k] \\ 1, \ \omega_n \in (\omega_k, \omega_c), n = (k+1) + \alpha N < c, N = 0, 1, 2, \ldots \\ 0, \ \omega_n \in (\omega_k, \omega_c), n \neq (k+1) + \alpha N < c, N = 0, 1, 2, \ldots \\ 1, \ \omega_n = \omega_c \end{cases} \quad (2.3.33)$$

It is indicated in Eq. (2.3.33) that in the frequency interval with frequency components less than or equal to the critical frequency $\omega_k$, the spectral densities exhibit larger values and are all included in the calculation, while in the frequency interval with frequency components larger than the critical frequency $\omega_k$ and less than the truncated frequency $\omega_c$, the spectral densities are utilized through spacing $\alpha$ frequency components.

Figure 2.2 shows the power spectral density and the pertinent spectral window function of a random excitation associated with specified parameters: the spacing number $\alpha = 10$, the truncated frequency point $c = 1201$. If the error between the original and the reconstructed power spectral densities is set to 1%, the point of the critical frequency is defined at $k = 96$. The introduction of this frequency point-reduced technique leads to that just 208 frequency points from the total 1201 points are required. The root-mean-square errors of the original and reconstructed power spectral densities and of the original and reconstructed random excitations are 0.2‰ and 0.3‰, respectively. It is thus revealed that merely using one-sixth of the frequency points can attain a satisfactory result by virtue of the weight-based point selection strategy.

Besides, from the perspective of the harmonic function in terms of random frequency and random phase, it can be proved that the simulated process exhibits an accurate power spectral density as the objective when the random frequency and random phase both follow the uniform distribution, and the random amplitude is positively proportional to the square root of power spectral density. A physical interpretation of the pseudo-excitation method is thus further noted that just traversing the frequency interval, the power spectral density of system response can be ready to be attained by virtue of the amplitudes of system responses under the random mono-harmonic excitation (Chen et al. 2011). It is indicated that only half of the

**Fig. 2.2** Power spectral
density and pertinent spectral
window function of random
excitation

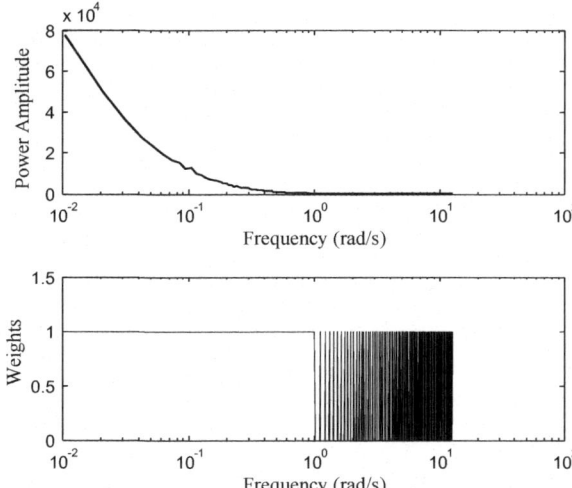

computational cost on the pseudo-excitation method is really required to gain the
power spectral density of structural responses.

## 2.3.2   Nonlinear Random Vibration

Without loss of generality, a nonlinear stochastic dynamical system is investigated,
of which the equation of motion is given by

$$\mathbf{M}\ddot{\mathbf{X}}(t) + \mathbf{f}(\mathbf{X}(t), \dot{\mathbf{X}}(t)) = \mathbf{F}(\mathbf{\Theta}, t) \tag{2.3.34}$$

where $\mathbf{f}(\cdot)$ is the $n$-dimensional column vector denoting nonlinear internal force.

The nonlinear internal force is assumed to be denoted by a polynomial function
of velocity and displacement. In fact, this is a weak hypothesis, and a large family
of dynamical systems can be represented by the formulation, such as the Duffing
oscillator with nonlinear stiffness force and the van der Pol oscillator with coupling
nonlinearities between stiffness and damping forces. The componental form of the
equation is then written as

$$\sum_{i=1}^{n} m_{ji}\ddot{x}_i(t) + \sum_{i=1}^{n}\sum_{k=0}^{q} \alpha_{ji,k}\dot{x}_i^{q-k}(t)x_i^{k}(t) = F_j(\mathbf{\Theta}, t) \tag{2.3.35}$$

where $j = 1, 2, \ldots, n$, $m_{ji}$ denotes the element of mass matrix; $\ddot{x}_i(t)$, $\dot{x}_i(t)$, $x_i(t)$
denote the acceleration, velocity, and displacement pertaining to the $i$th component,
respectively; $q$ denotes the highest order of the polynomial function of the internal

force; $\alpha_{ji,k}$ denotes the coefficient of the polynomial function. As the highest order $q$ is set as 1, Eq. (2.3.35) is reduced to a linear formulation:

$$\sum_{i=1}^{n} m_{ji}\ddot{x}_i(t) + \sum_{i=1}^{n} \alpha_{ji,0}\dot{x}_i(t) + \sum_{i=1}^{n} \alpha_{ji,1}x_i(t) = F_j(\mathbf{\Theta}, t) \tag{2.3.36}$$

where $\alpha_{ji,0}$, $\alpha_{ji,1}$ denote the coefficients relevant to damping force and the restoring force, respectively.

### 2.3.2.1 Polynomial Chaos Expansion

The solution of Eq. (2.3.35) can be represented by a truncated polynomial chaos expansion (PCE) (Ghanem and Spanos 1991), i.e.,

$$x_i(t) = \sum_{l=0}^{P} x_{il}(t)\Psi_l(\mathbf{\xi}) \tag{2.3.37}$$

where $\mathbf{\xi}$ is the $M$-dimensional row vector of Gaussian random variables; $P$ denotes the highest order of the polynomial chaos expansion; $\Psi_l(\mathbf{\xi})$ denotes the polynomial chaos with parameter of Gaussian random variables; $x_{il}(t)$ denotes the deterministic coefficient pertaining to the polynomial chaos which is often referred to as the random mode.

Substituting Eq. (2.3.37) into Eq. (2.3.35), then yields

$$\sum_{i=1}^{n}\sum_{l=0}^{P} m_{ji}\ddot{x}_{il}(t)\Psi_l(\mathbf{\xi}) + \sum_{i=1}^{n}\sum_{k=0}^{q} \alpha_{ji,k}\left(\sum_{l=0}^{P} \dot{x}_{il}(t)\Psi_l(\mathbf{\xi})\right)^{q-k}\left(\sum_{l=0}^{P} x_{il}(t)\Psi_l(\mathbf{\xi})\right)^{k}$$
$$= \sum_{l=0}^{P} F_{jl}(t)\Psi_l(\mathbf{\xi}) \tag{2.3.38}$$

Introducing a Galerkin projection technique (Ghanem and Spanos 1991), the polynomial chaos arises to pairwise orthogonal with respect to Gaussian measure, i.e.,

$$\langle \Psi_i \Psi_j \rangle = \langle \Psi_i^2 \rangle \delta_{ij} \tag{2.3.39}$$

where $\langle \cdot \rangle$ denotes the inner product; $\delta_{ij}$ denotes the Kronecker delta function with two variables, which is 1 if the variables are equal, and 0 otherwise:

$$\delta_{ij} = \begin{cases} 1, & i = j \\ 0, & i \neq j \end{cases} \tag{2.3.40}$$

Equation (2.3.38) is thus discretized into an equation set:

$$\sum_{i=1}^{n} m_{ji} \ddot{x}_{im}(t) + \sum_{i=1}^{n} \sum_{k=0}^{q} \sum_{l_1=0}^{P} \cdots \sum_{l_{q-k}=0}^{P} \sum_{l_{q-k+1}=0}^{P} \cdots \sum_{l_q=0}^{P} \frac{c_{l_1 \cdots l_{q-k} l_{q-k+1} \cdots l_q m}}{\langle \Psi_m^2 \rangle}$$

$$\alpha_{ji,k} \dot{x}_{il_1}(t) \cdots \dot{x}_{il_{q-k}}(t) x_{il_{q-k+1}}(t) \cdots x_{il_q}(t) = F_{jm}(t) \tag{2.3.41}$$

where $c_{l_1 \cdots l_{q-k} l_{q-k+1} \cdots l_q m} = \langle \Psi_{l_1} \cdots \Psi_{l_{q-k}} \Psi_{l_{q-k+1}} \cdots \Psi_{l_q} \Psi_m \rangle$, $m = 0, 1, 2, \ldots, P$. The coefficient $c_{l_1 \cdots l_{q-k} l_{q-k+1} \cdots l_q m}$ and $\langle \Psi_m^2 \rangle$ can be derived from a multiple-dimensional integral (Ghanem and Spanos 1991).

It is seen that the solution of Eq. (2.3.35) can be gained through solving the deterministic nonlinear equation set shown in Eq. (2.3.41) in conjunction with Eq. (2.3.37). Therefore, the standard schemes for solving the nonlinear equations can be readily employed. As to the $n$-dimensional stochastic dynamical system, the number of nonlinear equations is $n(P + 1)$.

Further, the mean-square response of the stochastic dynamical system is presented as follows:

$$E[x_i^2(t)] = \sum_{k=0}^{P} \sum_{l=0}^{P} x_{ik}(t) x_{il}(t) E[\Psi_k(\boldsymbol{\xi}) \Psi_l(\boldsymbol{\xi})] \tag{2.3.42}$$

It is indicated that the introduction of the PCE leads to that the solution of the pertinent complex stochastic system relies on the solving of a deterministic nonlinear equation set. The numerical procedure is definitive and simple. However, an eye-catching problem is that the number of expansion terms ($P + 1$) will increase dramatically along with the dimension of random vector $M$ and the expansion order $p$. Specifically, when a white noise-driven nonlinear system is investigated, the number of expansion terms is usually unacceptable (in this case, the dimension of random vector $M$ will be up to hundreds). In fact, the number of expansion terms, the dimension of random vector, and the expansion order have the functional relation as follows (Debusschere et al. 2004):

$$P + 1 = \frac{(p + M)!}{p! \, M!} \tag{2.3.43}$$

In order to reduce the computational cost, a family of adaptive polynomial chaos expansions has been proposed in recent years (Li and Ghanem 1998; Peng et al. 2010). The benefit of the adaptive polynomial chaos expansions lies in the significant term-number reduction of polynomial expansion through choosing the former random variables with larger contribution to system responses and considering the high-order modes of these random variables. These random variables are picked out according to their ranks in all the random parameters of the polynomial chaos expansion in terms of the maximum displacement norm or the maximum phase norm both in function of displacement and velocity. The adaptive polynomial chaos expansion on the solution of stochastic dynamical system; i.e., Eq. (2.3.37), can be written as

**Table 2.1** Term number of adaptive polynomial chaos expansion

| Order of expansion | $K = 2$ | $K = 4$ | $K = 12$ | $K = 24$ |
|---|---|---|---|---|
| $p = 1$ | 25 | 25 | 25 | 25 |
| $p = 3$ | 32 | 55 | 467 | 2,925 |
| $p = 5$ | 43 | 146 | 6,200 | 118,750 |
| $p = 7$ | 58 | 350 | 50,400 | 2,629,575 |

$$x(t) = \bar{x}(t) + \sum_{i=1}^{K} x_i(t)\Psi_i(\boldsymbol{\xi}) + \sum_{i=K+1}^{M} x_i(t)\Psi_i(\boldsymbol{\xi}) + \sum_{l=M+1}^{N} x_l(t)\Psi_l(\xi_i|_{i=1}^{K}), \quad N \leq P$$

$$(2.3.44)$$

where $\bar{x}(t)$ denotes the zero-order term, corresponding to the mean of the solution; $K$ denotes the number of the former random variables involving high-order random modes; $M$ denotes the total number of the random variables; $N$ denotes the term number of the adaptive polynomial chaos expansion.

It is seen from Eq. (2.3.44) that the right second and third parts of the expansion denote the first-order terms, corresponding to the variance of the solution; the fourth part of the expansion denotes the high-order terms.

Table 2.1 shows the term number of the adaptive polynomial chaos expansion changing with the number of random variable with high-order modes, and the order of expansion. It is seen that using the adaptive polynomial chaos expansion, the higher the expansion order, the more remarkable the term number is reduced, and the computational cost declines significantly. The term number of the seventh-order polynomial expansion reaches up to 2 million by the original polynomial chaos expansion.

It is revealed as well that the solving of the adaptive polynomial chaos expansion needs to initially assume the first $K$ random variables with larger contribution to the system response, and consider the high-order random modes of these variables into the equation set Eq. (2.3.41). The coefficients pertaining to the first-order polynomial chaos are then attained by solving the equation set. In the following step, the ranking of the $M$ random variables is carried out in terms of the maximum displacement norm or the maximum phase norm. The first $K$ random variables are picked out with high-order random modes. The coefficients pertaining to the second-order polynomial chaos at the second moment are gained. Likewise, until all the coefficients of the polynomial chaos in high-order expansion are solved.

### 2.3.2.2  Statistical Linearization Technique

An alternative method for random vibration analysis of nonlinear systems is the statistical linearization technique (Roberts and Spanos 1990). This method exhibits a hypothesis that the structural response is viewed as a stationary Gaussian process, thereby the equivalence between the linearized system and the original nonlinear system is attained by minimizing their differences in the sense of mean square. The random vibration analysis of nonlinear systems can then be carried out by the pertinent theory and methods to the random vibration of linear systems.

Therefore, the nonlinear multiple-degree-of-freedom system, shown in Eq. (2.3.34), can be substituted by a linearized system with equation of motion as follows:

$$\mathbf{M}\ddot{\mathbf{X}}(t) + \mathbf{C}_{eq}\dot{\mathbf{X}}(t) + \mathbf{K}_{eq}\mathbf{X}(t) = \mathbf{F}(\mathbf{\Theta}, t) \qquad (2.3.45)$$

where $\mathbf{C}_{eq}$, $\mathbf{K}_{eq}$ are the $n \times n$ equivalent damping and equivalent stiffness matrices, respectively.

Comparing Eqs. (2.3.34) and (2.3.45), and assuming that the linearized system and the original system have a same response, one can define the error vector between the internal forces of the two systems as follows:

$$\mathbf{e} = \mathbf{f}(\mathbf{X}(t), \dot{\mathbf{X}}(t)) - \mathbf{C}_{eq}\dot{\mathbf{X}}(t) - \mathbf{K}_{eq}\mathbf{X}(t) \qquad (2.3.46)$$

Minimization of the covariance matrix of the error vector, i.e.,

$$\frac{\partial E[\mathbf{e}\mathbf{e}^{T}]}{\partial \mathbf{C}_{eq}} = 0 \qquad (2.3.47a)$$

$$\frac{\partial E[\mathbf{e}\mathbf{e}^{T}]}{\partial \mathbf{K}_{eq}} = 0 \qquad (2.3.47b)$$

yields the basic equations:

$$\mathbf{C}_{eq}E[\dot{\mathbf{X}}\dot{\mathbf{X}}^{T}] + \mathbf{K}_{eq}E[\mathbf{X}\dot{\mathbf{X}}^{T}] = E[\mathbf{f}(\mathbf{X}, \dot{\mathbf{X}})\dot{\mathbf{X}}^{T}] \qquad (2.3.48a)$$

$$\mathbf{C}_{eq}E[\dot{\mathbf{X}}\mathbf{X}^{T}] + \mathbf{K}_{eq}E[\mathbf{X}\mathbf{X}^{T}] = E[\mathbf{f}(\mathbf{X}, \dot{\mathbf{X}})\mathbf{X}^{T}] \qquad (2.3.48b)$$

Given the joint probability density functions for solving the mathematical expectation of responses, shown in Eqs. (2.3.48a) and (2.3.48b), the equivalent damping and equivalent stiffness matrices can be readily attained. This treatment, however, often refers to an iteration procedure, as shown in Fig. 2.3, where the tolerant error can be set as the difference of response vectors or as the norm of the difference of mean-square response vectors between the sequential steps.

As to a single-degree-of-freedom system, the basic equations with respect to the equivalent damping and equivalent stiffness matrices are given as follows:

$$C_{eq}E[\dot{X}^{2}] + K_{eq}E[X\dot{X}] = E[f(X, \dot{X})\dot{X}] \qquad (2.3.49a)$$

**Fig. 2.3** Flowchart of statistical linearization technique

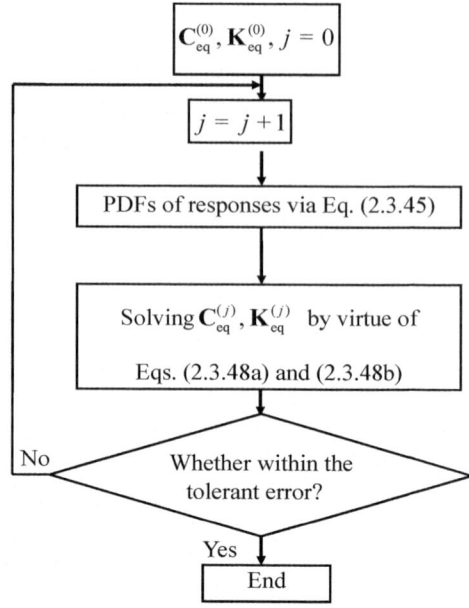

$$C_{eq}E[\dot{X}X] + K_{eq}E[X^2] = E[f(X, \dot{X})X] \qquad (2.3.49b)$$

thereby the solutions are derived as

$$c_{eq} = \frac{E[f(X, \dot{X})\dot{X}]E[X^2] - E[f(X, \dot{X})X]E[X\dot{X}]}{E[\dot{X}^2]E[X^2] - E^2[X\dot{X}]} \qquad (2.3.50a)$$

$$k_{eq} = \frac{E[f(X, \dot{X})X]E[\dot{X}^2] - E[f(X, \dot{X})\dot{X}]E[X\dot{X}]}{E[\dot{X}^2]E[X^2] - E^2[X\dot{X}]} \qquad (2.3.50b)$$

It is seen that the equivalent damping and equivalent stiffness matrices are both time-dependent if the system response is a nonstationary random process. In this case, the equivalent linearized system constructed by the statistical linearization technique is a time-dependent system, and the separation principle pertinent with system state estimation and control law design is not suitable for the classical linear quadratic Gaussian (LQG) control (Wonham 1968).

Moreover, the system velocity and displacement responses exhibit a certain orthogonality when the steady response of systems gives rise to stationary random processes, i.e., $E[X\dot{X}] = 0$. Equations (2.3.50a) and (2.3.50b) can thus be simplified to be

$$c_{eq} = \frac{E[f(X, \dot{X})\dot{X}]E[X^2]}{E[\dot{X}^2]E[X^2]} \qquad (2.3.51a)$$

$$k_{eq} = \frac{E[f(X, \dot{X})X]E[\dot{X}^2]}{E[\dot{X}^2]E[X^2]} \qquad (2.3.51b)$$

### 2.3.2.3  Fokker–Planck–Kolmogorov Equation

The mean-square solution of system response under random vibration just includes the former two-order moments information of stochastic dynamical systems, which is insufficient to represent the stochastic response as a complete probabilistic density function, especially for the nonlinear system, of which the probabilistic distribution is distinguished from the normal distribution. Therefore, seeking for the probability density of stochastic dynamical system has received extensive attention. Owing to the contributions from Fokker, Planck, and Kolmogorov, the probability density evolution equation related to random excitations were established in 1930s. This is the celebrated Fokker–Planck–Kolmogorov equation, i.e., FPK equation, in the classical random vibration theory.

Considering a random process $\mathbf{Z}(t)$, one has the Itô stochastic differential equation as follows:

$$d\mathbf{Z}(t) = \mathbf{A}(\mathbf{Z}, t)dt + \mathbf{B}(\mathbf{Z}, t)d\mathbf{w}(t) \qquad (2.3.52)$$

As for a random function $f(\mathbf{Z})$ in terms of random process $\mathbf{Z}(t)$, the Taylor series expansion is given by

$$
\begin{aligned}
df(\mathbf{Z}) &= \sum_{i=1}^{m} \frac{\partial f}{\partial z_i} dz_i + \frac{1}{2} \sum_{i=1}^{m} \sum_{j=1}^{m} \frac{\partial^2 f}{\partial z_i \partial z_j} dz_i dz_j + \cdots \\
&= \sum_{i=1}^{m} \frac{\partial f}{\partial z_i} \left[ A_i dt + \sum_{k=1}^{r} B_{ik} dw_k(t) \right] + \frac{1}{2} \sum_{i=1}^{m} \sum_{j=1}^{m} \left[ \frac{\partial^2 f}{\partial z_i \partial z_j} \sum_{k=1}^{r} B_{ik} dw_k(t) \sum_{s=1}^{r} B_{js} dw_s(t) \right] + \cdots
\end{aligned}
$$
$$(2.3.53)$$

Taking mathematical expectation on both sides of Eq. (2.3.53), and utilizing the product $E[(dw(t))^2] = Wdt$, the Taylor series expansion has a truncated formulation with respect to $dt$:

$$E[df(\mathbf{Z})] = E\left\{ \left[ \sum_{i=1}^{m} A_i \frac{\partial f}{\partial z_i} + \frac{1}{2} \sum_{i=1}^{m} \sum_{j=1}^{m} (\mathbf{BWB}^T)_{ij} \frac{\partial^2 f}{\partial z_i \partial z_j} \right] dt \right\} \qquad (2.3.54)$$

where $\mathbf{W}(t)$ is the $s \times s$ symmetric, and semi-positive spectral density matrix, shown in Eq. (2.2.4). It is noted as well $E[dw_k(t)] = 0$.

Noting the conditional probability density of $\mathbf{Z}(t)$ as $p_{\mathbf{Z}}(\mathbf{z}, t|\mathbf{z}_0, t_0)$, the derivative of left side of Eq. (2.3.54) is given by

$$\frac{dE[df(\mathbf{Z})]}{dt} = \frac{d}{dt}\int_{-\infty}^{\infty} f(\mathbf{z})p_{\mathbf{Z}}(\mathbf{z}, t|\mathbf{z}_0, t_0)d\mathbf{z} = \int_{-\infty}^{\infty} f(\mathbf{z})\frac{\partial p_{\mathbf{Z}}(\mathbf{z}, t|\mathbf{z}_0, t_0)}{\partial t}d\mathbf{z} \quad (2.3.55)$$

Meanwhile, the derivative of right side of Eq. (2.3.54) is given by

$$E\left\{\sum_{i=1}^{m}\frac{\partial f}{\partial z_i}A_i + \frac{1}{2}\sum_{i=1}^{m}\sum_{j=1}^{m}(\mathbf{BWB}^{\mathrm{T}})_{ij}\frac{\partial^2 f}{\partial z_i \partial z_j}\right\}$$

$$= \int_{-\infty}^{\infty}\left(\sum_{i=1}^{m}A_i(\mathbf{z}, t)\frac{\partial f(\mathbf{z})}{\partial z_i} + \frac{1}{2}\sum_{i=1}^{m}\sum_{j=1}^{m}(\mathbf{B}(\mathbf{z}, t)\mathbf{WB}^{\mathrm{T}}(\mathbf{z}, t))_{ij}\frac{\partial^2 f(\mathbf{z})}{\partial z_i \partial z_j}\right)p_{\mathbf{Z}}(\mathbf{z}, t|\mathbf{z}_0, t_0)d\mathbf{z}$$

$$(2.3.56)$$

Taking integration by parts, and noting that

$$A_i(\mathbf{z}, t)f(\mathbf{z})p_{\mathbf{Z}}(\mathbf{z}, t|\mathbf{z}_0, t_0)\big|_{z_i \to \pm\infty} = 0 \quad (2.3.57a)$$

$$\mathbf{B}(\mathbf{z}, t)\mathbf{WB}^{\mathrm{T}}(\mathbf{z}, t)\frac{\partial f(\mathbf{z})}{\partial z_i}p_{\mathbf{Z}}(\mathbf{z}, t|\mathbf{z}_0, t_0)\big|_{z_i \to \pm\infty} = \mathbf{0} \quad (2.3.57b)$$

$$f(\mathbf{z})\frac{\partial\{\mathbf{B}(\mathbf{z}, t)\mathbf{WB}^{\mathrm{T}}(\mathbf{z}, t)p_{\mathbf{Z}}(\mathbf{z}, t|\mathbf{z}_0, t_0)\}}{\partial z_i}\big|_{z_i \to \pm\infty} = \mathbf{0} \quad (2.3.57c)$$

Equation (2.3.56) is then transformed into

$$E\left\{\sum_{i=1}^{m}A_i\frac{\partial f}{\partial z_i} + \frac{1}{2}\sum_{i=1}^{m}\sum_{j=1}^{m}(\mathbf{BWB}^{\mathrm{T}})_{ij}\frac{\partial^2 f}{\partial z_i \partial z_j}\right\}$$

$$= \int_{-\infty}^{\infty}f(\mathbf{z})\left(-\sum_{i=1}^{m}\frac{\partial A_i(\mathbf{z}, t)p_{\mathbf{Z}}(\mathbf{z}, t|\mathbf{z}_0, t_0)}{\partial z_i} + \frac{1}{2}\sum_{i=1}^{m}\sum_{j=1}^{m}\frac{\partial^2\{(\mathbf{B}(\mathbf{z}, t)\mathbf{WB}^{\mathrm{T}}(\mathbf{z}, t))_{ij}\,p_{\mathbf{Z}}(\mathbf{z}, t|\mathbf{z}_0, t_0)\}}{\partial z_i \partial z_j}\right)d\mathbf{z}$$

$$(2.3.58)$$

Comparing Eq. (2.3.55) and Eq. (2.3.58), there is

$$\frac{\partial p_{\mathbf{Z}}(\mathbf{z}, t|\mathbf{z}_0, t_0)}{\partial t} = -\sum_{i=1}^{m}\frac{\partial A_i(\mathbf{z}, t)p_{\mathbf{Z}}(\mathbf{z}, t|\mathbf{z}_0, t_0)}{\partial z_i} + \frac{1}{2}\sum_{i=1}^{m}\sum_{j=1}^{m}\frac{\partial^2\{(\mathbf{B}(\mathbf{z}, t)\mathbf{WB}^{\mathrm{T}}(\mathbf{z}, t))_{ij}\,p_{\mathbf{Z}}(\mathbf{z}, t|\mathbf{z}_0, t_0)\}}{\partial z_i \partial z_j}$$

$$(2.3.59)$$

This equation is the so-called Fokker–Planck–Kolmogorov equation (FPK equation).

A family of particular nonlinear single-degree-of-freedom systems can be solved by virtue of Eq. (2.3.59). However, for the multi-degree-of-freedom system especially involved in the civil engineering, the solution of FPK equation is extremely difficult to be attained.

### 2.3.3 Generalized Probability Density Evolution Equation

Without loss of generality, a stochastic dynamical system under the random excitation can be represented by

$$\dot{\mathbf{Z}}(t) = \mathbf{g}[\mathbf{Z}(t), \mathbf{F}(\mathbf{\Theta}, t), t], \ \mathbf{Z}(t_0) = \mathbf{z}_0 \qquad (2.3.60)$$

where $\mathbf{F}(\cdot)$ is a column vector denoting the nonstationary and non-Gaussian random excitation; $\mathbf{\Theta}$ is a random parameter vector denoting the randomness inherent in the excitation.

As to the quantity of interest such as the system state or the control force $\mathbf{Z}^{\mathrm{T}} = \{Z_i\}_{i=1}^{m}$, the formal solution can be given by

$$\mathbf{Z}(t) = \mathbf{H}(\mathbf{\Theta}, \mathbf{Z}_0, t) \qquad (2.3.61)$$

where $\mathbf{H}$ is an $m$-dimensional column vector denoting arithmetic operator.

It is indicated in Eq. (2.3.61) that the randomness inherent in the random process $\mathbf{Z}(t)$ is completely represented by the random parameter vector $\mathbf{\Theta}$. In view of the probability preservation principle, the augmented system consisting of the quantity of interest and the random parameter vector, i.e., $(\mathbf{Z}(t), \mathbf{\Theta})$, thus sustains a preservative probability, that is

$$\frac{\mathrm{D}}{\mathrm{D}t} \int_{\Omega_t \times \Omega_{\mathbf{\Theta}}} p_{\mathbf{Z}\mathbf{\Theta}}(\mathbf{z}, \mathbf{\theta}, t) \, \mathrm{d}\mathbf{z}\mathrm{d}\mathbf{\theta} = 0 \qquad (2.3.62)$$

where $p_{\mathbf{Z}\mathbf{\Theta}}(\mathbf{z}, \mathbf{\theta}, t)$ denotes the joint probability density function of the augmented system $(\mathbf{Z}(t), \mathbf{\Theta})$; $\Omega_t$ denotes the time domain; $\Omega_{\mathbf{\Theta}}$ denotes the sample domain of random parameter vector $\mathbf{\Theta}$; $\mathrm{D}(\cdot)/\mathrm{D}t$ denotes the total derivative.

Extending Eq. (2.3.62), we have (Li and Chen 2009)

$$\frac{\mathrm{D}}{\mathrm{D}t} \int_{\Omega_t \times \Omega_{\mathbf{\Theta}}} p_{\mathbf{Z}\mathbf{\Theta}}(\mathbf{z}, \mathbf{\theta}, t) \, \mathrm{d}\mathbf{z}\mathrm{d}\mathbf{\theta}$$

$$= \frac{\mathrm{D}}{\mathrm{D}t} \int_{\Omega_{t_0} \times \Omega_{\mathbf{\Theta}}} p_{\mathbf{Z}\mathbf{\Theta}}(\mathbf{z}, \mathbf{\theta}, t) |J| \mathrm{d}\mathbf{z}\mathrm{d}\mathbf{\theta}$$

$$= \int_{\Omega_{t_0} \times \Omega_{\mathbf{\Theta}}} \left( |J| \frac{\mathrm{D}p_{\mathbf{Z}\mathbf{\Theta}}}{\mathrm{D}t} + p_{\mathbf{Z}\mathbf{\Theta}} \frac{\mathrm{D}|J|}{\mathrm{D}t} \right) \mathrm{d}\mathbf{z}\mathrm{d}\mathbf{\theta}$$

$$= \int_{\Omega_{t_0} \times \Omega_{\mathbf{\Theta}}} \left\{ |J| \left( \frac{\partial p_{\mathbf{Z}\mathbf{\Theta}}}{\partial t} + \sum_{j=1}^{m} \dot{Z}_j \frac{\partial p_{\mathbf{Z}\mathbf{\Theta}}}{\partial z_j} \right) + |J| p_{\mathbf{Z}\mathbf{\Theta}} \sum_{j=1}^{m} \frac{\partial \dot{Z}_j}{\partial z_j} \right\} \mathrm{d}\mathbf{z}\mathrm{d}\mathbf{\theta}$$

$$
= \int_{\Omega_{t_0} \times \Omega_\Theta} \left( \frac{\partial p_{Z\Theta}}{\partial t} + \sum_{j=1}^{m} \dot{Z}_j \frac{\partial p_{Z\Theta}}{\partial z_j} \right) |J| \mathrm{d}z \mathrm{d}\theta
$$

$$
= \int_{\Omega_t \times \Omega_\Theta} \left( \frac{\partial p_{Z\Theta}}{\partial t} + \sum_{j=1}^{m} \dot{Z}_j \frac{\partial p_{Z\Theta}}{\partial z_j} \right) \mathrm{d}z \mathrm{d}\theta \tag{2.3.63}
$$

where $|J|$ denotes the Jacobian determinant of the joint probability density function $p_{Z\Theta}(\mathbf{z}, \boldsymbol{\theta}, t)$.

Substituting Eq. (2.3.63) into Eq. (2.3.62) and noting the arbitrary characteristics on the integral domain $\Omega_t \times \Omega_\Theta$, one has

$$
\frac{\partial p_{Z\Theta}(\mathbf{z}, \boldsymbol{\theta}, t)}{\partial t} + \sum_{j=1}^{m} \dot{Z}_j(\boldsymbol{\theta}, t) \frac{\partial p_{Z\Theta}(\mathbf{z}, \boldsymbol{\theta}, t)}{\partial z_j} = 0 \tag{2.3.64}
$$

where $\dot{Z}_j(\boldsymbol{\theta}, t)$ denotes the velocity of $Z_j(t)$ at the condition of the sample $\{\boldsymbol{\Theta} = \boldsymbol{\theta}\}$, i.e., $\dot{Z}_j(\boldsymbol{\theta}, t) = \partial H_j(\boldsymbol{\theta}, t)/\partial t$.

Equation (2.3.64) is the so-called generalized probability density evolution equation (GDEE) (Li and Chen 2004a, b, 2008), of which the initial condition is given by

$$
p_{Z\Theta}(\mathbf{z}, \boldsymbol{\theta}, t)|_{t=0} = \delta(\mathbf{z} - \mathbf{z}_0) p_\Theta(\boldsymbol{\theta}) \tag{2.3.65}
$$

where $\mathbf{z}_0$ denotes the deterministic initial value of $\mathbf{Z}(t)$; $\delta(\cdot)$ denotes the Dirac delta function.

Solving the initial value problem of partial differential equation, see Eqs. (2.3.64) and (2.3.65), we can derive the joint probability density function $p_{Z\Theta}(\mathbf{z}, \boldsymbol{\theta}, t)$, and further gain the marginal probability density function of the quantity of interest:

$$
p_Z(\mathbf{z}, t) = \int_{\Omega_\Theta} p_{Z\Theta}(\mathbf{z}, \boldsymbol{\theta}, t) \mathrm{d}\boldsymbol{\theta} \tag{2.3.66}
$$

In general cases, the analytical solution of probability density function $p_Z(\mathbf{z}, t)$ is hardly to be derived. The introduction of numerical schemes is a practical choice. In view of the information propagation, a collection of representative points is first selected from the sample space of random variables. The probability density function of structural responses can be then obtained by virtue of the deterministic analysis and the finite difference method. The numerical procedure is detailed as follows (Li and Chen 2008):

*Step 1*: Partition of probability-assigned space $\Omega_\Theta$ of random variables and deriving the representative points $\boldsymbol{\theta}_q$'s, $q = 1, 2, \ldots, n_{\text{res}}$, where $n_{\text{res}}$ denotes the number of the representative points.

*Step 2*: Assigning sample value to the random variable, i.e., $\Theta = \theta_q$, and substituting them into Eq. (2.3.60), the numerical solution of the motion equation and its velocity derivation $\dot{Z}_j(\theta_q, t_m)$ can be gained. Herein, $t_m = m\Delta t$, $m = 0,1,2,\ldots$, denotes the time step.

*Step 3*: Integrating the solutions of all representative points $\theta_q :\Rightarrow \theta$, and substituting their velocity derivations into the generalized probability density evolution equation Eq. (2.3.64), the solution of the partial deferential equation $p_{Z\Theta}(z_{ji}, \theta_q, t_k)$ is obtained utilizing the finite difference method, where $z_{ji} = z_{j0} + i\Delta z_j$, $i = 0, \pm 1, \pm 2, \ldots$, $\Delta z_j$ denotes the spatial step; $t_k = k\Delta \hat{t}$, $k = 0,1,2,\ldots$, $\Delta \hat{t}$ denotes the time step.

*Step 4*: Integral of Eq. (2.3.66), then

$$p_Z(z_{ji}, t_k) = \sum_{q=1}^{n_{res}} p_{Z\Theta}(z_{ji}, \theta_q, t_k)S_q \qquad (2.3.67)$$

where $S_q$ denotes the area measure of the sub-domain represented by the representative point $\theta_q$.

The methods for the selection of representative points employed in *Step 1* refer to a series of schemes such as the dimension-reduction mapping method (Li and Chen 2006b), tangent spheres method (Chen and Li 2008), and the number theoretic method (Li and Chen 2007). The dynamic analysis involved in *Step 2* just resorts to as the conventional deterministic analysis. *Step 3* is carried out using a finite difference method (Li and Chen 2006a; Thomas 1995).

It is seen from Eq. (2.3.64) that the dimension $m$ of the generalized probability density evolution equation merely relies upon the physical quantity of interest, which is independent on the dimension $n$ of the stochastic dynamical system, as shown in Eq. (2.3.60), while the classical probability density evolution equations, such as the Liouville equation (Gardiner 1983), the Fokker–Planck–Kolmogorov equation (Kolmogorov 1931), and the Dostupov–Pugachev equation (Dostupov and Pugachev 1957), all have the same dimension as the stochastic system. This situation results in an extreme difficulty of solving the problem. In contrary, the generalized probability density evolution equation is a decoupled equation, and can be straightforwardly reduced into a one-dimensional version:

$$\frac{\partial p_{Z\Theta}(\mathbf{z}, \theta, t)}{\partial t} + \dot{Z}_j(\theta, t)\frac{\partial p_{Z\Theta}(\mathbf{z}, \theta, t)}{\partial z_j} = 0 \qquad (2.3.68)$$

Therefore, the physical quantities of concern can be explored individually.

The numerical scheme mentioned in this section for solving the generalized probability density evolution equation is an implementing means of the probability density evolution method (PDEM).

## 2.3.4 Historic Notes

It is generally recognized that the random vibration discipline originates from the research and application of stochastic dynamics that involves two logical clues (Li and Chen 2009). Einstein first explored the Brownian motion from a phenomenological perspective using the random process theory in 1905 (Einstein 1905), which was later developed by Fokker (Fokker 1914), Planck (Planck 1917), and mathematized by Kolmogorov (Kolmogorov 1931) that formed into the associated theory and methods with the FPK equation. From an almost coinstantaneous physical perspective, Langevin investigated the motion of a Brownian particle by the Newtonian equation (Langevin 1908), which was later developed by Wiener (Wiener 1923), Itô (Itô 1942) and Stratonovich (Stratonovich 1963) that underlined the formulation and solution schemes of the stochastic differential equation. Although the probabilistic description of structural vibration was pioneered in Rayleigh's investigation on the random flight in the early of twentieth century (Rayleigh 1919), the random vibration theory was widely applied in the engineering fields and gradually became to a new discipline until the middle of twentieth century. Since then, it has gained extensive progress from the primary linear random vibration analysis such as the random vibration with initial random conditions, the random vibration simultaneously involving the randomness inherent in external excitations and in structural parameters, to the nonlinear random vibration analysis (Crandall 1958; Crandall and Mark 1963; Lin 1967; Nigam 1983; Roberts and Spanos 1990; Lin and Cai 1995; Lutes and Sarkani 2004; Li and Chen 2009).

As to the classical linear random vibration analysis, an elegant theoretical formula and the pertinent numerical schemes have been formed by virtue of the statistical relation between the input and the output in temporal and frequency domains (Crandall 1958), e.g., the spectral transfer matrix method (Lutes and Sarkani 2004), the modal superposition method such as the complete quadratic combination (CQC) (Der Kiureghian 1981; Der Kiureghian and Neuenhofer 1992), the pseudo-excitation method (Lin et al. 2001; Li et al. 2004). However, the principle of superposition is not suitable for the nonlinear system. The temporal and frequency-domain methods prevailing in the linear random vibration analysis cannot deal with the problem of essentially nonlinear random vibrations. The classical Markov process method accommodates a few specific nonlinear systems but encounters the challenge as well in dealing with the general multi-degree-of-freedom and multidimensional systems. It is thus a preferable choice of deriving the approximate solution or the accurate stationary solution for the nonlinear random vibration analysis. In the past over 50 years, a collection of methods for nonlinear random vibration analysis were proposed, e.g., the statistical linearization technique (Caughey 1963) and the moment closure method (Stratonovich 1963) suitable for the weakly nonlinear systems; the extended statistical linearization technique (Beaman and Hedrick 1981), the equivalent nonlinear equation (Caughey 1986), and the Monte Carlo simulation (Shinozuka 1972) suitable for the strongly nonlinear systems. Meanwhile, the attempt of classical stochastic structure theory to the application of random vibration analysis was carried out. For instance, the perturbation expansion method was applied to deal with the random vibration analysis of low-order systems (Crandall 1963); the orthogonal function expansion was applied to deal with the random vibration analysis of white noise-driven Duffing oscillatory systems (Orabi and Ahmadi 1987); the polynomial chaos expansion was applied to deal with the random vibration analysis of stationary excitation-driven Duffing systems (Li and Ghanem 1998). One might realize that the mentioned methods cannot solve the problem of nonlinear random vibration of high-dimensional structural systems, not even to gain the complete probability density. In theory, the FPK equation is the most rigorous and most elegant method for the nonlinear random vibration analysis. As far as steady solution of stochastic dynamical systems is concerned, the method for solving high-dimensional FPK equation in Hamiltonian framework attained systematic progress since 1990s (Soize 1994; Zhu and Huang 1999; Er 2011). However, when the unsteady solution of stochastic dynamical systems is concerned, the computational complexity will increase exponentially with the dimension of systems. In this case, the solution is still hard to be derived even employing efficiently numerical schemes and advanced computational platforms. In recent years, the dimension reduction of FPK equation has been paid extensive attention (Chen and Yuan 2014; Chen and Rui 2018).

The probability density evolution method (PDEM) with the kernel generalized probability density evolution equation (GDEE) provided an efficient means for solving the stochastic dynamical system from a physical perspective. This method has been applied into the stochastic response analysis and reliability assessment of general nonlinear stochastic systems (Li and Chen 2004a, b, 2005, 2006a; Chen and Li 2005; Li and Chen 2008). The progress underlies the development of the physically based stochastic optimal control of structures.

## 2.4   Dynamic Reliability of Structures

The primary goal of structural analysis aims at the performance-based design and control of structures. If the random factors involved in the basic physical background are concerned, the logical manner of structural analysis is to carry out the reliability-based structural design and control.

As regards the assessment of dynamic reliability of structures as the first-passage failure criterion, the primary methods include the level-crossing process theory, the diffusion process theory and the probability density evolution method. Two families of criteria  are usually applied in the probability density evolution method, i.e., the absorbing boundary condition criterion and the equivalent extreme-value event criterion. Herein, the level-crossing process theory and the equivalent extreme-value event criterion-based probability density evolution method are introduced since the two methods are both widely used in practice.

**Fig. 2.4** Schematic of the
level-crossing process theory

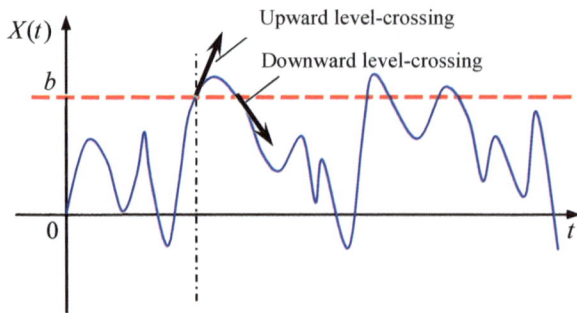

## 2.4.1 Level-Crossing Process Theory

The level-crossing process theory originated from Rice's researches on the digital
noise process in 1940s (Rice 1944, 1945). As to the $b$-level-crossing process, shown
in Fig. 2.4, the probability of occurring once upward level-crossing during the time
interval $t < \tau \leqslant t + \Delta t$ is given by

$$
\begin{aligned}
&\Pr\{N^+(t + \Delta t) - N^+(t) = 1\} \\
&= \Pr\{X(t + \Delta t) > b, X(t) < b\} \\
&= \Pr\{X(t) + \dot{X}(t)\Delta t > b, X(t) < b\}
\end{aligned}
\tag{2.4.1}
$$

where $N^+(t)$ denotes the total number of upward level-crossing during the time
interval $[0, t]$; $\Delta N^+ = N^+(t + \Delta t) - N^+(t)$ denotes the number of upward level-
crossing during the time interval $t < \tau \leqslant t + \Delta t$; $b$ denotes the threshold level.

The joint probability density function of random response processes $X(t)$, $\dot{X}(t)$ is
denoted by $p_{X\dot{X}}(x, \dot{x}, t)$. The probability of Eq. (2.4.1) can be derived by the integral
of the joint probability density function $p_{X\dot{X}}(x, \dot{x}, t)$ over the domain $(x + \dot{x}\Delta t >
b, x < b)$. One has

$$
\begin{aligned}
\Pr\{\Delta N^+ = 1\} &= \Pr\{X(t) + \dot{X}(t)\Delta t > b, X(t) < b\} \\
&= \int\limits_{x+\dot{x}\Delta t > b, x < b} p_{X\dot{X}}(x, \dot{x}, t)\mathrm{d}x\mathrm{d}\dot{x} \\
&= \int\limits_{x > b - \dot{x}\Delta t, x < b} p_{X\dot{X}}(x, \dot{x}, t)\mathrm{d}x\mathrm{d}\dot{x} \\
&= \int\limits_{0}^{\infty} \mathrm{d}\dot{x} \int\limits_{b-\dot{x}\Delta t}^{b} p_{X\dot{X}}(x, \dot{x}, t)\mathrm{d}x
\end{aligned}
$$

$$= \Delta t \int_0^\infty \dot{x} p_{X\dot{X}}(b, \dot{x}, t) \mathrm{d}\dot{x} + o(\Delta t) \tag{2.4.2}$$

Herein a formulation of the mean value theorem for integrals is applied:

$$\int_{b-\dot{x}\Delta t}^b p_{X\dot{X}}(x, \dot{x}, t)\mathrm{d}x = p_{X\dot{X}}(\tilde{x}, \dot{x}, t)[\dot{x}\Delta t] = \dot{x}p_{X\dot{X}}(b, \dot{x}, t)\Delta t + o(\Delta t) \tag{2.4.3}$$

where $\tilde{x} \in [b - \dot{x}\Delta t, b]$, and $o(\Delta t)$ denotes the infinitesimal of higher order with respect to $\Delta t$.

It is revealed in Eq. (2.4.3) that the probability of a level-crossing event in the time interval $\Delta t$ is proportional to $\Delta t$, and the longer the time interval $\Delta t$ arises, the larger of the possibility a level-crossing event occurs. Therefore, the probability of an upward level-crossing event in the unit time is given by

$$\alpha_b^+(t) = \lim_{\Delta t \to 0} \frac{\Pr\{\Delta N^+ = 1\}}{\Delta t} = \lim_{\Delta t \to 0} \frac{\Pr\{X(t) + \dot{X}(t)\Delta t > b, X(t) < b\}}{\Delta t}$$

$$= \int_0^\infty \dot{x} p_{X\dot{X}}(b, \dot{x}, t)\mathrm{d}\dot{x} \tag{2.4.4a}$$

Similarly, the probability of occurring a downward level-crossing event in the unit time is given by

$$\alpha_b^-(t) = \lim_{\Delta t \to 0} \frac{\Pr\{\Delta N^- = 1\}}{\Delta t} = \lim_{\Delta t \to 0} \frac{\Pr\{X(t) + \dot{X}(t)\Delta t < b, X(t) > b\}}{\Delta t}$$

$$= \lim_{\Delta t \to 0} \frac{1}{\Delta t} \int_{-\infty}^0 \mathrm{d}\dot{x} \int_b^{b-\dot{x}\Delta t} p_{X\dot{X}}(x, \dot{x}, t)\,\mathrm{d}x$$

$$= \int_{-\infty}^0 -\dot{x} p_{X\dot{X}}(b, \dot{x}, t)\,\mathrm{d}\dot{x}$$

$$= \int_{-\infty}^0 |\dot{x}| p_{X\dot{X}}(b, \dot{x}, t)\,\mathrm{d}\dot{x} \tag{2.4.4b}$$

Equations (2.4.4a) and (2.4.4b) are the celebrated Rice formulae. Meanwhile, it is revealed in Eqs. (2.4.2) and (2.4.4a) that

$$\Pr\{\Delta N^+ = 1\} = \alpha_b^+(t)\Delta t + o(\Delta t) \tag{2.4.5}$$

i.e., the probability of occurring an upward level-crossing event in the time interval $\Delta t$ is $\alpha^+(t)\Delta t$, which is a small quantity with a same order as $\Delta t$.

If there occur two upward level-crossing events in the time interval $\Delta t$: one occurs in the time interval $t < \tau_1 \leqslant t + \Delta t_1$; the other occurs in the time interval $t + \Delta t_1 < \tau_2 \leqslant t + \Delta t$, and meanwhile, the two upward level-crossing events are independent, one has

$$\Pr\{\Delta N^+ = 2\} = [\alpha_b^+(\tau_1)\Delta t_1] \cdot [\alpha_b^+(\tau_2)\Delta t_2] \leqslant \frac{1}{4}\alpha_b^+(\tau_1)\alpha_b^+(\tau_2)(\Delta t)^2 = o(\Delta t)$$

$$(2.4.6)$$

Therefore, the probability of occurring two upward level-crossing events is the infinitesimal of higher order with respect to $\Delta t$, which can be ignored safely in comparison with the probability of occurring one upward level-crossing event. Similarly, the probability of occurring three and more upward level-crossing events $\Pr\{\Delta N^+ \geqslant 3\}$ is the infinitesimal of higher order with respect to $\Delta t$ as well. There thus is

$$\Pr\{\Delta N^+ \geqslant 2\} = o(\Delta t) \tag{2.4.7}$$

Since $\sum\limits_{i=0}^{\infty} \Pr\{\Delta N^+ = i\} = 1$, there is

$$\Pr\{\Delta N^+ = 0\} = 1 - \Pr\{\Delta N^+ = 1\} - \sum_{i=2}^{\infty} \Pr\{\Delta N^+ = i\}$$

$$= 1 - \alpha_b^+(t)\Delta t + o(\Delta t) \tag{2.4.8}$$

Integrating Eqs. (2.4.6), (2.4.7) and (2.4.8), one can recognize that the level-crossing event in the time interval $\Delta t$ follows the Bernoulli distribution, i.e., 0–1 distribution, owing to the fact that the level-crossing event either occurs resulting in the structural damage or does not occur at all. Therefore, the mean of occurring level-crossing events is calculated by

$$E[\Delta N^+] = \Pr\{\Delta N^+ = 0\} \times 0 + \Pr\{\Delta N^+ = 1\} \times 1$$

$$= \Pr\{\Delta N^+ = 1\}$$

$$= \alpha_b^+(t)\Delta t + o(\Delta t) \tag{2.4.9}$$

which is denoted in formulation of the unit time as follows:

$$\lim_{\Delta t \to 0} \frac{E[\Delta N^+]}{\Delta t} = \lim_{\Delta t \to 0} \frac{\Pr\{\Delta N^+ = 1\}}{\Delta t} = \alpha_b^+(t) \tag{2.4.10}$$

It is indicated that $\alpha_b^+(t)$ not only denotes the probability of occurring one upward level-crossing event in the unit time, but also denotes the mean of occurring one upward level-crossing event in the unit time. The two symbols $\alpha_b^+(t)$, $\alpha_0^-(t)$ are thus both termed as the expected level-crossing rate or the averaged level-crossing rate.

If the random process $X(t)$ is assumed to be a stationary Gaussian process with zero mean, the expected level-crossing rate can be derived as a specified solution from Eqs. (2.4.4a) and (2.4.4b):

$$\alpha_b^+(t) = \alpha_b^-(t) = \frac{1}{2\pi} \frac{\sigma_{\dot{X}}}{\sigma_X} \exp\left(-\frac{b^2}{2\sigma_X^2}\right) \tag{2.4.11}$$

When the threshold level $b = 0$, the expected level-crossing rate can be further denoted by

$$\alpha_0^+(t) = \alpha_0^-(t) = \frac{1}{2\pi} \frac{\sigma_{\dot{X}}}{\sigma_X} \tag{2.4.12}$$

where $\sigma_X$, $\sigma_{\dot{X}}$ denote the standard deviations of the random process $X(t)$ and its velocity argument $\dot{X}(t)$, respectively.

Therefore, the critical step on the dynamic reliability assessment of structures by virtue of the level-crossing process theory is to derive the expected level-crossing rate according to the second-order statistics of random processes. The reliability of structural damage as the first-passage failure criterion is then derived by Eq. (2.4.13):

$$R(t) = L_0 \exp\left(-\int_0^t \lambda(\tau)d\tau\right) \tag{2.4.13}$$

where $L_0 = R(0)$ denotes the dynamic reliability of structures at the initial instant of time; $\lambda(\tau)$ denotes the risk-rate function which has a relation with the expected level-crossing rate $\alpha_b^+(t)$ in: (i) the single-wall relevant case, $\lambda(t) = \alpha_b^+(t)$; (ii) the double-wall relevant case, $\lambda(t) = \alpha_b^+(t) + \alpha_{-b}^-(t)$; (iii) the circle-wall relevant case, $\lambda(t) = \alpha_b^+[1 - \exp(-\alpha_{b,A}^+/\alpha_b^+)]$. In these symbols, the superscript "+" denotes upward level-crossing, the superscript "-" denotes downward level-crossing, the subscripts "$b$" and "$-b$" denote the thresholds of upward and downward level-crossings, respectively, and $\alpha_{b,A}^+$ denotes the expected level-crossing rate of envelop processes.

## 2.4.2 Equivalent Extreme–Value Event Criterion

As indicated in the level-crossing process theory, the dynamic reliability assessment still remains the situation utilizing the second-order statistics of structural responses, i.e., moment-based reliability methods, owing to the limitation of the classical random vibration theory. The critical step of the classical dynamic reliability analysis relies upon the Rice formulae (Rice 1944, 1945), thereby the relation between the expected level-crossing rate and the variance of structural responses can be readily derived in terms of the Gaussian stationarity assumption of structural responses. The functional relation between dynamic reliability and response variance is then established. In this transform, however, the assumption of Gaussian stationarity of structural responses and the assumption of Poisson (Coleman 1959) or Markov (Chandiramani 1964) behaviors of the level-crossing events often fail for the non-Gaussian noise-driven stochastic systems. Moreover, the system reliability or even global reliability considered from the component reliability would result in extremely complicated issues due to the relevance among failure modes and the curse of dimensionality. Following the solving procedure, however, of classical system reliability methods such as the Cornell bound method (Cornell 1967), the narrow bound method (Ditlevsen 1979), and the branch and bound method (Murotsu et al. 1984) tends to be trapped in relevance analysis.

Toward the perspective of the physical stochastic system, however, one might recognize that solving the system reliability along the nonlinearity development of structural systems not only bypasses the relevance analysis, also gains the solution of system reliability of structures straightforwardly. An elegant scheme, entitled as equivalent extreme-value event criterion, was proposed in recent years that includes the relevance information with respect to the complicated failure events represented by the segmented or partitioned limit state functions (Li et al. 2007). This progress shows that by virtue of the equivalent extreme-value event criterion, the system reliability of structures can be readily solved utilizing the probability density evolution method.

The definition of dynamic reliability of structures is given by

$$R(T) = \Pr\{X(\mathbf{\Theta}, t) \in \Omega_s, 0 \leq t \leq T\} \tag{2.4.14}$$

where $\Pr\{\cdot\}$ denotes the probability of random event; $X(\mathbf{\Theta}, t)$ denotes the random event; $\Omega_s$ denotes the safe domain.

The dynamic reliability assessment of structures involves the relevance analysis of various random variables. Assuming that the structural system relies upon two dependent random variables $X_1$, $X_2$, and their joint probability density function is denoted by $p_{X_1 X_2}(x_1, x_2)$, one has the failure probability of the structural system as follows:

$$\Pr\{(X_1 > a) \cap (X_2 > a)\} = \Pr\{X_{\min} > a\} \tag{2.4.15}$$

$$\Pr\{(X_1 > a) \cup (X_2 > a)\} = \Pr\{X_{\max} > a\} \tag{2.4.16}$$

where $X_{\min} = \min\{X_1, X_2\}$, $X_{\max} = \max\{X_1, X_2\}$.

For the extended cases, $X$ is an $n \times m$ matrix of random variables, and there is

$$\Pr\left\{ \bigcup_{i=1}^{n} \left( \bigcap_{j}^{m} (X_{ij} > a) \right) \right\} = \Pr\{X_{\text{eq}} > a\} \tag{2.4.17}$$

where $X_{eq}(\boldsymbol{\Theta}, t) = \max\limits_{1 \le i \le n}\left\{\min\limits_{1 \le j \le m} (X_{ij})\right\}$, and $X_{eq} > a$ denotes the equivalent extreme-value event of failure random event $\bigcup\limits_{i=1}^{n}\left(\bigcap\limits_{j}^{m}(X_{ij} > a)\right)$.

It is indicated in Eq. (2.4.17) that the equivalent extreme-value event criterion reveals a fact that the various complicated failure events can be integrated into a simple extreme-value event, where all the relevances among these random events are included.

The dynamic reliability of structures can be derived from the integral of probability density function of extreme-value variables of structural responses, thereby an extreme-value variable is defined:

$$W_{eq}(\boldsymbol{\Theta}, T) = \operatorname*{ext}_{t \in [0,T]} (X_{eq}(\boldsymbol{\Theta}, t)) \tag{2.4.18}$$

and a pseudo-random process is introduced:

$$Z(\tau) = \varphi(W_{eq}(\boldsymbol{\Theta}, T), \tau) \tag{2.4.19}$$

$$Z(\tau)|_{\tau=\tau_0} = 0, \ Z(\tau)|_{\tau=\tau_c} = W_{eq}(\boldsymbol{\Theta}, T) \tag{2.4.20}$$

Thereotically, the functional formulation of the pseudo-random process $\varphi(\cdot)$ can be arbitrary only if the initial condition, shown in Eq. (2.4.20), is satisfied. A simple function such as $Z(\tau) = W(\boldsymbol{\Theta}, T)\tau/\tau_c$ can be utilized; and more robust formulation refers to the function $Z(\tau) = W(\boldsymbol{\Theta}, T)\sin(\bar{\omega}\tau/\tau_c)$, where the frequency argument can be set as $\bar{\omega} = 0.5\pi, 2.5\pi, \ldots, (2n + 0.5)\pi$.

It is ready to understand that the state quantity $Z(\tau)$ and the random parameter $\boldsymbol{\Theta}$ constitute a conservative system. According to the probability preservation principle of random events, the generalized probability density evolution question with respect to the state quantity $Z(\tau)$ can be readily presented as follows:

$$\frac{\partial p_{Z\Theta}(z, \boldsymbol{\theta}, \tau)}{\partial \tau} + \dot{\varphi}(W_{eq}(\boldsymbol{\theta}, T), \tau)\frac{\partial p_{Z\Theta}(z, \boldsymbol{\theta}, \tau)}{\partial z} = 0 \tag{2.4.21}$$

where $\tau$ denotes the generalized time. The corresponding initial condition is given by

$$p_{Z\Theta}(z, \boldsymbol{\theta}, \tau_0) = \delta(z - z_0)p_{\Theta}(\boldsymbol{\theta}) \tag{2.4.22}$$

Using the numerical procedure shown in Sect. 2.3.3, the probability density function of extreme-value variable of system quantity $Z(\tau)$ can be readily attained. Therefore, the dynamic reliability of structures in the provided time length $T$ is given by

$$R(T) = \Pr\{W_{eq}(\boldsymbol{\Theta}, T) \in \Omega_s\} = \int_{\Omega_s} P_Z(z, \tau_c)\, dz \tag{2.4.23}$$

**Fig. 2.5** Cumulative
distribution functions
(CDFs) of extreme-value and
equivalent extreme-value
(EEV) interstory drifts

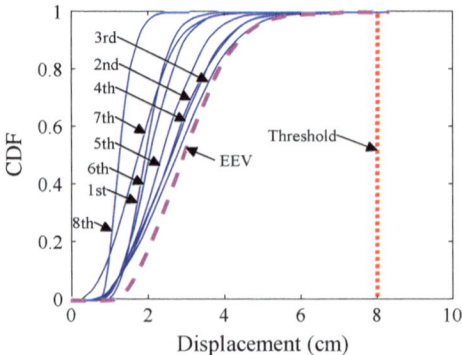

where $P_Z(z, \tau_c)$ denotes the probability density of the pseudo-random process $Z(\tau)$
at the time instant $\tau = \tau_c$.

Figure 2.5 shows cumulative distribution functions (CDFs) of extreme-value and
equivalent extreme-value interstory drifts of a randomly base-excited eight-story
frame structure with hysteretic components. The structural parameters are listed as
follows: $m_1 = m_2 = 1.0 \times 10^5$ kg, $m_3 = m_4 = 0.9 \times 10^5$ kg, $m_5 = m_6 = 0.9 \times
10^5$ kg, $m_7 = m_8 = 0.8 \times 10^5$ kg; $k_1 = k_2 = 36$ kN/mm, $k_3 = k_4 = 32$ kN/mm,
$k_5 = k_6 = 32$ kN/mm, $k_7 = k_8 = 28$ kN/mm; the heights of the interstories are
all 4.0 m. Rayleigh's damping $\mathbf{C} = a\mathbf{M} + b\mathbf{K}$ is employed, where $\mathbf{M}$, $\mathbf{K}$ denote
mass and stiffness matrices, respectively, $a = 0.01$, $b = 0.005$. The damping ratio of
the first vibrational mode is 1.05%. An extended Bouc–Wen model for describing
the behaviors of hysteretic components of the structure is employed; see Appendix
A. Parameters of the extended Bouc–Wen model are valued by $\alpha = 0.01$, $A =
1.0$, $\beta = 140.0$, $\gamma = 20.0$, $n = 1.0$, $\delta_v = 0.002$, $\delta_\eta = 0.001$, $q = 0.25$, $\zeta_s =
0.95$, $p = 2000$, $\psi = 0.2$, $\delta_\psi = 0.005$, $\lambda = 0.1$, respectively. A threshold of
interstory drift 8 cm is marked as well which distinguishes the boundary of the safe
domain $\Omega_s$. It is readily seen that the reliability of the top interstory drift is maximum
since the corresponding curve of cumulative distribution function mostly lies in the
safe domain and is far from the threshold. Meanwhile, the reliability of the third
interstory drift is minimum since the corresponding curve of cumulative distribution
function mostly approaches to the threshold except the equivalent extreme-value
(EEV) interstory drift. It is seen as well that as far as the interstory drift is concerned,
all the component reliabilities are larger than the system reliability. This result is
different from that of systems meeting with the weakest chain assumption where
the system reliability defined by Eq. (2.4.17) accurately equals the reliability of
the weakest component. In fact, the seismic structure is not only a spatial system
but also a temporal system, which exhibits complex failure correlated modes. The
failure modes of structural components are not completely correlated as a series
system. The classical weakest chain assumption thus does not remain herein. In
summary, the definition of equivalent extreme-value event and its integration with
PDEM, without any assumptions, provide a feasible means for accurately solving
the system reliability of engineering structures.

## 2.5 Modeling of Random Dynamic Excitations

Performance-based design and control of structures not only relies upon the structural model and the computational method but also relies upon the rationality of the modeling of random dynamic excitations of structures. Classical random process theory usually employs the power spectral density to describe the random excitations, such as the Kanai–Tajimi spectrum (Kanai 1957; Tajimi 1960) used in the earthquake engineering community, the Davenport spectrum (Davenport 1961) used in the wind engineering community, and the Pierson–Moskowitz spectrum (Pierson and Moskowiz 1964) used in the marine engineering community. One might recognize that the power spectral density denotes the second-order statistics of stationary processes in essence, which hardly reveals, however, the complete probabilistic information of original random processes. Moreover, the measure on the power spectral density of random excitations cannot be accurately delivered to the stochastic response through nonlinear structural systems, not mentioned to carry out the logical control of structural performance. However, a family of physically motivated random excitation models has been developed in recent years by exploring the physical mechanism of engineering excitations (Li 2006; 2008). For illustrative purposes, the modeling of random seismic ground motion and of spatial fluctuating wind-velocity field are investigated herein, and the pertinent theory and methods are introduced.

### 2.5.1 Random Seismic Ground Motion

It is well understood that the behaviors of seismic ground motions rely upon a series of critical factors such as the fault mechanism, propagation medium, and properties of the local site (Boore 2003). Due to the uncontrollability of these factors, the observed seismic ground motion arises to have a significant randomness. An efficient means for exploring the seismic wave and its propagation is to establish a wave equation with boundary conditions in conjunction with the seismic source motion (Aki and Richards 1980).

#### 2.5.1.1 Spectral Transfer Function

Assuming that the propagation medium is homogenous, elastic, and time-independent, the one-dimensional seismic ground motion field is governed by a wave equation as follows (Wang and Li 2011):

$$\sum_{j=0}^{n} \sum_{k=0}^{m} a_{jk} \frac{\partial^{j+k}}{\partial x^j \partial t^k} u(x, t) = 0 \qquad (2.5.1)$$

where $a_{jk}$ is a medium-relevant parameter; $u(x, t)$ denotes the wave displacement of seismic ground motion. The initial and boundary conditions are given by

$$u(0, t) = u_0(t), \quad \left.\frac{\partial^i u(x, t)}{\partial t^i}\right|_{t \to 0} = 0, \quad \left.\frac{\partial^i u(x, t)}{\partial t^i}\right|_{t \to +\infty} = 0, \ i = 0, 1, \ldots, n$$

$$(2.5.2)$$

By virtue of the Fourier transform, the partial differential equation shown in Eq. (2.5.1) can be transformed into an ordinary differential equation, of which the solution has a formulation as follows:

$$U(x, \omega) = \sum_{j=0}^{n} b_j(\omega) \exp(-ik_j(\omega)x) \qquad (2.5.3)$$

where $k_j(\omega)$ is the eigenvalue of wave displacement, which relies upon the propagation medium; $b_j(\omega)$ denotes the synthetic effect of seismic source and propagation path.

Inverse Fourier transform on the wave displacement $U(x, \omega)$, yields

$$u(x, t) = \frac{1}{2\pi} \sum_{j=0}^{n} \int_{-\infty}^{\infty} B_j(\omega, x) \exp[i\omega(t - \frac{x}{c_j(\omega)})] d\omega \qquad (2.5.4)$$

where $c_j(\omega) = \omega / \mathrm{Re}[k_j(\omega)]$; $\mathrm{Re}[\cdot]$ denotes real component.

Equation (2.5.4) can be further expanded as

$$u(x, t) = \frac{1}{2\pi} \int_{-\infty}^{\infty} A(b_0(\omega), \ldots, b_n(\omega); k_0(\omega), \ldots, k_n(\omega); \omega, x)$$

$$\cdot \cos[\omega t + \Phi(b_0(\omega), \ldots, b_n(\omega); k_0(\omega), \ldots, k_n(\omega); \omega, x)] d\omega \qquad (2.5.5)$$

It is indicated that the seismic ground motion field can be represented as a formulation of superposition harmonics, of which the amplitude and phase both are influenced by the boundary condition and the characteristics of propagation medium.

Assuming that the specific engineering site is far from the seismic source and the fault develops extensively fast, the dislocation process of seismic source can be viewed as irrelevance with the behaviors of the propagation path of seismic wave. Meanwhile, the scale of the local engineering site is far less than that of the propagation path of seismic wave, and the frequency scatter effect of local site on the seismic ground motion can be ignored safely. The amplitude spectrum $A(\omega, x)$ and the phase angle $\Phi(\omega, x)$ in Eq. (2.5.5) can be thus written in a separation formulation (Wang and Li 2011):

$$u(x, t) = \frac{1}{2\pi} \int\limits_{-\infty}^{\infty} A_s(\alpha_1, \ldots, \alpha_s, \omega) H_{Ap}(\beta_1, \ldots, \beta_h, \omega, x) H_{As}(\gamma_1, \ldots, \gamma_l, \omega)$$

$$\cdot \cos[\omega t + \Phi_s(\alpha_1, \ldots, \alpha_s, \omega) + H_{\Phi p}(\beta_1, \ldots, \beta_h, \omega, x) + H_{\Phi s}(\gamma_1, \ldots, \gamma_l, \omega)] \, d\omega$$

$$(2.5.6)$$

where $A_s(\cdot)$ denotes the amplitude spectrum of seismic source displacement; $H_{Ap}(\cdot)$ denotes the amplitude spectrum transfer function of propagation path; $H_{As}(\cdot)$ denotes the amplitude spectrum transfer function of local site; $\Phi_s(\cdot)$ denotes the phase spectrum of seismic source displacement; $H_{\Phi p}(\cdot)$ denotes the phase spectrum transfer function of propagation path; $H_{\Phi s}(\cdot)$ denotes the phase spectrum transfer function of local site; $\alpha_i$, $\beta_i$, $\gamma_i$ denote the physical parameters associated with the seismic source model, the propagation path model and the local site model, respectively.

Equation (2.5.6) is the so-called Fourier spectrum transfer function of seismic ground motions, which reveals the physical law governing the behaviors of seismic ground motions. Considering the randomness inherent in the seismic source, the propagation path and the local site, a physically motivated model of random seismic ground motion can be represented as follows (Wang and Li 2011):

$$a(R, t) = \ddot{u}(R, t) = -\frac{1}{2\pi} \int\limits_{-\infty}^{\infty} \omega^2 A_s(\boldsymbol{\alpha}_E, \omega) H_{Ap}(\boldsymbol{\beta}_E, \omega, R) H_{As}(\boldsymbol{\gamma}_E, \omega)$$

$$\cdot \cos[\omega t + \Phi_s(\boldsymbol{\alpha}_E, \omega) + H_{\Phi p}(\boldsymbol{\beta}_E, \omega, R) + H_{\Phi s}(\boldsymbol{\gamma}_E, \omega)] \, d\omega$$

$$(2.5.7)$$

where $\boldsymbol{\alpha}_E = (\alpha_1, \ldots, \alpha_s)$ is a $s$-dimensional vector of random parameters denoting the randomness inherent in seismic source; $\boldsymbol{\beta}_E = (\beta_1, \ldots, \beta_h)$ is a $h$-dimensional vector of random parameters denoting the randomness inherent in propagation path; $\boldsymbol{\gamma}_E = (\gamma_1, \ldots, \gamma_l)$ is a $l$-dimensional vector of random parameters denoting the randomness inherent in local site randomness; $R$ is the separation between seismic source and the local site, which is a constant.

### 2.5.1.2 Seismic Source Model

Seismic source models in seismology are mainly classified into the kinematic models and dynamic models (Aki and Richards 1980). The former describes the kinematic characteristics of seismic source and focuses on the modeling of motion amplitude of seismic source. The latter describes the dynamic characteristics of seismic source and focuses on the modeling of dislocation and dynamic development of seismic source. The kinematic model of seismic source is widely used in the earthquake engineering community. The most celebrated spectral models pertaining to the kinematics of seismic source are the $\omega^{-3}$ model based on the Haskell rectangular dislocation mechanism of seismic source (Haskell 1964, 1966), the $\omega^{-2}$ model based on the

Haskell rectangular dislocation mechanism of seismic source (Aki 1967), and the Brune source model based on the Brune circle dislocation mechanism of seismic source (Brune 1970). Among these models, the Brune source model has the benefits of less parameters and solid physical background, in which the fault surface is assumed to be circular and the dislocation distributes uniformly on the fault surface, and the shear stress wave caused by the shear stress drop propagates perpendicular to the dislocation surface. The Fourier amplitude spectrum and the Fourier phase spectrum on the Brune source model are thus denoted as follows (Brune 1970):

$$A_s(\alpha_E, \omega) = \frac{A_0}{\omega\sqrt{\omega^2 + \left(\frac{1}{\tau}\right)^2}}, \quad \Phi_s(\alpha_E, \omega) = \arctan\left(\frac{1}{\omega\tau}\right) \tag{2.5.8}$$

where $\alpha_E = (A_0, \tau)$ denotes the random vector of physical parameters relevant to the seismic source; $A_0$ denotes the amplitude parameter which is a random variable pertaining to intensity of seismic source; $\tau$ denotes the source parameter which is a random variable pertaining to the characteristics of seismic source.

### 2.5.1.3 Propagation Path Effect

Physical factors influencing the amplitude and phase of seismic waves propagating in the earth medium are mainly the geometric spreading, reflection, and refraction at the surfaces between the layers and the attenuation caused by internal friction in the medium (Aki and Richards 1980). The geometric spreading effect just contributes to the amplitude of seismic waves other than the shape of the amplitude spectrum. Since the statistical modeling involves a scale normalization on the real records of seismic ground motion, the geometric spreading effect originated from the propagation path can be ignored safely. Meanwhile, the influence of propagation path on the shape of the amplitude spectrum of seismic waves mainly comes from the attenuation effect of medium damping. The amplitude spectrum transfer function is thus written by

$$H_{Ap}(\omega, R) = \exp(-KR\omega) \tag{2.5.9}$$

where $K$ is a parameter denoting the attenuation effect of medium damping.

The influence of the propagation path on the phase arises to be complicated, which is caused by the reflection and refraction at the layer surfaces, and the attenuation effect of medium damping. The associated frequency scatter effect can be hardly represented by a general expression. Herein an empirical relation between wave number and frequency is utilized to denote the transfer function of phase spectrum (Wang and Li 2011):

$$H_{\Phi p}(\omega, R) = -Rd\ln\left[(a + 0.5)\omega + b + \frac{1}{4c}\sin(2c\omega)\right] \tag{2.5.10}$$

where $a$, $b$, $c$, and $d$ are empirical parameters which are defined by the realistic relation between wave number and frequency.

### 2.5.1.4 Local Site Effect

The local engineering site exhibits a sound filtering effect on the seismic wave propagating from the bedrock. Therefore, the local site effect is largely independent of the propagation path effect. A logical means is to separate the two effects from the modeling of seismic ground motion.

The local site is usually modeled as an equivalent single-degree-of-freedom (SDOF) system (Kanai 1957). The transfer function relevant to the filtering effect of local site is then given by

$$H_{As}(\pmb{\gamma}_E, \omega) = \sqrt{\frac{1 + 4\zeta_g^2(\omega/\omega_g)^2}{[1 - (\omega/\omega_g)^2]^2 + 4\zeta_g^2(\omega/\omega_g)^2}} \qquad (2.5.11)$$

Since the geometric scale of the local site is far smaller than that of the propagation path, a straightforward spreading effect of seismic waves is considered. Assuming that the influence of the local site on the phase variation is small and can be ignored safely, there is

$$H_{\Phi s}(\omega) = 0 \qquad (2.5.12)$$

where $\pmb{\gamma}_E = (\zeta_g, \omega_g)$ denotes the random vector relevant to the local site; $\zeta_g$ denotes the equivalent damping ratio of local site; $\omega_g$ denotes the equivalent circular frequency.

According to the Fourier spectrum transfer function of the random seismic ground motion shown in Eq. (2.5.7), and the pertaining components, i.e., Eqs. (2.5.8)–(2.5.12), a complete stochastic function model of seismic ground motion is derived as follows:

$$a_R(t) = -\frac{1}{2\pi} \int_{-\infty}^{\infty} A_R(\pmb{\xi}_E, \omega) \cos[\omega t + \Phi_R(\pmb{\xi}_E, \omega)] \, d\omega \qquad (2.5.13)$$

$$A_R(\pmb{\xi}_E, \omega) = \frac{A_0 \omega e^{-K\omega R}}{\sqrt{\omega^2 + (1/\tau)^2}} \cdot \sqrt{\frac{1 + 4\zeta_g^2(\omega/\omega_g)^2}{[1 - (\omega/\omega_g)^2]^2 + 4\zeta_g^2(\omega/\omega_g)^2}} \qquad (2.5.14)$$

$$\Phi_R(\pmb{\xi}_E, \omega) = \arctan\left(\frac{1}{\omega\tau}\right) - Rd \ln\left[(a + 0.5)\omega + b + \frac{1}{4c}\sin(2c\omega)\right] \qquad (2.5.15)$$

where $\pmb{\xi}_E = (\pmb{\alpha}_E, \pmb{\beta}_E, \pmb{\gamma}_E) = (A_0, \tau, \zeta_g, \omega_g)$ is the vector of random parameters pertaining to the seismic wave propagation from the seismic source to the local site.

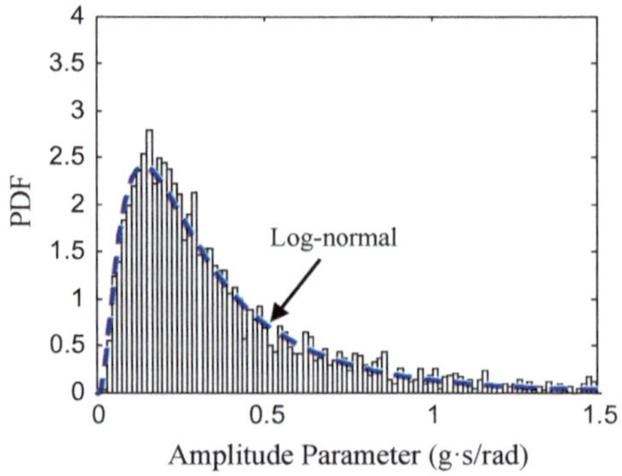

**Fig. 2.6** Statistical histogram of amplitude parameter

**Fig. 2.7** Statistical histogram of source parameter

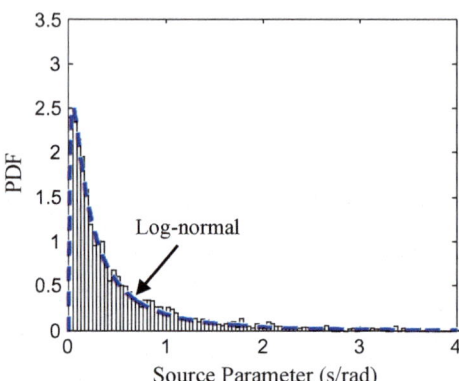

A superposition scheme of narrowband harmonics is introduced (Wong and Tri-funac 1979), and a collection of seismic ground motion samples can be simulated. Utilizing the acceleration records of strong seismic ground motions collected from the NGA database of the Pacific Earthquake Engineering Research Center (PEER NGA Database), the identification and modeling of random parameters are carried out. These acceleration records are selected as the following rules: (i) the moment magnitude of earthquakes is not less than 4; (ii) the peak ground acceleration is not less than $0.35g$. Total 4,438 acceleration records are selected. In order to retain the consistency on the data processing, all the records are scaled to a same peak ground acceleration $0.1g$, of which the time interval of sampling is scaled to 0.02 s and the upper bound of frequency is set as 25 Hz.

Figures 2.6 and 2.7 show the statistical histograms of the amplitude parameter $A_0$ and the source parameter $\tau$. It is readily seen that both the two random parameters approximately follow the log-normal distribution:

$$f(x) = \frac{1}{\sqrt{2\pi}\sigma x}e^{-\frac{(\ln x - \mu)^2}{2\sigma^2}}, \quad x \geq 0 \tag{2.5.16}$$

**Fig. 2.8** Statistical histogram of equivalent damping ratio pertaining to site class II

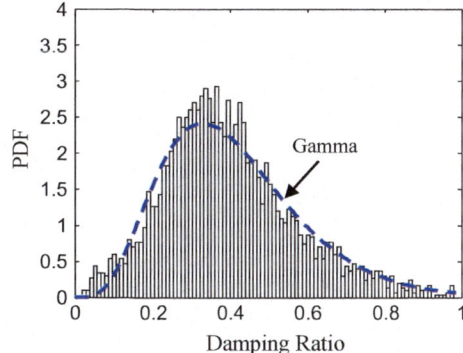

**Fig. 2.9** Statistical histogram of equivalent circular frequency pertaining to site class II

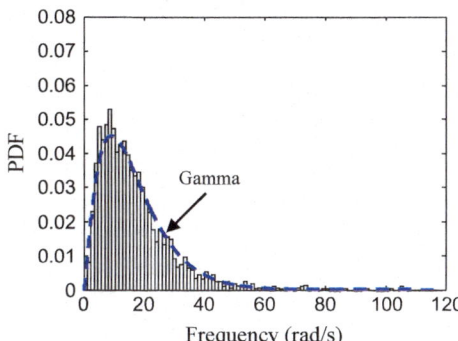

where $\mu$ and $\sigma$ denote the mean and standard deviation of the normal distribution, respectively.

By virtue of the maximum likelihood estimation, the probability density function of random parameters can be readily obtained: the mean and standard deviation of the amplitude parameter $A_0$ are $-1.2712g$ and $0.8267g$, respectively, and the relevant significance level is 0.05; and mean and standard deviation of the source parameter $\tau$ are $-1.2403$ s/rad and $1.3436$ s/rad, respectively, and the relevant significance level is 0.05 as well.

According to the Chinese Code for Seismic Design of Building Structures (GB50011-2010), the acceleration records of seismic ground motions are grouped in terms of the height of overlying strata and the equivalent shear velocity of site soil. The record numbers corresponding to the site classes I, II, III, and IV are 652, 3047, 671, and 68, respectively. For illustrative purposes, the statistical histograms of equivalent damping ratio $\zeta_g$ and equivalent circular frequency $\omega_g$ pertaining to site class II are provided; see Figs. 2.8 and 2.9. It is seen that both the two parameters approximately follow the Gamma distribution:

$$f(x; k, \theta) = x^{k-1} \frac{e^{-x/\theta}}{\theta^k \Gamma(k)}, \quad x \geq 0 \qquad (2.5.17)$$

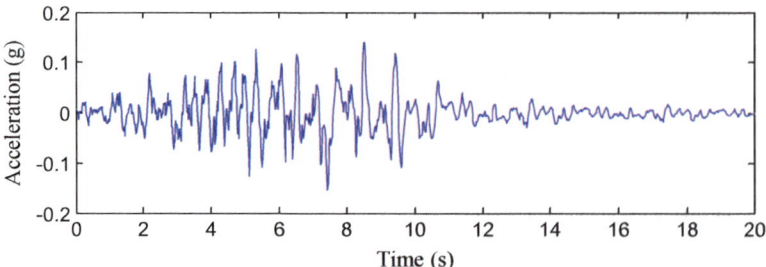

**Fig. 2.10** Seismic ground motion with means of random parameters pertaining to site class II

where $k$ denotes the shape parameter; $1/\theta$ denotes the scale parameter; $\Gamma(\cdot)$ denotes Gamma function $\Gamma(k) = \int_0^\infty (t^{k-1}/e^t)\,\mathrm{d}t$.

Similarly, the probability density functions of random parameters are identified using the maximum likelihood estimation: the shape parameter and scale parameter of the equivalent damping ratio of local site are 5.1326 and 0.08, respectively, and the significance level is 0.05; the shape parameter and scale parameter of the equivalent circular frequency of local site are 2.2415 and 7.4136 rad/s, respectively, and the significance level is 0.05 as well.

Utilizing the superposition scheme of narrowband harmonics, the simulation of random seismic ground motion is carried out. Figure 2.10 shows the seismic ground acceleration with the means of random parameters pertaining to site class II. It is revealed that the seismic ground acceleration exhibits a remarkable nonstationarity.

Further, the number theoretic method is employed to perform the partition of high-dimensional probability-assigned space spanned by the physical random parameters (Li and Chen 2009). A total of 309 samples of seismic ground motions are simulated. For validating purposes, the simulation of random seismic ground motion on the site class II in Shanghai is carried out and the peak ground acceleration of frequently occurring seismic ground motions is set as $0.035g$ according to the Chinese code (GB50011-2010). Figure 2.11 shows the mean response spectrum of simulated random seismic ground acceleration, the mean response spectrum of recorded seismic ground acceleration, and the design response spectrum provided in the code. It is seen that the simulated random seismic ground acceleration shows a consistency with the recorded seismic ground acceleration and the design provisions in the sense of mean.

### 2.5.2  Fluctuating Wind-Velocity Field

In the wind engineering community, the investigation of wind field mainly focuses on the modeling of fluctuating wind velocity and spatial coherence analysis. The

**Fig. 2.11** Comparison of mean response spectra of simulated and recorded seismic ground accelerations and design response spectrum

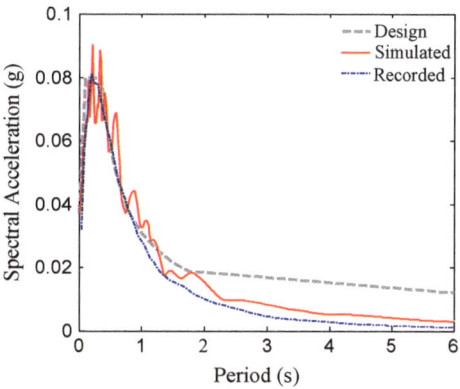

fluctuating wind velocity at an arbitrary spatial point $j$ can be represented as its relevance with the reference point $i$:

$$u_j(\boldsymbol{\xi}_W, t) = \mathrm{Re}\left(\sqrt{T} \int_0^{F_s} |F_i(\boldsymbol{\alpha}_W, f)| e^{i\varphi_i(\boldsymbol{\beta}_W, f) + \Delta\phi_{ij}(\boldsymbol{\gamma}_W, f)} \mathrm{d}f\right) \quad (2.5.18)$$

where $\boldsymbol{\xi}_W = (\boldsymbol{\alpha}_W, \boldsymbol{\beta}_W, \boldsymbol{\gamma}_W)$ is the vector of random parameters pertaining to the fluctuating wind-velocity field; $T$ denotes the time length of wind-velocity processes; $F_s$ denotes the upper bound of the sampling frequency $f$; $|F_i(\boldsymbol{\alpha}_W, f)|$ denotes the amplitude spectrum of the Fourier expansion which describes the energy distribution of fluctuating wind-velocity processes at the reference point $i$; $\varphi_i(\boldsymbol{\beta}_W, f)$ denotes the phase spectrum of the Fourier expansion which governs the shape of fluctuating wind-velocity processes at the reference point $i$. The relation between the shapes of fluctuating wind-velocity processes at spatial points $i$ and $j$ can be represented by the phase-delay spectrum $\Delta\phi_{ij}(\boldsymbol{\gamma}_W, f)$. Therefore, integrating the Fourier amplitude spectrum, the Fourier phase spectrum and the phase-delay spectrum, a complete modeling of spatial wind-velocity field can be implemented.

### 2.5.2.1 Modeling of Fourier Amplitude Spectrum

Atmosphere flows in the nature almost exist in the form of turbulence (Monin and Yaglom 1971). Atmosphere turbulence can be viewed as the consisting of a series of eddies with diverse scales. The largest eddy is directly resulted from the instability or the boundary condition of the main flow. The large-scale eddies are broken into the small-scale eddies, and the small-scale eddies then are broken into less-scale eddies. Moreover, the large-scale eddies gain the kinetic energy from the interaction between turbulence and main flow, and then progressively transfer the energy to small-scale eddies. The kinetic energy eventually dissipates at the less-scale eddies due to viscous effects of flows. This is the so-called energy cascade process of atmosphere turbulence. In this process, the small-scale eddies which exhibit high frequencies and large wave numbers eventually attain to a certain statistical equilibrium state. This equilibrium state does not rely upon the external conditions that motivate atmosphere turbulence, but forms into the so-called local homogeneous isotropic turbulence.

According to the Navier–Stokes equation, the dynamic equation of arguments relevant to the fluctuating velocity at the two spatial points of homogeneous shear turbulence is given by

$$\frac{\partial Q_{i,j}}{\partial t} + \left( 2Q_{i,j} + \xi \frac{\partial Q_{i,j}}{\partial \xi} \right) \frac{\partial \overline{U_i}}{\partial x_j} = S_{i,j} + 2v \frac{\partial^2 Q_{i,j}}{\partial \xi^2} \tag{2.5.19}$$

where $Q_{i,j}$ denotes the velocity-difference tensor between spatial points $i$ and $j$; $S_{i,i}$ denotes the three-order velocity tensor; $\xi$ denotes the position-difference tensor between the spatial points $i$ and $j$; $v$ denotes the coefficient of kinematic viscosity of flow.

Introducing the Fourier transform, the velocity-relevant equation shown in Eq. (2.5.19) can be transferred into the energy-spectrum equation:

$$\varepsilon = 2v \int_0^k k^2 E(k) \mathrm{d}k - \int_0^k F(k) \mathrm{d}k - \frac{\mathrm{d}\overline{U_i}}{\mathrm{d}x_j} \int_k^\infty \zeta(k) \mathrm{d}k \tag{2.5.20}$$

where $\varepsilon = -\overline{u_i u_j} \mathrm{d}\overline{U_i}/\mathrm{d}x_j$ denotes the total turbulence production deduced from the main flow, which can be viewed as the total dissipation of the turbulence; $2v \int_0^k k^2 E(k)\mathrm{d}k$ denotes the viscous dissipation of turbulence in the wave number range $[0, k]$; $-\int_0^k F(k)\mathrm{d}k$ denotes the transfer term of turbulence energy from the small-wave-number eddies to the large-wave-number eddies in the range $(0, k]$; $-\mathrm{d}\overline{U_i}/\mathrm{d}x_j \int_k^\infty \zeta(k)\mathrm{d}k$ denotes the turbulence production in the wave number range $(k, \infty)$.

When the wave number range is considered as $0 < k < k_c$, in which $k_c$ denotes the cutoff wave number, the eddy amount involved in the main flow has a same order as that involved in the turbulence, and the interaction between the turbulence and the main flow occupies a leading position in comparison with the viscous dissipation and eddy transfer. The right former two terms of Eq. (2.5.20) can be ignored safely, and the solution has the formulation as follows:

$$E(k) = \frac{1}{\alpha'} \frac{\varepsilon}{\frac{\mathrm{d}\overline{U_i}}{\mathrm{d}x_j}} k^{-1} \tag{2.5.21}$$

where $\alpha'$ denotes a constant.

When the wave number range is considered as $k_c \leq k \ll k_d$, in which $k_d$ denotes the wave number of energy-dissipation-scale eddies, the eddy amount involved in the main flow is far less than that involved in the turbulence, and the interaction between

**Fig. 2.12** Schematic of relevance between energy spectrum and multiscale eddies in turbulence

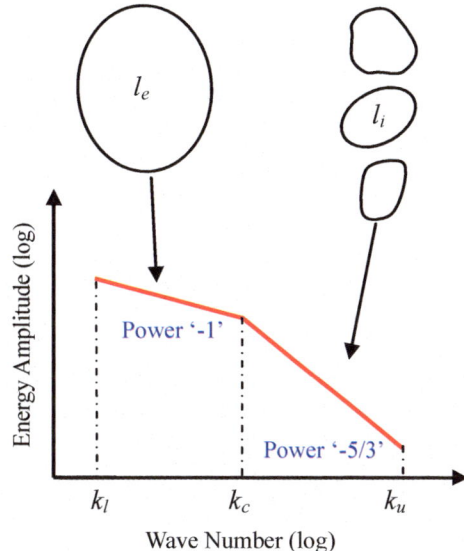

the turbulence and the main flow is very small. The right third term of Eq. (2.5.20) can be ignored safely, and the solution is then given by

$$E(k) = \left(\frac{8}{9\alpha''}\right)^{2/3} \varepsilon^{2/3} k^{-5/3} \tag{2.5.22}$$

where $\alpha''$ denotes a constant.

Therefore, the energy spectrum of atmosphere turbulence can be divided into three sub-domains with different physical meanings according to energy transfer process of atmosphere flow (Kaimal and Finnigan 1994). It is shown in Fig. 2.12 that the energy-containing sub-domain with wave number range $k_l \leq k < k_c$ corresponds to an energy-spectrum distribution submitted to the rule of power "$-1$" with respect to the wave number $k$, in which the eddies gain the energy from the main flow so as to yield turbulent kinetic energy. The inertial sub-domain with wave number range $k_c \leq k < k_u$ corresponds to an energy-spectrum distribution submitted to the rule of power "$-5/3$" with respect to the wave number $k$, in which the turbulent kinetic energy remains unchanged and is just transferred from the large-scale eddies to the small-scale eddies. The energy-dissipation sub-domain conforms with a wave number range $k_u \leq k < \infty$, in which the turbulent kinetic energy dissipates to internal energy due to the viscosity inherent in the flow. In the civil engineering community, the former two sub-domains are of concern since they occupy a majority of turbulent energy and cover the predominant frequencies of most engineering structures.

On this basis, the stochastic Fourier amplitude spectrum is proposed as follows (Li et al. 2012):

$$|F(\alpha_W, k)| = \begin{cases} \sqrt{\alpha_1}\frac{u_*(\overline{U}_{10}, z_0)}{(\kappa z k_c)^{1/3}}k^{-1/2} & (k_l < k < k_c) \\ \sqrt{\alpha_1}\frac{u_*(\overline{U}_{10}, z_0)}{(\kappa z)^{1/3}}k^{-5/6} & (k \geq k_c) \end{cases} \qquad (2.5.23)$$

where $\alpha_W = (\overline{U}_{10}, z_0)$ is the vector of random parameters pertaining to stochastic Fourier amplitude spectrum; $\alpha_1$ denotes the Kolmogorov constant pertaining to the one-dimensional turbulent energy spectrum; $k_l$ denotes the lower bound of the wave number; $k_c$ denotes the cutoff wave number between the energy-containing sub-domain and the inertial sub-domain; $\kappa$ denotes the von Karman constant; $u_*(\cdot)$ denotes the shear wave velocity, $u_* = \overline{U}_{10}\kappa / \ln(10/z_0)$; $z$ denotes the spatial height.

In view of the theoretical framework of the physical stochastic system, the elementary random variables in the stochastic Fourier amplitude spectrum merely are related to two basic parameters, i.e., the mean wind velocity $\overline{U}_{10}$ in the time interval 10 min at the standard height 10 m, and the surface roughness length $z_0$ that describes the characterizes of the local site of wind-field measurement.

For validating purposes, the elementary random variables in the stochastic Fourier amplitude spectrum are identified utilizing the wind data collected at a certain bridge in Hongkong. The wind-velocity indicators locate at the heights 30 m and 50 m separated from the water surface, of which the sampling frequency is set as 4 Hz. The statistical results show that the 10-min mean wind velocity $\overline{U}_{10}$ at the standard height 10 m follows the extreme-value type I distribution, of which the position parameter is 5.065 m/s and the scale parameter is 0.953 m/s. The surface roughness length $z_0$ follows the log-normal distribution, of which the log-mean is $-1.5795$ and the log-standard deviation is 1.4090. The 10-min mean wind velocity at the target height is derived from that at the standard height 10 m and the surface roughness length in terms of the log-law formula of wind profile. Figure 2.13 shows the comparison between the modeling Fourier amplitude spectrum and the measured Fourier amplitude spectrum, involving the mean and standard deviation spectra. It is readily seen that both the mean and standard deviation show a consistency between the modeling and measured Fourier amplitude spectra.

### 2.5.2.2  Modeling of Fourier Phase Spectrum

The phase spectrum in the spectral representation for simulating wind field is usually assumed to consist of a series of independent random initial phases which follows the uniform distribution in the frequency domain $[0, 2\pi)$ (Shinozuka and Jan 1972). This treatment results in a large number of random variables which often attains to 400 even 600. It seriously declines the computational efficiency of randomly wind-induced vibration analysis of structures due to the requirement of large-size samples in high-dimensional space of random variables. Moreover, the classical spectral representation exclusively neglects the correlation among the phases of various frequencies.

It is well known that the wind flow is the motion of various particles passing through a certain spatial point. According to the Taylor's hypothesis, if the wind-

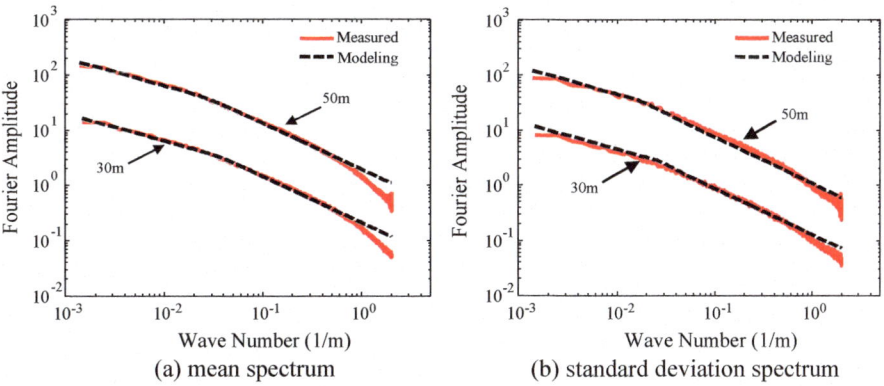

**Fig. 2.13** Comparison between modeling and measured Fourier amplitude spectra

velocity indicator moves along the main flow with a same speed as the mean wind velocity, the velocity of a same particle recorded at different spatial positions denotes the fluctuating wind velocity. In this case, the fluctuating wind velocity can be viewed as the longitudinal vibration velocity of the same particle in the kinetic reference coordinate moving as the mean wind velocity. In physics, the vibration velocity of the particle can be viewed as the superposition of a series of eddy vibrations with different scales and different frequencies. The characteristic velocity of eddies is thus defined as (Hinze 1975):

$$v(f) = \sqrt{|F(f)|^2 \Delta f} \qquad (2.5.24)$$

It is indicated that the ratio between the moving distance of eddies with characteristic velocity $v(f)$ in time interval $[t_0, t_1]$ and the perimeter of the eddy just denotes the period number of the eddy changes in the time interval, where each period corresponds to a $2\pi$ phase variation. Thus, the phase variation $\Delta\varphi(f)$ of eddies with different scales and different frequencies in the time interval $\tau$ can be given by

$$\Delta\varphi(f, \tau) = 2\pi \frac{v(f)\tau}{2\pi l(f)} = v(f)k(f)\tau \qquad (2.5.25)$$

where $\tau = t_1 - t_0$; $l(f)$ denotes the wavelength of the eddy; $k(f)$ denotes the wave number which is reciprocal to the wave length. The wave number and the natural frequency have the relation as follows:

$$k(f) = 2\pi \frac{f}{\overline{U}} \qquad (2.5.26)$$

where $\overline{U}$ denotes the mean wind velocity of spatial points.

Differential of Eq. (2.5.25) with respect to the time $t$, the phase evolution velocity of the eddies with different frequencies is thus given by

$$\Delta\dot{\varphi}(f) = v(f)k(f) \tag{2.5.27}$$

It is seen that owing to the difference from vibration velocity, the phase evolution velocity of the eddies with different frequencies is different, which is related to the energy amplitude and the scales of the eddies. Generally, the eddies with low frequencies and large scales exhibit a slow phase variation, while the eddies with high frequencies and small scales exhibit a rapid phase variation. One might imagine that the real fluctuating wind velocity can be viewed as the superposition of a collection of harmonics, i.e., eddies, after evolution over time length $T_e$ from the same initial phase. A simple treatment is that the same initial phase is set as zero. Therefore, the recorded wind velocity can be viewed as the evolutionary result of eddies with the initial zero phase over time length $T_e$ (Li et al. 2013). Here, the time length $T_e$ is termed as the zero-phase evolution time. In the next step, one can build the phase spectrum model based on the zero-phase evolution time:

$$\varphi(\boldsymbol{\beta}_W, f) = v(f)k(f)T_e \tag{2.5.28}$$

where $\boldsymbol{\beta}_W = (T_e)$ is the vector of random parameters denoting the randomness inherent in the Fourier phase spectrum.

It is indicated in Eq. (2.5.28) that the stochastic Fourier phase spectrum just relies upon the random variable $T_e$. By comparison with the traditional spectral representation method, the introduction of zero-phase evolution time correlates the phases of frequency points, and significantly reduces the number of random variables. This scheme efficiently implements the reconstruction of Fourier phase spectrum of fluctuating wind velocity.

### 2.5.2.3   Modeling of Phase-Delay Spectrum

Coherence function is a critical argument for representing wind field structure. In fact, the correlation between fluctuating wind velocities at the spatial points $i$ and $j$ can be represented by the difference between the Fourier phase spectra of the fluctuating wind velocities. Here, a phase-delay spectrum is defined as follows:

$$\Delta\phi(f) = \varphi_j(f) - \varphi_i(f) \tag{2.5.29}$$

The phase-delay spectrum and the coherence function exhibit the following relationship:

$$\gamma(f) = \left|E\left[e^{i\Delta\phi(f)}\right]\right| = |E[\cos(\Delta\phi(f)) + i\sin(\Delta\phi(f))]| \tag{2.5.30}$$

If the spatial point $i$ is defined as the reference point, and the phase spectrum of the fluctuating wind velocity at the reference point $\varphi_i(f)$ and the pertinent phase-delay spectrum between spatial points $i$ and $j$ $\Delta\phi(f)$ are provided, the fluctuating wind velocity at the any spatial point can be readily derived by integrating the Fourier amplitude spectrum into Eq. (2.5.18). Besides, the spatial wind field represented by Eq. (2.5.18) is a random field, since the Fourier amplitude spectrum and the Fourier phase spectrum both are in formula of stochastic functions.

The main factors associated with the phase-delay spectrum include: (i) the natural frequency $f$ which has a positive relation with the phase delay; (ii) the spatial separation along horizontal and vertical dimensions $r_y$, $r_z$, respectively, which has a positive correlation with the phase delay as well; (iii) the mean wind velocity $\overline{U}$, which has a negative relation with the phase delay; (iv) the shear ratio $d\overline{U}/dz$, which has a positive relation with the phase delay since a high friction upon the flow near to the ground tends to cause significant wave differences. According to the dimension analysis, the horizontal and vertical phase-delay spectra can be defined as follows (Yan et al. 2013):

$$\Delta\phi_y(\gamma_W, f) = \frac{\eta_y r_y \left(f d\overline{U}/dz\right)^{0.5}}{\overline{U}} \tag{2.5.31}$$

$$\Delta\phi_z(\gamma_W, f) = \frac{\eta_z r_z \left(f d\overline{U}/dz\right)^{0.5}}{\overline{U}} \tag{2.5.32}$$

where $\eta_y$, $\eta_z$ are the amplification coefficients along the horizontal and vertical dimensions, respectively. It is seen from the formula of shear ratio of main flow $d\overline{U}/dz = \overline{U}_{10}/z \ln(10/z_0)$ that the phase-delay spectrum does not introduce new random variables, and the vector of random parameters still remains as $\gamma_W = (\overline{U}_{10}, z_0)$.

In summary, the model for representing fluctuating wind field involves three random variables, i.e., the 10-min mean wind velocity at the height 10 m, the surface roughness length and the zero-phase evolution time. The vector of random parameters in Eq. (2.5.18) is thus denoted by $\xi_W = (\alpha_W, \beta_W, \gamma_W) = (\overline{U}_{10}, z_0, T_e)$. The basic procedure for simulation of fluctuating wind velocities at spatial points involves four steps: (i) constructing the phase spectrum at the reference point; (ii) generating the phase-delay spectrum of target points separated from the reference point; (iii) gaining the phase spectrum of target points; and (iv) deriving the fluctuating wind velocities of target points by virtue of inverse Fourier transform through integrating with the Fourier amplitude spectra at the target points.

A first large-scale platform for strong wind measurement in a certain area of East China was established in 2006. Four anemometer towers P1, P2, P3, and P4 are involved in the platform, and their separations are 40 m, 80 m, and 120 m, respectively, as shown in Fig. 2.14. Total 10 supersonic anemometers were deployed at the heights 10, 20, 28, and 43 m of the tower P1, and at the heights 10 and 20 m of the towers P2, P3, and P4. Meanwhile, for validating purposes, a horizontal mechanical anemometer was deployed at the height 10 m of the tower P1, which is 1 m separation from the

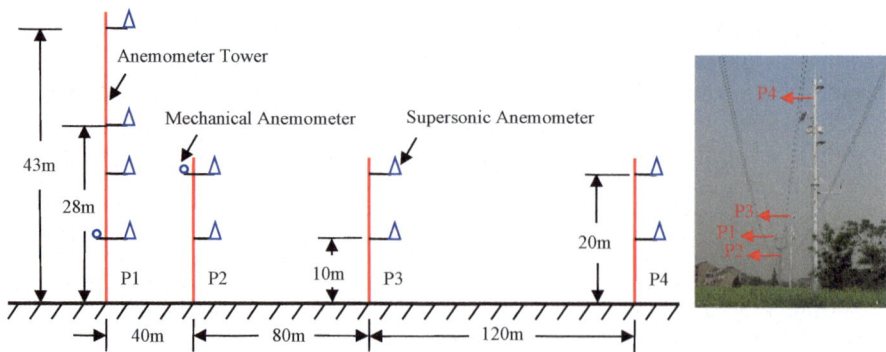

**Fig. 2.14** Platform for strong wind measurement at a certain area of East China

supersonic anemometer; and a vertical mechanical anemometer was deployed at the height 20 m of the tower P2.

By virtue of the wind data, the three elementary random variables of the Fourier spectrum model of fluctuating wind velocity are identified. Statistical results show that the 10-min mean wind velocity $\overline{U}_{10}$ at the standard height 10 m follows the extreme-value type I distribution, of which the position parameter is 5.1746 m/s, and the scale parameter is 0.7475 m/s. According to the power law of wind profile, the 10-min mean wind velocity at other three heights follows the extreme-value type I distribution as well, of which the statistical parameters are given as follows: the position and scale parameters at the height 20 m are 6.3349 m/s and 0.8286 m/s, at the height 28 m, 6.7712 m/s and 0.8402 m/s, and at the height 43 m, 7.5151 m/s and 1.0337 m/s. The surface roughness length $z_0$ follows the log-normal distribution, of which the log-mean is $-1.2155$, and log-standard deviation is 1.0052. The zero-phase evolution time $T_e$ follows the Gamma distribution, of which the scale parameter is $0.82 \times 10^9$s, and the shape parameter is 1.1.

According to the procedure for simulation of the fluctuating wind velocity at spatial points, and taking the standard height 10 m of the tower P1 as the reference point, the fluctuating wind velocities at the heights 20 m, 28 m, 43 m of the tower P1, and at the height 20 m of the towers P2, P3 and P4 are simulated. The amplification factors of the phase-delay spectrum $\eta_y$, $\eta_z$ are set as 35 and 80, respectively. Figure 2.15 shows the simulated and measured fluctuating wind velocities at these target points. It is readily seen that the simulated results have a good consistency with the measured results.

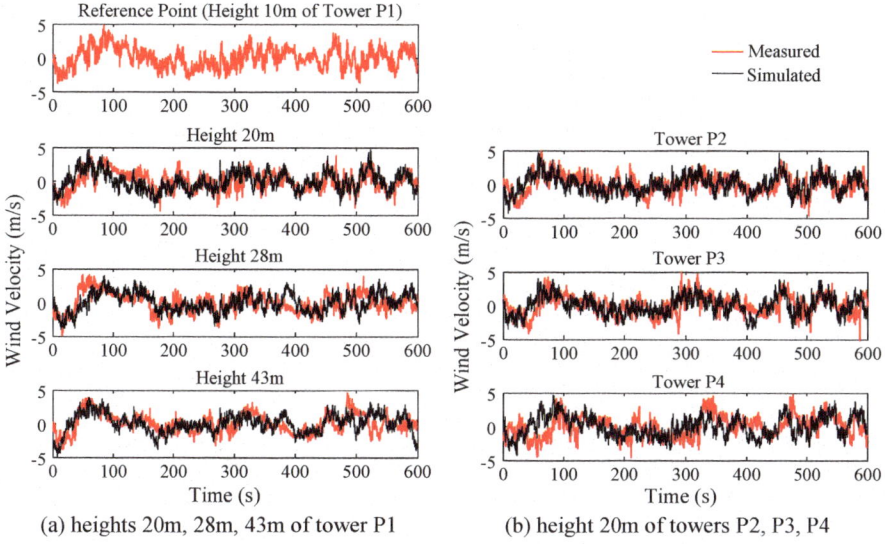

**Fig. 2.15** Comparison between simulated and measured fluctuating wind velocities at target points

# References

Aki K (1967) Scaling law of seismic spectrum. J Geophys Res 72(4):1217–1231

Aki K, Richards PG (1980) Quantitative seismology theory and methods. W.H. Freeman and Company, San Francisco

Athans M, Falb P (1966) Optimal control: an introduction to the theory and its applications. McGraw Hill, New York

Beaman JJ, Hedrick JK (1981) Improved statistical linearization for analysis and control of nonlinear stochastic-systems I: an extended statistical linearization technique. J Dyn Syst Meas Control 103(1):14–21

Bellman R (1957) Dynamic programming. Princeton University Press, Princeton

Boore DM (2003) Simulation of ground motion using the stochastic method. Pure Appl Geophys 160:635–676

Brune JN (1970) Tectonic stress and the spectra of seismic shear waves from earthquakes. J Geophys Res 75(26):4997–5009

Caughey TK (1963) Equivalent linearization techniques. J Acoust Soc Am 35:1706–1711

Caughey TK (1986) On the response of non-linear oscillators to stochastic excitation. Probab Eng Mech 1:2–4

Chandiramani KI (1964) First passage problem probability for a linear oscillator. Doctoral Thesis, Department of Mechanical Engineering, MIT, Cambridge, MA

Chatfield C (1989) The analysis of time series-an introduction, 4th edn. Chapman and Hall, London

Chen JB, Li J (2005) Dynamic response and reliability analysis of non-linear stochastic structures. Probab Eng Mech 20:33–44

Chen JB, Li J (2008) Strategy for selecting representative points via tangent spheres in the probability density evolution method. Int J Numer Methods Eng 74(13):1988–2014

Chen JB, Peng YB, Li J (2011) A note on the pseudo-excitation method. Chinese J Comput Mech 28(2):163–167 (in Chinese)

Chen JB, Rui ZM (2018) Dimension-reduced FPK equation for additive white noise excited non-linear structures. Probab Eng Mech 53:1–13

Chen JB, Yuan SR (2014) Dimension reduction of the FPK equation via an equivalence of probability flux for additively excited systems. J Eng Mech 140(11):04014088

Coleman JJ (1959) Reliability of aircraft structures in resisting chance failure. Oper Res 7(5):639–645

Cornell CA (1967) Bounds on the reliability of structural systems. ASCE J Struct Div 93(1):171–200

Crandall SH (1958) Random vibration. Technology Press of MIT, Wiley, New York

Crandall SH (1963) Perturbation techniques for random vibration of nonlinear systems. J Acoust Soc Am 35:1700–1705

Crandall SH, Mark W (1963) Random vibration in mechanical systems. Academic Press, New York

Davenport AG (1961) The spectrum of horizontal gustiness near the ground in high winds. Q J R Meteorol Soc 87:194–211

Debusschere BJ, Najm HN, Pébay PP, Knio OM, Ghanem R, Le Maître OP (2004) Numerical challenges in the use of polynomial chaos representations for stochastic processes. SIAM J Sci Comput 26(2):698–719

Der Kiureghian A (1981) A response spectrum method for random vibration analysis of MDF systems. Earthq Eng Struct Dyn 9:419–435

Der Kiureghian A, Neuenhofer A (1992) Response spectrum method for multi-support seismic excitations. Earthq Eng Struct Dyn 21:713–740

Ditlevsen O (1979) Narrow reliability bounds for structural systems. ASCE J Struct Mech 7(4):453–472

Dostupov BG, Pugachev VS (1957) The equation for the integral of a system of ordinary differential equations containing random parameters. Autom i Telemekhanika 18:620–630

Einstein A (1905) In: Furth R (1956) Investigations on the theory of the Brownian movement. Dover Publications, New York

Er GK (2011) Methodology for the solutions of some reduced Fokker-Planck equations in high dimensions. Ann Phys 523(3):247–258

Fang T, Zhang TS, Wang ZN (1991) Complex modal analysis of nonstationary random process. In: Proceedings of the 9th international modal analysis conference (IMAC), Florence, Italy, 15–18 Apr 1991

Florentin JJ (1961) Optimal control of continuous time, Markov, stochastic systems. J Electron Control 10:473–488

Fokker AD (1914) Die mittlere Energie rotierender elektrischer Dipole im Strahlungsfeld. Ann der Phys (Leipz) 43:810–820 (in German)

Gardiner CW (1983) Handbook of stochastic methods for physics, chemistry and the natural sciences, 2nd edn. Springer, Berlin

Ghanem R, Spanos PD (1991) Stochastic finite elements: a spectral approach. Springer, New York

Haskell NA (1964) Total energy and energy spectral density of elastic wave radiation from propagating faults. Bull Seismol Soc Am 54(6):1811–1841

Haskell NA (1966) Total energy and energy spectral density of elastic wave radiation from propagating faults. Part II. A statistical source model. Bull Seismol Soc Am 56(1):125–140

Hinze JQ (1975) Turbulence. McGraw-Hill, New York

Housner GW, Bergman LA, Caughey TK, Chassiakos AG, Claus RO, Masri SF, Skelton RE, Soong TT, Spencer BF Jr, Yao James TP (1997) Structural control: past, present, and future. ASCE J Eng Mech 123(9):897–971

Itô K (1942) Differential equations determining a Markoff process. Zenkoku Sizyo Sugaku Danwakasi 1077

Kaimal JC, Finnigan JJ (1994) Atmospheric boundary layer flows: their structure and measurement. Oxford University Press, New York

Kanai K (1957) Semi-empirical formula for the seismic characteristics of the ground. Bull Earthq Res Inst 35:309–325 (University of Tokyo, Japan)

Kolmogorov AN (1931) Über die analytischen Methoden in der Wahrscheinlichkeitsrechnung. Math Ann 104:415–458 (in German)

Kushner HJ (1962) Optimal stochastic control. IRE Trans Autom Control AC-7:120–122

Lanczos C (1970) The variational principles of mechanics, 4th edn. Dover, New York

Langevin P (1908) Sur la théorie du mouvement brownien. Comptes Rendus de l'Académie des Sci 146:530–533 (in German)

Li J (2006) Key scientific issues in structural engineering. Tongji University, Shanghai (in Chinese)

Li J, Chen JB (2004a) Probability density evolution method for dynamic reliability analysis of stochastic structures. J Vib Eng 17(2):121–125 (in Chinese)

Li J, Chen JB (2004b) Probability density evolution method for dynamic response analysis of structures with uncertain parameters. Comput Mech 34:400–409

Li J, Chen JB (2005) Dynamic response and reliability analysis of structures with uncertain parameters. Int J Numer Methods Eng 62:289–315

Li J, Chen JB (2006a) The dimension-reduction strategy via mapping for probability density evolution analysis of nonlinear stochastic systems. Probab Eng Mech 21(4):442–453

Li J, Chen JB (2006b) The probability density evolution method for dynamic response analysis of non-linear stochastic structures. Int J Numer Methods Eng 65:882–903

Li J, Chen JB (2007) The number theoretical method in response analysis of nonlinear stochastic structures. Comput Mech 39(6):693–708

Li J, Chen JB (2008) The principle of preservation of probability and the generalized density evolution equation. Struct Saf 30:65–77

Li J, Chen JB (2009) Stochastic dynamics of structures. Wiley, Singapore

Li R, Ghanem R (1998) Adaptive polynomial chaos expansions applied to statistics of extremes in nonlinear random vibration. Probab Eng Mech 13(2):125–136

Li QS, Zhang YH, Wu JR, Lin JH (2004) Seismic random vibration analysis of tall buildings. Eng Mech 26:1767–1778

Li J, Chen JB, Fan WL (2007) The equivalent extreme-value event and evaluation of the structural system reliability. Struct Saf 29(2):112–131

Li J, Peng YB, Yan Q (2013) Modeling and simulation of fluctuating wind speeds using evolutionary phase spectrum. Probab Eng Mech 32:48–55

Li J, Yan Q, Chen JB (2012) Stochastic modeling of engineering dynamic excitations for stochastic dynamics of structures. Prob Eng Mech 27(1):19–28

Liberzon D (2012) Calculus of variations and optimal control theory: A concise introduction. Princeton University Press, Princeton

Lin YK (1967) Probabilistic theory of structural dynamics. McGraw-Hill, New York

Lin YK, Cai GQ (1995) Probabilistic structural dynamics: advanced theory and applications. McGraw-Hill, New York

Lin JH, Zhao Y, Zhang YH (2001) Accurate and highly efficient algorithms for structural stationary/non-stationary random responses. Comput Methods Appl Mech Eng 191:103–111

Lutes LD, Sarkani S (2004) Random vibrations, analysis of structural and mechanical systems. Elsevier, Amsterdam

Monin AS, Yaglom AM (1971) Statistical fluid mechanics: mechanics of turbulence, vol 1. The MIT Press, Cambridge

Murotsu Y, Okada H, Taguchi K, Grimmelt M, Yonezawa M (1984) Automatic generation of stochastically dominant failure modes of frame structures. Struct Saf 2(1):17–25

Naidu DS (2003) Optimal control systems. CRC Press, Boca Raton

Nigam N (1983) Introduction to random vibrations. The MIT Press, Cambridge

Orabi II, Ahmadi G (1987) A functional series expansion method for response analysis of non-linear systems subjected to random excitations. Int J Non-Linear Mech 22(6):451–465

Peng YB, Ghanem R, Li J (2010) Polynomial chaos expansions for optimal control of nonlinear random oscillators. J Sound Vib 329(18):3660–3678

Pierson WJ Jr, Moskowitz L (1964) A proposed spectral form for fully developed wind seas based on the similarity theory of S. A. Kitaigorodskii. J Geophys Res 69(24):5181–5190

Planck M (1917) Uber einen Satz der statistichen Dynamik und eine Erweiterung in der Quantumtheorie. Sitzungberichte der Preussischen Akadademie der Wissenschaften 324–341 (in German)

Rayleigh L (1919) On the problem of random vibrations, and of random flights in one, two, or three dimensions. Philos Mag 37:321–347

Rice S (1944, 1945) Mathematical analysis of random noise. Bell Syst Tech J 23(3):282–332, 24(1):46–156, Reprinted in: Wax N (ed) (1954) Selected papers on noise and stochastic process. Dover Publications, Inc., New York, pp 133–294

Roberts JB, Spanos PD (1990) Random vibration and statistical linearization. Wiley, West Sussex

Shinozuka M (1972) Monte Carlo simulation of structural dynamics. Comput Struct 2:855–874

Shinozuka M, Jan CM (1972) Digital simulation of random processes and its applications. J Sound Vib 25(1):111–128

Soize C (1994) The Fokker-Planck equation for stochastic dynamical systems and its explicit steady state solutions. Utopia Press, Singapore

Sperb RP (1981) Maximum principles and their applications. Academic Press, New York

Stratonovich RL (1963) Topics in the theory of random noise. Gordon and Breach, New York

Tajimi H (1960) A statistical method of determining the maximum response of a building structure during an earthquake. In: Proceedings of 2nd world conference on earthquake engineering. Tokyo and Kyoto, vol 2, pp 781–798

Thomas JW (1995) Numeral partial differential equations: finite difference methods. Springer, New York

Wang D, Li J (2011) Physical random function model of ground motions for engineering purposes. Sci China Technol Sci 54(1):175–182

Wiener N (1923) Differential space. J Math Phys 58:131–174

Wiener N (1964) Time series. The MIT Press, Cambridge

Wong HL, Trifunac MD (1979) Generation of artificial strong motion accelerograms. Earthq Eng Struct Dyn 77:509–527

Wonham WM (1968) On separation theory of stochastic control. SIAM J Control 2:312–326

Yan Q, Peng YB, Li J (2013) Scheme and application of phase delay spectrum towards spatial stochastic wind fields. Wind Struct 16(5): 433–455

Yong JM, Zhou XY (1999) Stochastic controls: Hamiltonian systems and HJB equations. Springer, New York

Zhou XY, Yu RF, Dong D (2004) Complex mode superposition algorithm for seismic responses of non-classically damped linear MDOF system. J Earthq Eng 8(4):597–641

Zhu WQ, Huang ZL (1999) Stochastic Hopf bifurcation of quasi-nonintegrable-Hamiltonian systems. Int J Non-Linear Mech 34:437–447

# Chapter 3
# Physically Based Stochastic Optimal Control

## 3.1 Preliminary Remarks

The notion of stochastic optimal control as currently defined has its roots in statistical methods for dealing with certain tracking and signal estimation problems arising from the existence of uncertainties inherent either in the measurement or in the excitation that drives the evolution of systems, which involve prediction, filtering, and data smoothing. The pioneering work on these problems was done by the mathematician Wiener, who is accredited as the founder of control theory (Wiener 1949). A large number of research efforts were devoted to estimation problems of practical interest in electronics, communications and control engineering. An important attempt was the filtering and prediction theory by Kalman and Bucy in the early 1960s (Bucy and Kalman 1961). Almost in the same period, the introduction of the state-space method (Kalman 1960a, b), the developments of the stochastic maximum principle (Kushner 1962), and the stochastic dynamic programming (Florentin 1961) in the context of Itô calculus received great attention. The stochastic optimal control theorem was then developed into a rather integrated system in the early 1970s (Åström 1970). Thereafter, the duality methods, as a major branch of the stochastic optimal control theory, also known as the Martingale approach, have been paid extensive attention in recent years because they offered powerful tools for the study of some classes of stochastic optimal control problems (Josa-Fombellida and Rincón-Zapatero 2007).

In the classical stochastic optimal control theory, the random disturbance specifying external excitations and measurement noise is typically assumed to be the additive white Gaussian noise or the filtered white Gaussian noise, and the pertinent schemes, such as the linear quadratic Gaussian (LQG) control and the covariance control, which aim to seek the optimal control gain in an admissible set by minimizing or maximizing the cost function of system state and control force (Stengel 1986). The application of the classical stochastic optimal control theory in the civil engineering has attained an extensive progress. For instance, Yang applied the LQG control into the active optimal control of engineering structures under random excitations

© Springer Nature Singapore Pte Ltd. and Shanghai Scientific and Technical Publishers 2019
Y. Peng and J. Li, *Stochastic Optimal Control of Structures*,
https://doi.org/10.1007/978-981-13-6764-9_3

(Yang 1975). Chang and Yu developed an optimal pole assignment method in response to the vibration control of a single-degree-of-freedom system subjected to the white-noise excitation, i.e., using the control gain with the minimum variance, the pole of the closed-loop system could be transferred to the prespecified complex plane region (Chang and Yu 1998). Ho and Ma proposed a synthesis method combining the LQG and input estimation schemes, which was demonstrated to be better than the pure LQG control by a numerical simulation of active vibration control of lumped-mass systems (Ho and Ma 2007). Bani-Hani and Alawneh developed a set of posttensioning system with active prestress for the vibration control of bridges and utilized the LQG control in design of constant and variant control gains (Bani-Hani and Alawneh 2007). Kohiyama and Yoshida proposed a parameter design method for the LQG control so as to reduce the displacement and acceleration of computational facilities under the strong earthquakes (Kohiyama and Yoshida 2014).

The classical stochastic optimal control theory, however, implies an assumption of weak excitations in essence (Zhu 2006). Actually, as seen in the history of stochastic optimal control, the stochastic dynamics underlies its elementary substance, but the present theoretical frame of the stochastic dynamics is exclusively based on the white or filtered white noises and the Itô calculus (Lin and Cai 1995; Øksendal 2005). Therefore, the applicability of the classical stochastic optimal control theory in the vibration control of civil engineering structures still remains open since the practical excitations are nonstationary and non-Gaussian processes, such as seismic ground motions, high winds, and huge waves (Sun 2006). As an insight into this challenge, this chapter is devoted to developing a methodology of stochastic optimal control for response reduction of structures with actively closed-loop control systems, integrating the physically motivated random excitation model and the probability density evolution theory. The pertinent topics include the definition of control law of stochastic optimal control of structures using Pontryagin's maximum principle, the parameter design and optimization of controllers. Since the concern of the methodology lies upon the probability density evolution of structural systems during the control process, it is also referred to as the probability density evolution method (PDEM)-based stochastic optimal control.

## 3.2  Performance Evolution of Controlled Systems

As mentioned in the previous chapters, the probability density evolution method provides the theoretical foundation for the accurate analysis and design of stochastic dynamical systems. Naturally, this method can be extended to stochastic optimal control of stochastic dynamical systems so as to circumvent the dilemma that the classical stochastic optimal control confronted with.

Without loss of generality, the state equation of controlled systems subjected to random excitations is written as

$$\dot{\mathbf{Z}} = \mathbf{L}[\mathbf{Z}, \mathbf{U}, \mathbf{\Theta}, t] \tag{3.2.1}$$

where $\mathbf{Z}(t)$ is the $2n$-dimensional column vector denoting system state; $\mathbf{U}(t)$ is the $r$-dimensional column vector denoting control force; $\mathbf{\Theta}$ is the random vector characterizing the randomness inherent in the system; and $\mathbf{L}[\cdot]$ denotes the $2n$-dimensional vector of operator.

It is noted that the intervention of the control force necessarily affects the evolution trajectory of the system state, and the control force, on the contrary, needs to be regulated by the instantaneous system state in terms of the control law in feedback logic. In most cases, Eq. (3.2.1) is a well-posed equation and the system state $\mathbf{Z}(t)$ can be determined uniquely, which is a function of $\mathbf{\Theta}$ and might be assumed to take the following form:

$$\mathbf{Z}(t) = \mathbf{H_Z}(\mathbf{\Theta}, t) \tag{3.2.2}$$

At the present stage, the explicit expression of the formal function $\mathbf{H_Z}(\cdot)$ is not requisite and the sufficient condition is just its existence and uniqueness. Likewise, the control force $\mathbf{U}(t)$ is also a function of $\mathbf{\Theta}$ and can be assumed to take the following form:

$$\mathbf{U}(t) = \mathbf{H_U}(\mathbf{\Theta}, t) \tag{3.2.3}$$

The velocities of $\mathbf{Z}(t)$ and $\mathbf{U}(t)$ can be thus assumed to take the following forms:

$$\dot{\mathbf{Z}}(t) = \mathbf{h_Z}(\mathbf{\Theta}, t) \tag{3.2.4}$$

$$\dot{\mathbf{U}}(t) = \mathbf{h_U}(\mathbf{\Theta}, t) \tag{3.2.5}$$

If the probability density function of a component of $\mathbf{Z}(t)$, denoted as $Z(t)$, without risk of confusion, is of interest, i.e.,

$$\dot{Z}(t) = h_Z(\mathbf{\Theta}, t) \tag{3.2.6}$$

The augmented system $(Z(t), \mathbf{\Theta})$ sustains a conservative probability since all the randomness involved in this system comes from $\mathbf{\Theta}$. There thus has

$$\frac{\mathrm{D}}{\mathrm{D}t} \int_{\Omega_t \times \Omega_{\mathbf{\Theta}}} p_{Z\mathbf{\Theta}}(z, \mathbf{\theta}, t) \mathrm{d}z \mathrm{d}\mathbf{\theta} = 0 \tag{3.2.7}$$

where $\Omega_t, \Omega_{\mathbf{\Theta}}$ are the distribution domain of $t, \mathbf{\Theta}$, respectively; $p_{Z\mathbf{\Theta}}(z, \mathbf{\theta}, t)$ is the joint probability density function of $(Z(t), \mathbf{\Theta})$. Through some mathematical manipulations, it follows (Li and Chen 2009)

$$\frac{D}{Dt} \int\limits_{\Omega_t \times \Omega_\Theta} p_{Z\Theta}(z, \boldsymbol{\theta}, t) dz d\boldsymbol{\theta} = \int\limits_{\Omega_t \times \Omega_\Theta} \left( \frac{\partial p_{Z\Theta}}{\partial t} + h_Z \frac{\partial p_{Z\Theta}}{\partial z} \right) dz d\boldsymbol{\theta} \qquad (3.2.8)$$

Combining Eqs. (3.2.7) and (3.2.8) and considering the arbitrary characteristics on the integral domain $\Omega_t \times \Omega_\Theta$, we have

$$\frac{\partial p_{Z\Theta}(z, \boldsymbol{\theta}, t)}{\partial t} + \dot{Z}(\boldsymbol{\theta}, t) \frac{\partial p_{Z\Theta}(z, \boldsymbol{\theta}, t)}{\partial z} = 0 \qquad (3.2.9)$$

Equation (3.2.9) is the so-called generalized probability density evolution equation (GDEE) for the augmented system $(Z(t), \Theta)$.

Likewise, for the component of control force $U(t)$, we have

$$\frac{\partial p_{U\Theta}(u, \boldsymbol{\theta}, t)}{\partial t} + \dot{U}(\boldsymbol{\theta}, t) \frac{\partial p_{U\Theta}(u, \boldsymbol{\theta}, t)}{\partial u} = 0 \qquad (3.2.10)$$

The pertinent instantaneous PDFs of $Z(t)$ and $U(t)$ can be obtained by solving a family of partial differential equations with provided initial conditions as follows:

$$p_{Z\Theta}(z, \boldsymbol{\theta}, t)|_{t=0} = \delta(z - z_0) p_\Theta(\boldsymbol{\theta}) \qquad (3.2.11)$$

$$p_{U\Theta}(u, \boldsymbol{\theta}, t)|_{t=0} = \delta(u - u_0) p_\Theta(\boldsymbol{\theta}) \qquad (3.2.12)$$

where $\delta(\cdot)$ is the Dirac delta function; $z_0, u_0$ are determinative initial values of $Z(t), U(t)$, respectively. We then have

$$p_Z(z, t) = \int\limits_{\Omega_\Theta} p_{Z\Theta}(z, \boldsymbol{\theta}, t) d\boldsymbol{\theta} \qquad (3.2.13)$$

$$p_U(u, t) = \int\limits_{\Omega_\Theta} p_{U\Theta}(u, \boldsymbol{\theta}, t) d\boldsymbol{\theta} \qquad (3.2.14)$$

where the joint PDFs $p_{Z\Theta}(z, \boldsymbol{\theta}, t)$ and $p_{U\Theta}(u, \boldsymbol{\theta}, t)$ are the solutions of Eqs. (3.2.9) and (3.2.10), respectively.

It is noted that the GDEE reveals the intrinsic relation of stochastic control system and deterministic control system via the realization of random vector $\boldsymbol{\theta}$, which underlies the realizability of probability-density-based optimal control for high-dimensional stochastic systems driven by practical nonstationary and non-Gaussian random excitations (Li and Chen 2008). One might recognize from Eqs. (3.2.9), (3.2.10), (3.2.13), and (3.2.14) that the kernel of implementing the probability-density-based optimal control is solving the physical quantity change $\dot{Z}(\boldsymbol{\theta}, t)$, $\dot{U}(\boldsymbol{\theta}, t)$ of systems with respect to the realization of random vector $\boldsymbol{\theta}$. Distinguished from the classical stochastic optimal control such as the LQG, the optimal control methodol-

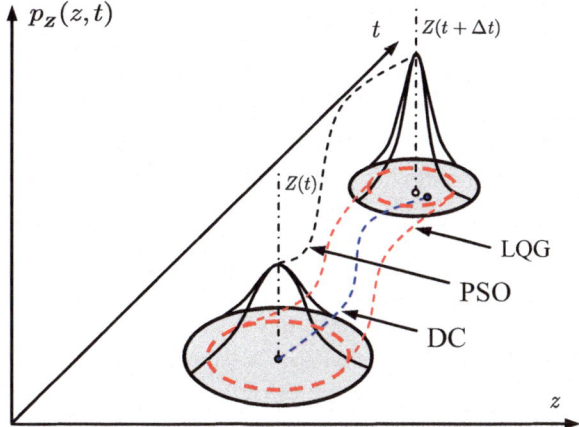

**Fig. 3.1** Schematic of deterministic control (DC), LQG control, and physically based stochastic optimal (PSO) control

ogy on the basis of the GDEEs is termed as the physically based stochastic optimal control.

Figure 3.1 shows the differences from the deterministic control (DC), the LQG control, and the physically based stochastic optimal (PSO) control upon tracing the state evolution of controlled systems. It is ready to see that the trajectory of the deterministic control system is from point to point, which obviously lacks the ability of governing the system performance due to the randomness inherent in external excitations and measurement noise. The trajectory of the control system by the LQG is from circle to circle. It is noted that the classical stochastic optimal control is essentially a control scheme based on the second-order statistics, which just holds the system performance in the sense of mean-square quantities, and is incapable of attaining the complete probability information. The trajectory of the control system by the PSO, however, is from domain to domain, which can readily complement the accurate control of the system performance in the sense of probability density of quantities, owing to the advantage that the system quantities of interest are governed by the GDEEs, i.e., Eqs. (3.2.9) and (3.2.10).

## 3.3 Scheme of Stochastic Optimal Control

### 3.3.1 Closed-Loop Control Systems

Consider an $n$-degree-of-freedom linear structural system with active control devices and subjected to random excitations. The equation of motion is given by

$$\mathbf{M}\ddot{\mathbf{X}}(t) + \mathbf{C}\dot{\mathbf{X}}(t) + \mathbf{K}\mathbf{X}(t) = \mathbf{B}_s\mathbf{U}(t) + \mathbf{D}_s\mathbf{F}(\mathbf{\Theta}, t) \tag{3.3.1}$$

where $\mathbf{X}(t)$ is the $n$-dimensional column vector denoting system displacement; $\mathbf{U}(t)$ is the $r$-dimensional column vector denoting control force; $\mathbf{F}(\cdot)$ is the $p$-dimensional column vector denoting random excitations, which can be represented by a stochastic function in terms of the orthogonal decomposition of random processes (Li and Liu 2006) or introducing the physical mechanism of random processes (Li and Ai 2006); $\mathbf{M}$, $\mathbf{C}$, and $\mathbf{K}$ are $n \times n$ mass, damping, and stiffness matrices, respectively; $\mathbf{B}_s$ is the $n \times r$ matrix denoting the location of control devices; and $\mathbf{D}_s$ is the $n \times p$ matrix denoting the location of external excitations.

It should be noted that the random vector $\mathbf{\Theta}$ is considered to represent the randomness inherent in external excitations, and the measurement noise is ignored in this study since that the uncertainty arising from the measurement noise is more controllable compared with that arising from the random excitation. Meanwhile, although it is somewhat cumbersome, the notation $\mathbf{\Theta}$ underlies the fact that a random process is a function defined over the space of events of which $\mathbf{\Theta}$ is an element. Having noted this, the symbol $\mathbf{\Theta}$ would be dropped in the following development when the random nature of a certain quantity is obvious from the context except in the case of special denotation for a key quantity.

In the state space, Eq. (3.3.1) becomes

$$\dot{\mathbf{Z}}(t) = \mathbf{A}\mathbf{Z}(t) + \mathbf{B}\mathbf{U}(t) + \mathbf{D}\mathbf{F}(\mathbf{\Theta}, t) \tag{3.3.2}$$

with the initial condition

$$\mathbf{Z}(t_0) = \mathbf{z}_0 \tag{3.3.3}$$

where $\mathbf{A}$ is the $2n \times 2n$ system matrix; $\mathbf{B}$ is the $2n \times r$ matrix denoting the location of control devices, and $\mathbf{D}$ is the $2n \times p$ matrix denoting the location of external excitation, respectively,

$$\mathbf{Z}(t) = \begin{bmatrix} \mathbf{X}(t) \\ \dot{\mathbf{X}}(t) \end{bmatrix}, \mathbf{A} = \begin{bmatrix} \mathbf{0} & \mathbf{I} \\ -\mathbf{M}^{-1}\mathbf{K} & -\mathbf{M}^{-1}\mathbf{C} \end{bmatrix}, \mathbf{B} = \begin{bmatrix} \mathbf{0} \\ \mathbf{M}^{-1}\mathbf{B}_s \end{bmatrix}, \mathbf{D} = \begin{bmatrix} \mathbf{0} \\ \mathbf{M}^{-1}\mathbf{D}_s \end{bmatrix} \tag{3.3.4}$$

Equation (3.3.2) is numerically tractable using time integration methods, thereby any system quantities of interest such as displacement, velocity, acceleration, and control force can be readily derived.

The stochastic optimal control involves maximizing or minimizing a specified cost function. The generalized form of cost function is typically a quadratic combination of displacement, velocity, acceleration, and control force (Yang et al. 1994). Considering the linear quadratic regulator (LQR) as the control logic of the PSO, a standard quadratic cost function is given by (Soong 1990)

$$J_1(\mathbf{Z}, \mathbf{U}, \mathbf{\Theta}) = \frac{1}{2}\mathbf{Z}^{\mathrm{T}}(t_f)\mathbf{S}(t_f)\mathbf{Z}(t_f) + \frac{1}{2}\int_{t_0}^{t_f} [\mathbf{Z}^{\mathrm{T}}(t)\mathbf{Q}_{\mathbf{Z}}\mathbf{Z}(t) + \mathbf{U}^{\mathrm{T}}(t)\mathbf{R}_{\mathbf{U}}\mathbf{U}(t)]\mathrm{d}t$$

$$(3.3.5)$$

where $\mathbf{Q}_{\mathbf{Z}}$ is a $2n \times 2n$ positive semi-definite weighting matrix with respect to system state; $\mathbf{R}_{\mathbf{U}}$ is an $r \times r$ positive definite weighting matrix with respect to control force; $t_0$ is the initial time; and $t_f$ is the terminal time which is usually larger than the duration of the external excitation. As should be noted, the cost function of the classical LQG is defined as ensemble average on the right terms of Eq. (3.3.5), which is a deterministic function; its minimization aims to attain the optimal gain with minimum cost under the assumption of white Gaussian noise as the external excitation and the given parameters of control law. In this case, the optimal gain relies upon second-order statistics of system quantities, which allows for a mean-square solution of control system, but the probability distribution of the system state pertaining to structural reliability is still unknown. The cost function of the PSO, i.e., Eq. (3.3.5), however, is a random function; its minimization aims to derive a stochastic optimal gain with parameters of control law of which the design and optimization through cost-effect analysis over realizations of random vector can attain the desired probability distribution of the system state. The proposed procedure is suitable for the optimal control of general stochastic systems, without assumption of white Gaussian noise as the external excitation.

In brief, the procedure involves a two-step optimization; see Fig. 3.2. In the first step, for each realization (sample) $\mathbf{\theta}$ of the random vector $\mathbf{\Theta}$, the minimization of the cost function Eq. (3.3.5) is carried out to build a functional mapping from the set of parameters of control law to the set of control gains. In the second step, the optimal parameters of control law to be used are obtained by optimizing the control gain as a probabilistic criterion pertaining to the structural performance objective.

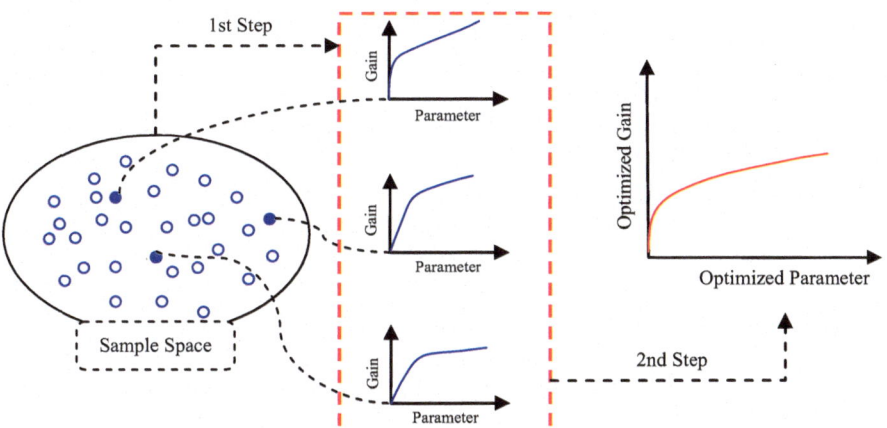

**Fig. 3.2** Two-step optimization of physically based stochastic optimal control

Therefore, viewed from representative realizations, the minimization of the cost function $J_1$ leads to a conditional extreme-value problem. Introducing the costate vector $\boldsymbol{\lambda}(t) \in \mathbb{R}^n$ and utilizing the Lagrange multiplier method, we have

$$J_1(\mathbf{Z},\mathbf{U},\boldsymbol{\lambda},\mathbf{F}, \boldsymbol{\Theta}) = \frac{1}{2}\mathbf{Z}^{\mathrm{T}}(t_f)\mathbf{S}(t_f)\mathbf{Z}(t_f) + \int_{t_0}^{t_f} [H(\mathbf{Z},\mathbf{U},\boldsymbol{\lambda},\mathbf{F}, \boldsymbol{\Theta},t) - \boldsymbol{\lambda}^{\mathrm{T}}(t)\dot{\mathbf{Z}}(t)]\mathrm{d}t$$

(3.3.6)

where the Hamiltonian function is given by

$$H(\mathbf{Z},\mathbf{U},\boldsymbol{\lambda},\mathbf{F}, \boldsymbol{\Theta},t) = \frac{1}{2}[\mathbf{Z}^{\mathrm{T}}(t)\mathbf{Q}_{\mathbf{Z}}\mathbf{Z}(t) + \mathbf{U}^{\mathrm{T}}(t)\mathbf{R}_{\mathbf{U}}\mathbf{U}(t)] + \boldsymbol{\lambda}^{\mathrm{T}}(t)[\mathbf{A}\mathbf{Z}(t) + \mathbf{B}\mathbf{U}(t) + \mathbf{D}\mathbf{F}(\boldsymbol{\Theta}, t)]$$

(3.3.7)

The necessary condition for the minimization of the cost function $J_1(\mathbf{Z},\mathbf{U},\boldsymbol{\lambda},\mathbf{F}, \boldsymbol{\Theta})$ is deduced from the celebrated Pontryagin's maximum principle that the system state $\mathbf{Z}^*(t)$ denotes the optimal trajectory if the control force $\mathbf{U}^*(t)$ is referred to as an optimal control, and there must exist a costate $\boldsymbol{\lambda}^*(t)$ that allows for the Euler–Lagrange equation, as shown in Eqs. (2.2.13)–(2.2.15), in the presence of the random excitation. Then, we have

$$\frac{\partial H}{\partial \mathbf{U}} = \mathbf{R}_{\mathbf{U}}\mathbf{U}(t) + \mathbf{B}^{\mathrm{T}}\boldsymbol{\lambda}(t) = \mathbf{0}$$

(3.3.8)

which yields

$$\mathbf{U}(t) = -\mathbf{R}_{\mathbf{U}}^{-1}\mathbf{B}^{\mathrm{T}}\boldsymbol{\lambda}(t)$$

(3.3.9)

The costate equation Eq. (2.2.14) then turns to be

$$\dot{\boldsymbol{\lambda}}(t) = -\left(\frac{\partial H}{\partial \mathbf{Z}}\right)^{\mathrm{T}} = -\mathbf{Q}_{\mathbf{Z}}\mathbf{Z}(t) - \mathbf{A}^{\mathrm{T}}\boldsymbol{\lambda}(t)$$

(3.3.10)

As to a closed-open-loop control system with the state feedback and the input feedback simultaneously (Yang et al. 1987), the linear mapping between the costate $\boldsymbol{\lambda}(t)$ and the state $\mathbf{Z}(t)$, and the random excitation $\mathbf{F}(\boldsymbol{\Theta}, t)$ is deduced as (for the details; see Appendix B)

$$\boldsymbol{\lambda}(t) = \mathbf{P}(t)\mathbf{Z}(t) + \mathbf{S}_{\mathbf{F}}(t)\mathbf{F}(\boldsymbol{\Theta}, t)$$

(3.3.11)

where $\mathbf{P}(t)$, $\mathbf{S}_{\mathbf{F}}(t)$ are undetermined matrices with the terminal conditions

$$\mathbf{P}(t_f) = \mathbf{S}_{\mathbf{F}}(t_f) = \mathbf{0}$$

(3.3.12)

Substituting Eq. (3.3.11) into Eq. (3.3.9), one could obtain the control law

$$\mathbf{U}(t) = -\mathbf{R}_U^{-1}\mathbf{B}^T\mathbf{P}(t)\mathbf{Z}(t) - \mathbf{R}_U^{-1}\mathbf{B}^T\mathbf{S}_F(t)\mathbf{F}(\Theta, t) \tag{3.3.13}$$

Introducing Eq. (3.3.11) into Eq. (3.3.10) yields

$$\dot{\mathbf{P}}(t)\mathbf{Z}(t) + \mathbf{P}(t)\dot{\mathbf{Z}}(t) + \dot{\mathbf{S}}_F(t)\mathbf{F}(\Theta, t) + \mathbf{S}_F(t)\dot{\mathbf{F}}(\Theta, t)$$
$$= -[\mathbf{Q}_Z + \mathbf{A}^T\mathbf{P}(t)]\mathbf{Z}(t) - \mathbf{A}^T\mathbf{S}_F(t)\mathbf{F}(\Theta, t) \tag{3.3.14}$$

Namely,

$$[\dot{\mathbf{P}}(t) + \mathbf{P}(t)\mathbf{A} + \mathbf{A}^T\mathbf{P}(t) - \mathbf{P}(t)\mathbf{B}\mathbf{R}_U^{-1}\mathbf{B}^T\mathbf{P}(t) + \mathbf{Q}_Z]\mathbf{Z}(t)$$
$$= [\dot{\mathbf{S}}_F(t) + \mathbf{A}^T\mathbf{S}_F(t) - \mathbf{P}(t)\mathbf{B}\mathbf{R}_U^{-1}\mathbf{B}^T\mathbf{S}_F(t) + \mathbf{P}(t)\mathbf{D}]\mathbf{F}(\Theta, t) - \mathbf{S}_F(t)\dot{\mathbf{F}}(\Theta, t)$$
$$\tag{3.3.15}$$

Equation (3.3.15) is the so-called differential Riccati equation and $\mathbf{P}(t)$ denotes the Riccati matrix.

It is indicated in Eq. (3.3.15) that the control law of a continuous time system involving the input feedback must be computed in real time according to the measured data since $\mathbf{P}(t)$, $\mathbf{S}_F(t)$ are both coupled with $\mathbf{F}(\Theta, t)$, $\mathbf{Z}(t)$. As mentioned previously, a critical task included in the proposed control scheme is the determination of probabilistic criterion, which relies upon the structural performance objective and naturally considers the influence of the random excitation. Therefore, the excitation-relevant term can be removed safely from the expression of control law. This treatment leads to a closed-loop control with the state feedback. The Riccati equation of the closed-loop control is then written as

$$\dot{\mathbf{P}}(t) = -\mathbf{P}(t)\mathbf{A} - \mathbf{A}^T\mathbf{P}(t) + \mathbf{P}(t)\mathbf{B}\mathbf{R}_U^{-1}\mathbf{B}^T\mathbf{P}(t) - \mathbf{Q}_Z \tag{3.3.16}$$

It is indicated in previous studies that the Riccati matrix $\mathbf{P}(t)$ remains the steady solution in a long interval after the initial time $t_0$, and comes into the transient solution rapidly until to zero near the final time $t_f$ (Athans and Falb 1966). The starting time of the transient solution moves forward to $t_f$ when the final time $t_f \rightarrow \infty$. Consequently, for the infinite-time control system, the Riccati matrix $\mathbf{P}(t)$ equals to its steady solution $\mathbf{P}$, and Eq. (3.3.16) thus becomes a matrix algebraic equation

$$\mathbf{P}\mathbf{A} + \mathbf{A}^T\mathbf{P} - \mathbf{P}\mathbf{B}\mathbf{R}_U^{-1}\mathbf{B}^T\mathbf{P} + \mathbf{Q}_Z = 0 \tag{3.3.17}$$

According to Eq. (3.3.13), the control law of closed-loop control is thus given by

$$\mathbf{U}(\Theta, t) = -\mathbf{G}_Z\mathbf{Z}(\Theta, t) \tag{3.3.18}$$

where $\mathbf{G}_Z$ denotes the gain matrix of state-feedback control

$$\mathbf{G}_Z = \mathbf{R}_U^{-1}\mathbf{B}^T\mathbf{P} \tag{3.3.19}$$

**Fig. 3.3** Schematic of probability density of system state at typical instant of time with and without controls

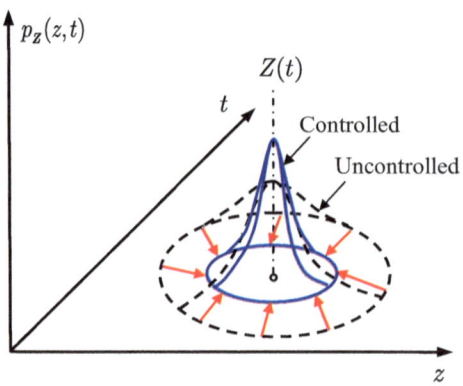

Substituting the physical solutions of state quantity $\mathbf{Z}(t)$ and control force $\mathbf{U}(t)$ in the equation of motion of controlled structural system; see Eq. (3.3.2), into the generalized probability density evolution equations Eqs. (3.2.9) and (3.2.10), one can readily derive the probability density evolution process of system quantities of concern. Fig. 3.3 shows the schematic of probability density of system state at typical instant of time with and without controls.

### 3.3.2  Parameter Optimization of Control Law

The optimal control involves minimizing or maximizing a cost function in terms of system state and control force, either the deterministic cost function included in the classical LQG control or the stochastic cost function included in the proposed PSO control. The control effectiveness relies upon the derived control law pertaining to the structural performance objective. A critical step of designing control system is the determination of parameters of control law. It is seen from Eqs. (3.3.17)–(3.3.19) that the effort of designing the linear quadratic regulator (LQR) ought to be paid on the choice of cost-function weights $\mathbf{Q_Z}$ and $\mathbf{R_U}$. A number of strategies for the choice of cost-function weights were developed in the context of the classical LQG, such as the statistical moment evaluation based on the mathematical expectation of quantities of interest (Zhang and Xu 2001), the system robustness analysis in the sense of optimal probability (Stengel et al. 1992), and the weighting matrices comparison in the context of Hamiltonian theoretical framework (Zhu et al. 2001). In the context of the PSO, a strategy for cost-function weights choice is developed which is referred to as the system second-order statistics evaluation (Li et al. 2010), where the pertinent performance function involving evaluation and constraint quantities is proposed as follows:

$$J_2 = \bigcup_{j=1}^{N} F[\tilde{W}_j] \,\bigg|\, \bigcup_{k=1}^{M} \{F[\tilde{V}_k] \le \tilde{V}_{k,\mathrm{con}}\} \tag{3.3.20}$$

where $\tilde{W}_j = \max_t[\max_i|W_{ji}(\Theta, t)|]$ denotes the $j$th component of the equivalent extreme-value vector to be evaluated; $\tilde{V}_k = \max_t[\max_i|V_{ki}(\Theta, t)|]$ denotes the $k$th component of the equivalent extreme-value vector used as the constraint; $\tilde{V}_{k,con}$ denotes the threshold of the $k$th constraint; and the hat "~" on symbols indicates the equivalent extreme-value vector (Li et al. 2007); $\cup$ denotes the union operator; $N, M$ denote the number of evaluation and constraint quantities, respectively; and $F[\cdot]$ is the quantile function denoting confidence level.

Therefore, the probabilistic criterion in terms of the system second-order statistics evaluation is defined as follows:

$$\{\mathbf{Q}_Z^*, \mathbf{R}_U^*\} = \arg\min_{\mathbf{Q}_Z, \mathbf{R}_U}\{J_2\} \tag{3.3.21}$$

The employment of the probabilistic criterion of Eq. (3.3.21) aims to seek the optimal cost-function weights $\mathbf{Q}_Z^*, \mathbf{R}_U^*$, under the condition of the quantile of the constraint less than its threshold, such that the quantile of the evaluation quantity is minimized. Herein, the evaluation quantity could be recognized as the extreme value of a structural response, e.g., interstory drift, interstory velocity, story acceleration, interstory shear force, and control force.

In this sense, the cost-function weights can be employed as (Soong 1990)

$$\mathbf{Q}_Z = q\begin{bmatrix} \mathbf{I} & \mathbf{0} \\ \mathbf{0} & \mathbf{I} \end{bmatrix}, \mathbf{R}_U = r\mathbf{I} \tag{3.3.22}$$

where $q, r$ are coefficients of weighting matrices pertaining to system state and control force, respectively. The ratio between the two coefficients denotes the trade-off between the effect (mitigation ratio) and the cost.

It is worth noting that the abovementioned procedure underlies a heuristic algorithm of defining the cost-function weights $\mathbf{Q}_Z$ and $\mathbf{R}_U$. The details of the procedure will be presented in the numerical examples as shown in the following section.

## 3.4 Numerical Examples

### 3.4.1 Controlled Single-Story Building Structure

A planar single-story shear frame attached with an active tendon system as sketched in Fig. 3.4 is considered here, which is subjected to the horizontal random seismic ground motion $\ddot{x}_g(\Theta, t)$. The properties of the system are as follows: the mass of the story is $m = 1 \times 10^5$ kg; the circular frequency of the uncontrolled structural system is $\omega_0 = 11.22$ rad/s; the control force of the actuator is denoted by $f(t)$; $\alpha$ represents the inclination angle of the tendon with respect to the base and the acting force $u(t)$

**Fig. 3.4** Sketch of
single-story shear frame with
active tendon system

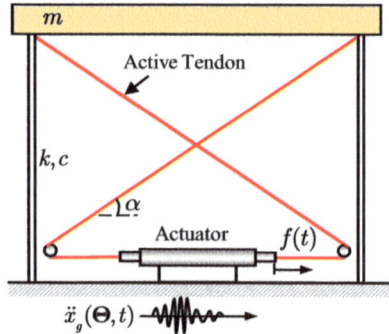

on the structure is simulated; and the damping ratio is set as 0.05. The interstory drift
is considered as the constraint and the evaluation quantities include the interstory
drift, the story acceleration, and the control force. The quantile function is defined as
mean plus three times of standard deviation of the equivalent extreme-value variables.
The threshold of interstory drift is specified to be 10 mm. The stochastic optimal
control aims to assure the structural safety through controlling the interstory drift, to
accommodate the structural habitability through controlling the story acceleration,
and to satisfy with the system workability through controlling the output of the active
tendon. Numerical simulation of structural responses employs a transfer function
method, i.e., the S-transform of linear time-invariant (LTI) systems (Mathews and
Fink 2003).

The physically motivated random seismic ground motion model addressed in
Sect. 2.5.1 is employed to represent the random seismic ground acceleration. The conditional ground motion pertaining to the background of seismic hazards is introduced
such that the influences of seismic source and wave propagation can be integrated
as the input at the bedrock. The local site is viewed as a single-degree-of-freedom
system; see Fig. 3.5. The physical relation between the surface ground motion and

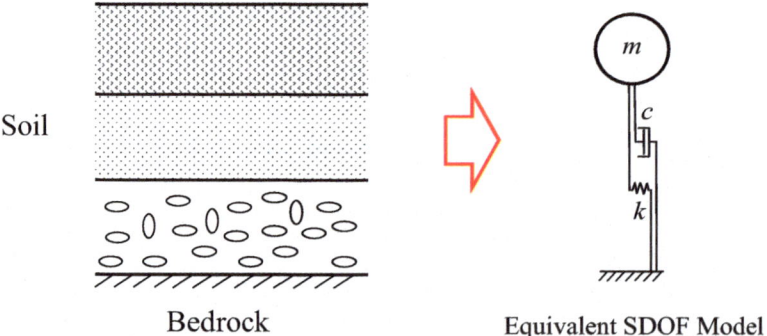

**Fig. 3.5** Equivalent single-degree-of-freedom model of local site

bedrock ground motion, predominant circular frequency, and equivalent damping ratio of local site has a theoretical formulation as follows (Li and Ai 2006):

$$\ddot{X}_g(\Theta, \omega) = \frac{\Theta_{\bar{\omega}_0}^2 + 2i\,\Theta_{\bar{\zeta}}\Theta_{\bar{\omega}_0}\omega}{\Theta_{\bar{\omega}_0}^2 - \omega^2 + 2i\,\Theta_{\bar{\zeta}}\Theta_{\bar{\omega}_0}\omega} \cdot \ddot{U}_b(\Theta_b, \omega) \tag{3.4.1}$$

where $\ddot{X}_g(\Theta, \omega)$, $\ddot{U}_b(\Theta_b, \omega)$ are the frequency-domain expressions of ground motions at the surface of engineering site and at the bedrock, respectively. The ground motion at the bedrock is mathematically assumed to be a band-limited white noise, of which the Fourier amplitude is defined based on the background of seismic hazards; $\Theta = (\Theta_{\bar{\omega}_0}, \Theta_{\bar{\zeta}}, \Theta_b)$ is the random vector characterizing the randomness inherent in the ground motion at the surface of engineering site, which is used to model the randomness inherent in systems, of which $\Theta_{\bar{\omega}_0}$, $\Theta_{\bar{\zeta}}$ are the elementary random variables denoting the uncertainty of the site soil, i.e., the predominant circular frequency of engineering site $\bar{\omega}_0$ and the equivalent damping ratio $\bar{\zeta}$; $\Theta_b = \{\Theta_{b,i}\}_{i=1}^{s_b}$ is the random vector characterizing the randomness inherent in the ground motion at the bedrock coming from the properties of seismic sources and wave propagation; $s_b$ denotes the number of random variables; $\omega$ is the circular frequency; and i is the imaginary unit.

The time history of the random ground motion then could be attained by the inverse Fourier transform:

$$\ddot{x}_g(\Theta, t) = \frac{1}{2\pi} \int_{-\infty}^{\infty} \ddot{X}_g(\Theta, \omega) e^{i\omega t} d\omega \tag{3.4.2}$$

The local site is assumed to have the properties of site class III and exhibit seismic fortification intensity 8 in terms of the Chinese Code for Seismic Design of Building Structures (GB50011-2010). Following the basic principle of stochastic modeling, the probabilistic structures and distribution parameters of the elementary random variables, i.e., the predominant circular frequency $\bar{\omega}_0$ and the equivalent damping ratio $\bar{\zeta}$ can be derived by the data fitting of recorded seismic accelerations using the least squares method. Numerical results show that the predominate circular frequency $\bar{\omega}_0$ and the equivalent damping ratio $\bar{\zeta}$ both follow the lognormal distribution, of which the mean and coefficient of variation of $\bar{\omega}_0$ are 12 rad/s, 0.42, respectively; the mean and coefficient of variation of $\bar{\zeta}$ are 0.1, 0.35, respectively. Considering the seismic hazard with return period 50 years, i.e., frequently occurring earthquake and peak ground acceleration $0.11g$, the Fourier amplitude of ground motion at the bedrock is set as $0.20$ m/s$^2$. Meanwhile, the initial phase angle in the inverse Fourier transform for simulating seismic ground accelerations is assumed to follow the normal distribution, of which mean and coefficient of variation are $\pi$, 1.2, respectively. Utilizing the tangent spheres method to carry out the partition of probability-assigned space, 221 representative points and the pertinent time histories of seismic ground accelerations are generated (Chen et al. 2007; Chen and Li

2008). Sampling frequency and duration of the simulated ground motions are 50 Hz, 20.48 s, respectively.

In order to reveal the nonstationary intensity of seismic ground motions, the following uniform modulation function is used (Li and Chen 2009):

$$f(t) = \begin{cases} t^2/4, & t \le t_a \\ 1, & t_a < t \le t_b \\ e^{-0.8(t-t_b)}, & t_b < t \le T \end{cases} \qquad (3.4.3)$$

where $t_a$ and $t_b$ are set as 2 s and 16 s, respectively and $T$ denotes the duration of the ground motion.

Statistical moments of the random seismic ground acceleration are shown in Fig. 3.6. It is seen that the amplitude of the mean (approximate 0.06 m/s$^2$) is around 8% of the amplitude of the standard deviation (approximate 0.8 m/s$^2$), indicating that the physically motivated random ground motion model exhibits the property of zero mean. Time history of a representative seismic ground acceleration is shown in Fig. 3.7. It is recognized that the random seismic ground acceleration exhibits remarkable nonstationary behaviors both in temporal and frequency domains. Two recorded seismic ground accelerations from the same site class, labeled as EL270 and EMC90, are shown in Fig. 3.8, of which the peak ground acceleration (PGA) is scaled to 0.1$g$. For comparative purposes, the acceleration response spectra of the random seismic ground motion, representative, and recorded seismic ground motions are pictured; see Fig. 3.9. It is shown clearly that the mean plus standard deviation of acceleration response spectrum derived from the random seismic ground motion accommodates the acceleration response spectra derived from the recorded seismic ground motion, indicating the moderation of the selected seismic ground motions and the rationality of the physically motivated random seismic ground motion model. This knowledge owes to the fact that the physically motivated random seismic ground

**Fig. 3.6** Mean and standard deviation of random seismic ground acceleration

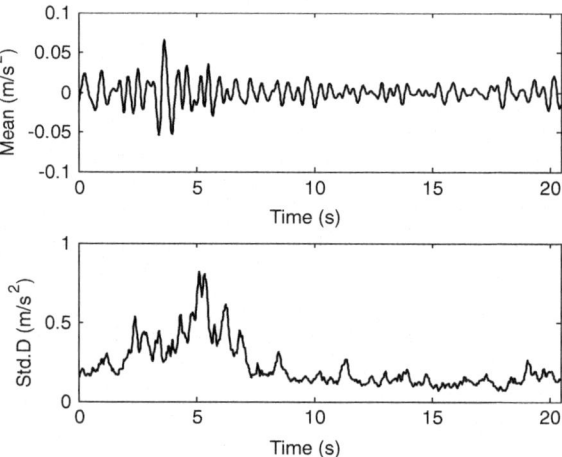

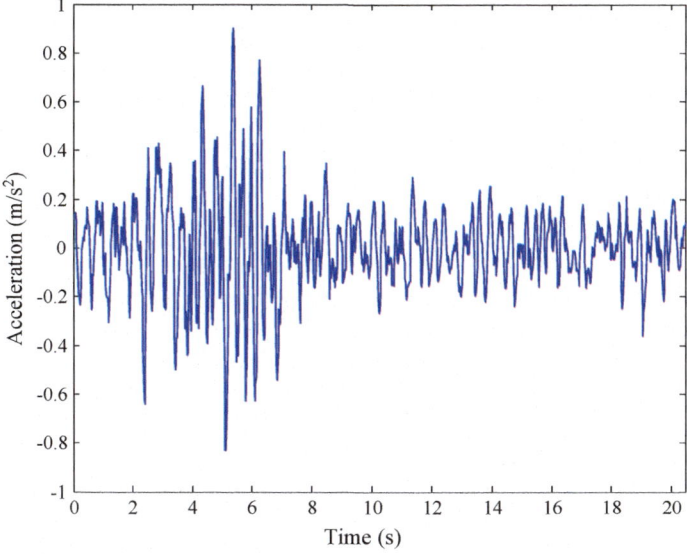

**Fig. 3.7** Time history of representative seismic ground acceleration

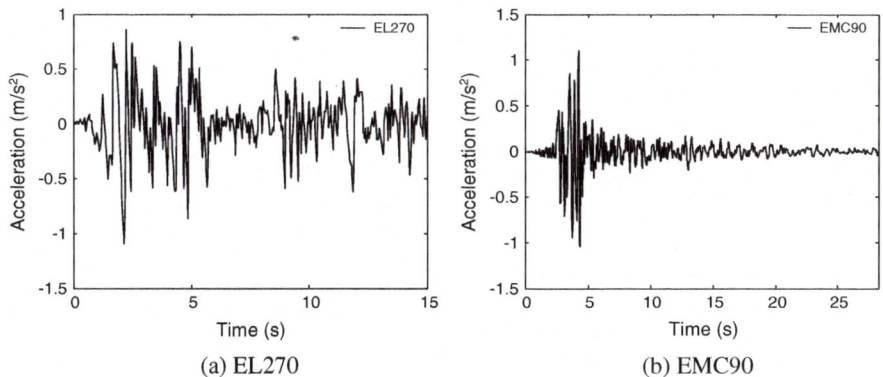

(a) EL270                               (b) EMC90

**Fig. 3.8** Time histories of recorded seismic ground accelerations from the same site class

motion model is a self-contained sample set with respect to simulated seismic ground motions; the selected seismic ground motions, in some sense, can be viewed as elements of another sample set of recorded seismic ground motions which has the same background of seismic hazards to the self-contained sample set.

In order to reveal the influence of cost-function weights on the stochastic optimal control, the relations between the equivalent extreme-value displacement, the equivalent extreme-value acceleration, the equivalent extreme-value control force, and the ratio of coefficients of weighting matrices $q/r$ are presented in Fig. 3.10, where $q$ is set as 100. It is clearly seen that these relation curves are different both from the

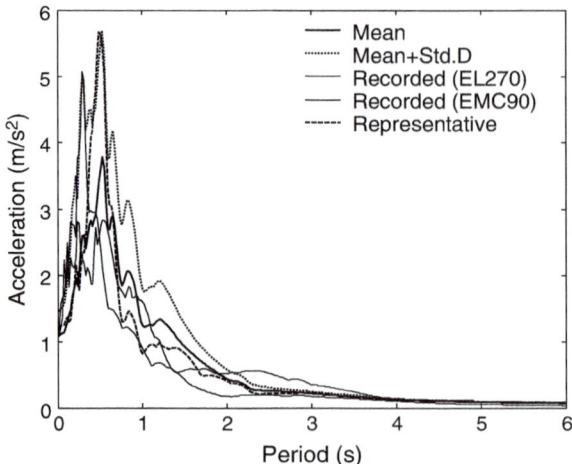

**Fig. 3.9** Acceleration response spectra of random seismic ground motion, representative, and recorded seismic ground motions

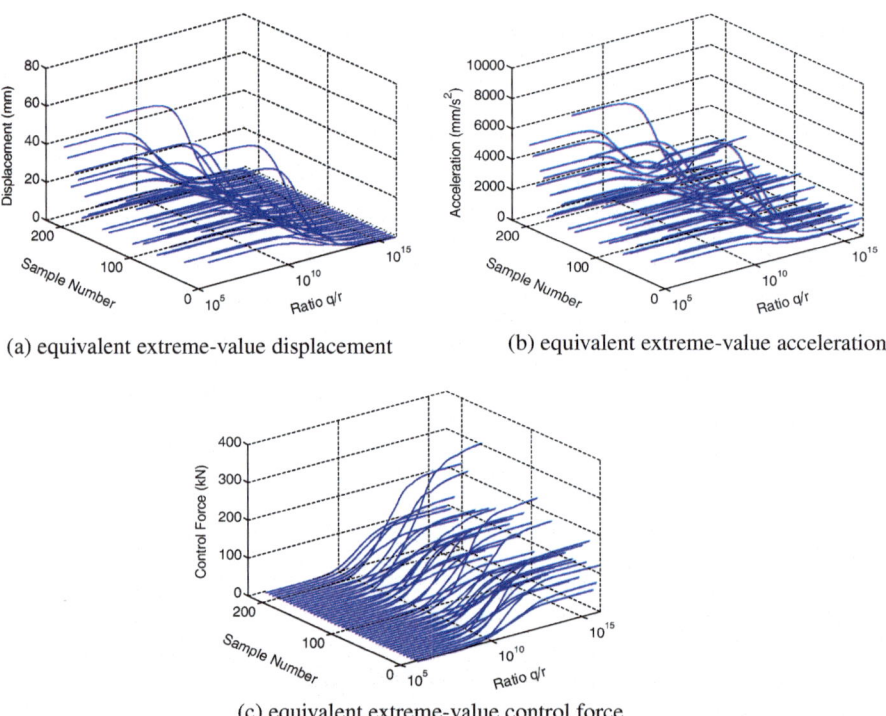

(a) equivalent extreme-value displacement

(b) equivalent extreme-value acceleration

(c) equivalent extreme-value control force

**Fig. 3.10** Relation between equivalent extreme-value quantities and ratio of coefficients of weighting matrices

structural responses and seismic ground motions, e.g., concerning the same seismic ground motion, the relation between control force and the ratio $q/r$ changes from the structural displacement, velocity, and acceleration; concerning the same structural response, the relation between control force and the ratio $q/r$ changes from seismic ground motions. The optimal ratio $q^*/r^*$, in other words, nominally arises to be different from samples of the random seismic ground motion and exhibits a certain randomness. Moreover, there exists a relevance between control effectiveness and control cost; the definition of the optimal ratio $q^*/r^*$ shall consider the trade-off between system quantities of interest. One might wonder, however, which ratio of coefficients of weighting matrices for the stochastic optimal control exhibits the optimality in a global sense?

In fact, the control law involves a deterministic gain matrix even in the stochastic optimal control, which relies upon the structural performance objective and can be derived as a probabilistic criterion, e.g., the system second-order statistics evaluation. Figure 3.11 shows the relation between the mean of equivalent extreme -value displacement, equivalent extreme-value velocity, equivalent extreme-value

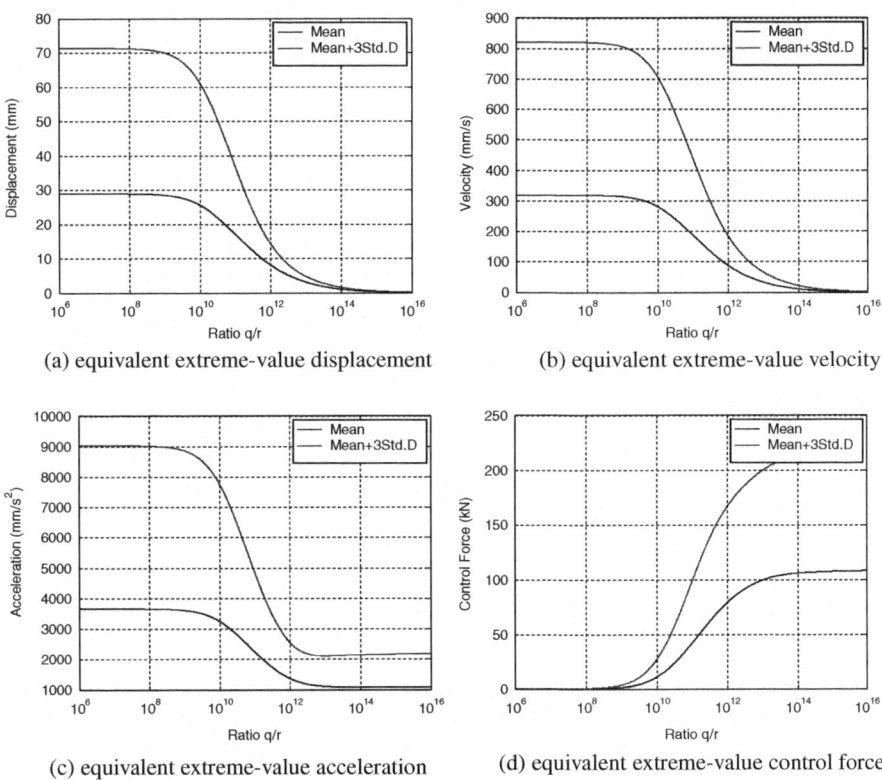

(a) equivalent extreme-value displacement

(b) equivalent extreme-value velocity

(c) equivalent extreme-value acceleration

(d) equivalent extreme-value control force

**Fig. 3.11** Relation between mean, quantile of quantities, and ratio of coefficients of weighting matrices

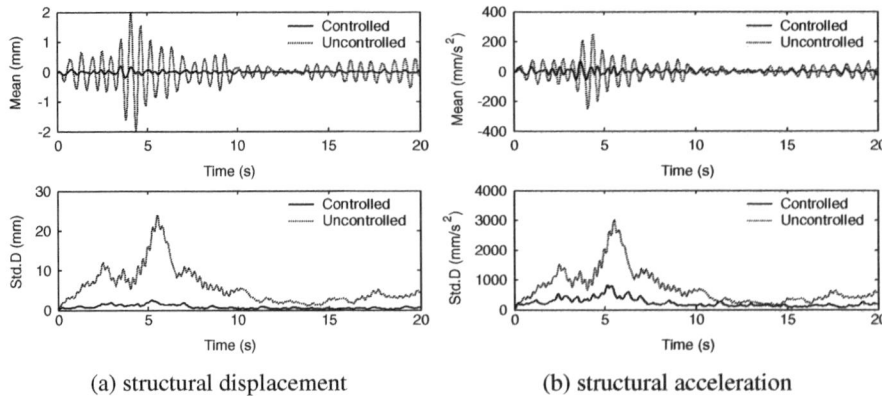

**Fig. 3.12** Time histories of mean and standard deviation of structural responses with and without controls

acceleration, equivalent extreme-value control force, and the ratio $q/r$. The quantiles of these quantities are shown as well. It is seen that: (i) as ratio $q/r \geq 2 \times 10^{12}$, the quantile of the displacement is completely within its threshold as a constraint; (ii) as ratio $q/r \geq 2 \times 10^{14}$, the standard deviation of the displacement is minimum and the mean decreases slowly, whereas the standard deviation of the acceleration increases and the mean of acceleration decreases very gently; the mean and standard deviation of the control force, however, increase significantly; (iii) as ratio $q/r = 8 \times 10^{12}$, the standard deviation of the acceleration attains minimum, although the standard deviation of the displacement is not minimum; meanwhile, the means of the acceleration and the displacement decrease evidently. The benefit at this ratio, moreover, lies in that the mean and standard deviation of the control force are much less than those at the ratio $q/r \geq 2 \times 10^{14}$. Considering the trade-off between the quantities of interest, it is thus reasonable to take the optimal ratio $q^*/r^* = 8 \times 10^{12}; q^* = 80, r^* = 10^{-11}$ in the numerical case.

Time histories of the mean and standard deviation of structural displacement and those of structural acceleration with and without controls are shown in Fig. 3.12. It is seen that the structural responses with control are reduced significantly in comparison with those without control. Moreover, the structural responses decrease significantly in the time interval with larger amplitudes; see the interval from 2 s to 8 s, which indicates that the stochastic optimal control aims at enhancing the structural robustness in a global sense. The amplitudes of standard deviations of the displacement and the acceleration with control are reduced by 5 and 3 times than those without control, respectively. It is also seen from Fig. 3.12 that the amplitude of the mean is around 8% smaller than that of the standard deviation. It is understood that the linear structural system is driven by the random seismic ground motion with zero mean.

Figure 3.13 shows the PDFs of the displacement of the controlled and uncontrolled structures at typical instants of time 4 s, 7 s, and 10 s. It is seen that the variation of the structural displacement with control is reduced significantly by comparison with

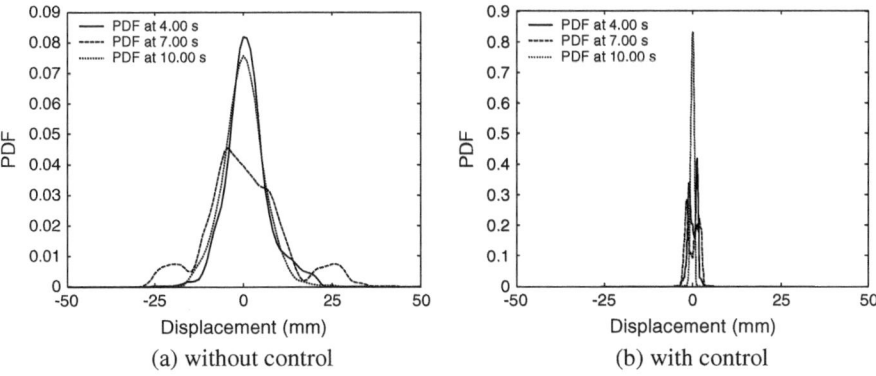

**Fig. 3.13** PDFs of structural displacement at typical instants of time

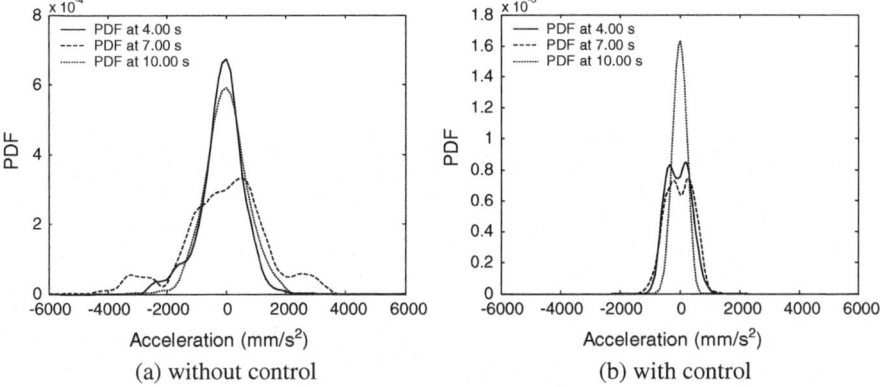

**Fig. 3.14** PDFs of structural acceleration at typical instants of time

that without control. The PDFs of the structural acceleration at typical instants of time show similar properties to the structural displacement, as shown in Fig. 3.14. It is revealed that the seismic performance of the structure has been enhanced greatly after the stochastic optimal control is applied. For details of time-varying probabilistic information, the probability densities of structural displacement and structural acceleration at typical time interval from 4 s to 10 s are shown in Figs. 3.15 and 3.16, respectively.

The mean and standard deviation of control force and PDFs at typical instants of time are shown in Fig. 3.17. It is seen that the shapes of the mean and standard deviation of control force and the pertinent curves of the probability density function exhibit certain similarities to the structural displacement and structural acceleration. These similarities are resulted from the cause that the optimal control force is pursuing the structural response in real time, which is a weighted combination of structural displacement and velocity as the relevant elements of the gain matrix (Chung et al.

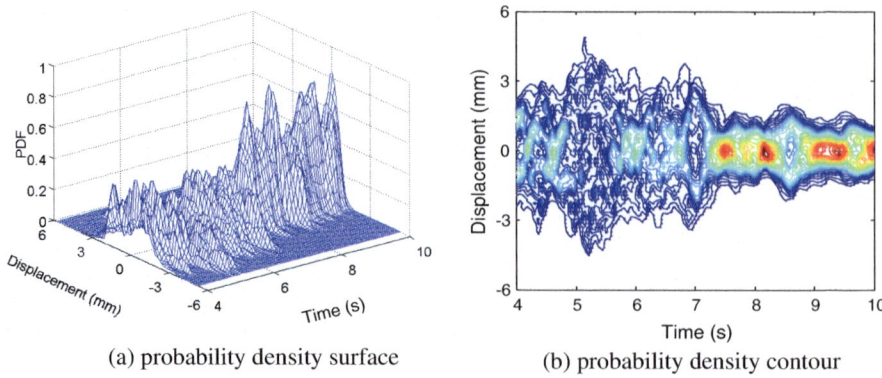

(a) probability density surface                (b) probability density contour

**Fig. 3.15** Probability density of structural displacement at typical time interval

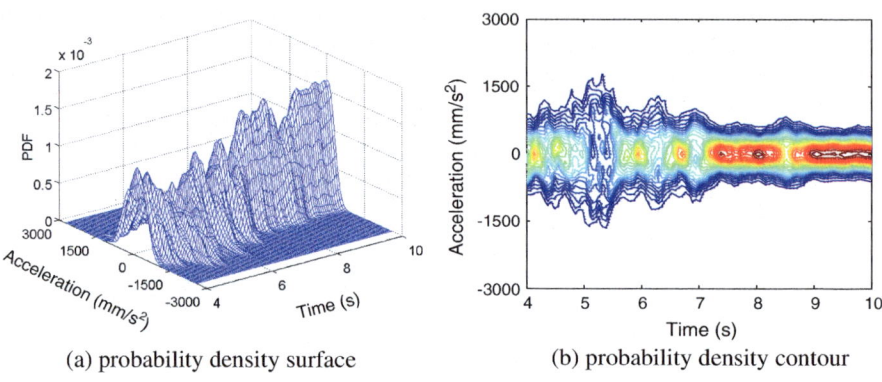

(a) probability density surface                (b) probability density contour

**Fig. 3.16** Probability density of structural acceleration at typical time interval

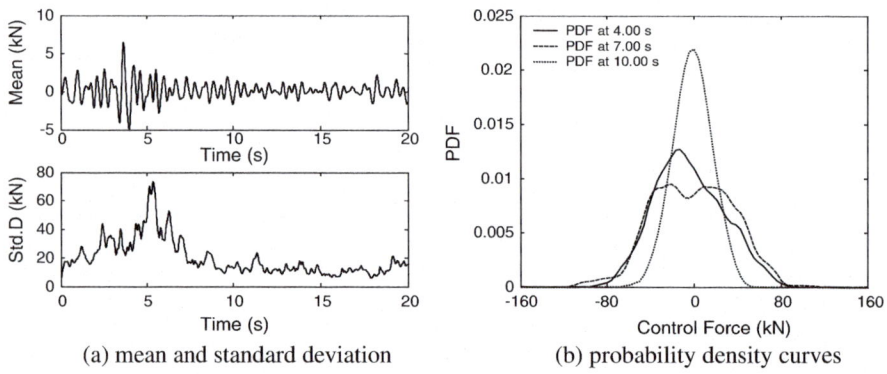

(a) mean and standard deviation                (b) probability density curves

**Fig. 3.17** Mean and standard deviation of control force and PDFs at typical instants of time

1988). However, the generalized probability density evolution equation is essentially a nonlinear first-order partial differential equation, which results in the differences of PDFs between system state and control force even if the control force is linearly mapped from the system state.

## 3.4.2  Controlled Multiple-Story Building Structure

An eight-story shear frame attached with fully distributed active tendon systems is studied. The properties of the uncontrolled structure are taken from the publication by Yang et al. (Yang et al. 1987). The story mass is $m_i = 3.456 \times 10^5$ kg; interstory stiffness is $k_i = 3.404 \times 10^2$ kN/mm; and internal damping coefficient of each story is $c_i = 2.937$ kNs/mm, which corresponds to a 2% damping ratio for the first vibration mode of the entire structure. The external damping is assumed to be zero. The calculated natural frequencies are 5.79, 17.18, 27.98, 37.82, 46.38, 53.36, 58.53, and 61.69 rad/s, respectively. The constraint quantity, evaluation quantity, and control objective are the same as those of the case shown in Sect. 3.4.1. The quantile function is defined as the mean plus one time of standard deviation. The threshold of interstory drift is set as 15 mm. The random seismic ground motion model is employed, of which the peak ground acceleration is $0.3g$.

Figure 3.18 shows the relation between the mean, quantile of equivalent extreme-value displacement, equivalent extreme-value acceleration, equivalent extreme-value control force, and the ratio of coefficients of weighting matrices. It is shown that (i) as ratio $q/r \geq 4 \times 10^{13}$, the quantile of the displacement is completely within its threshold as a constraint; (ii) as ratio $q/r = 1 \times 10^{14}$, the standard deviation of the displacement is minimum and the mean approaches to its minimum, which are both decreasing significantly; and the mean of control force increases constantly, of which the standard deviation, however, provisionally possesses a small value. Therefore, it is reasonable to take the optimal ratio $q^*/r^* = 1 \times 10^{14}$; $q^* = 100, r^* = 10^{-12}$ in this numerical case.

Figure 3.19 shows the time histories of mean and standard deviation of the first and eighth interstory drifts with and without controls. It is seen that the interstory drifts are reduced significantly when the structure is under control. The amplitudes of interstory drifts with control are nearly 4 times smaller than those without control. Similar to the displacement of the single-story structural system with control shown in Sect. 3.4.1, the time interval with larger amplitudes gets an obvious improvement. The interstory drift with control exhibits almost same mitigation ratio along the story level of the structure. Story acceleration, however, does not exhibit this behavior. Shown in Fig. 3.20 is the time histories of the mean and standard deviation of the first and eighth story accelerations with and without controls. It is seen that the first story acceleration remains nearly unchanged; while the eighth story acceleration is improved significantly. The two story accelerations, however, have almost the same variation, owing to the fact that the objective of the stochastic optimal control is

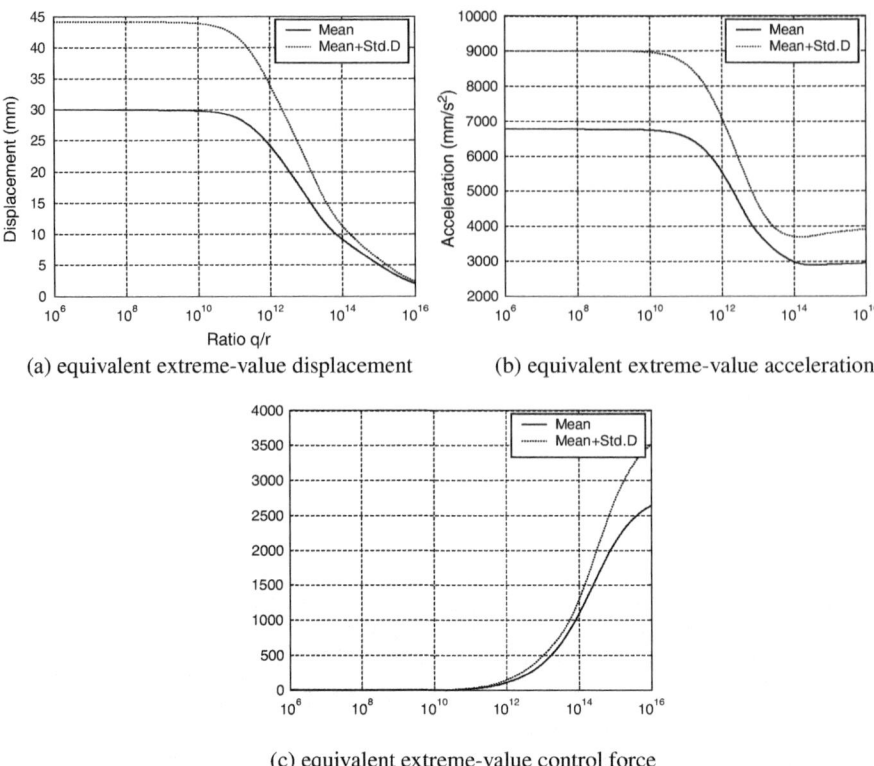

(a) equivalent extreme-value displacement          (b) equivalent extreme-value acceleration

(c) equivalent extreme-value control force

**Fig. 3.18** Relation between mean, quantile of quantities, and ratio of coefficients of weighting matrices

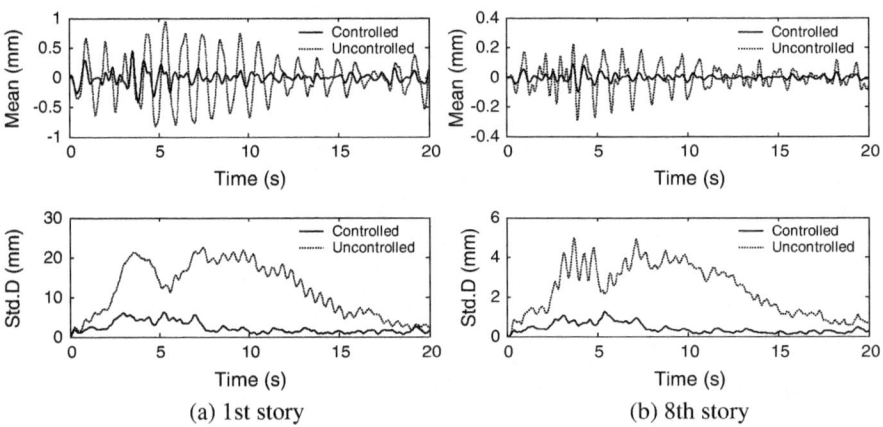

(a) 1st story                              (b) 8th story

**Fig. 3.19** Time histories of mean and standard deviation of interstory drifts of structure with and without controls

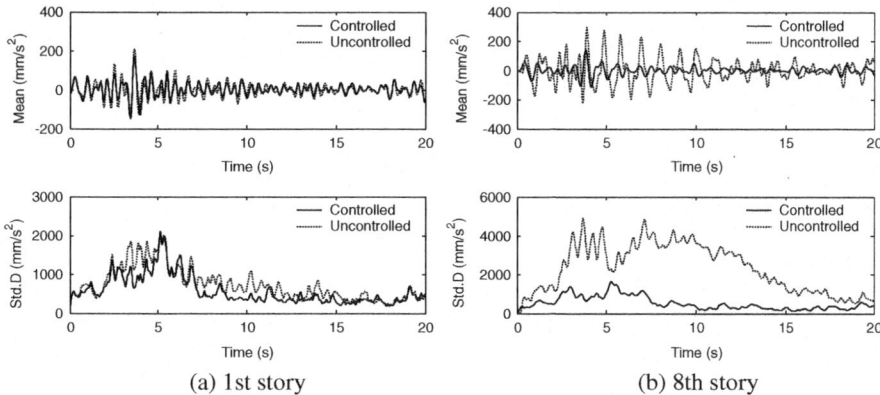

**Fig. 3.20** Time histories of mean and standard deviation of story accelerations of structure with and without controls

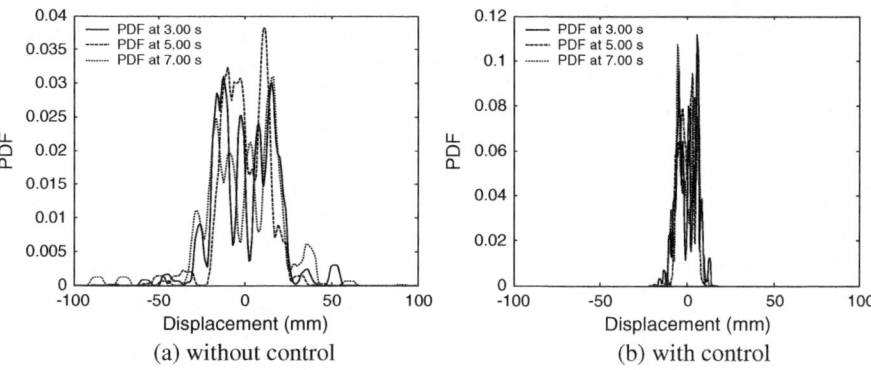

**Fig. 3.21** PDFs of first interstory drift at typical instants of time

to optimize the structural performance in the sense of the trade-off between system quantities of interest.

The PDFs of the first and eighth interstory drifts at typical instants of time 3 s, 5 s, and 7 s with and without controls are shown in Figs. 3.21 and 3.22. By comparison with the cases without control, the distribution range of the PDFs of interstory drifts with control becomes narrower and the shape of the PDFs arises to be more irregular. It is indicated that the shear frame structure with control does not move as a similar profile to that without control since the introduction of the control force leads to a change of contribution from vibrational modes to structural responses. Similar control effectiveness is shown in the PDFs of the first and eighth story accelerations at typical instants of time with and without controls; see Figs. 3.23 and 3.24. The probability densities of the first interstory drift and story acceleration at typical time interval from 3 s to 7 s are pictured in Figs. 3.25 and 3.26, respectively.

Time histories of the mean and standard deviation of the first and eighth interstory control forces are shown in Fig. 3.27. It is readily seen that the first interstory control

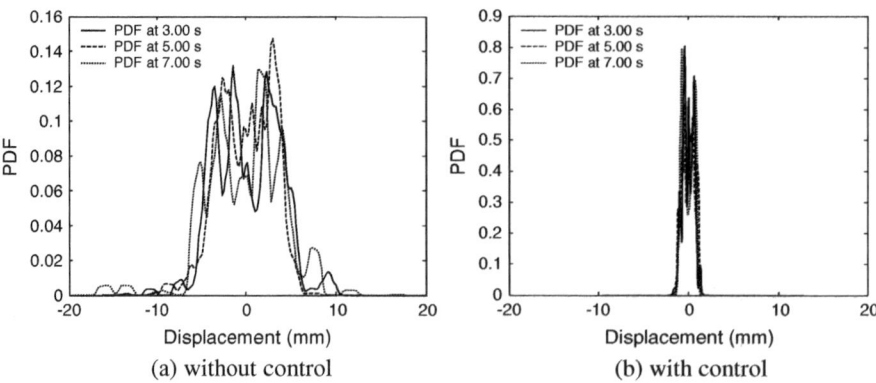

**Fig. 3.22**  PDFs of eighth interstory drift at typical instants of time

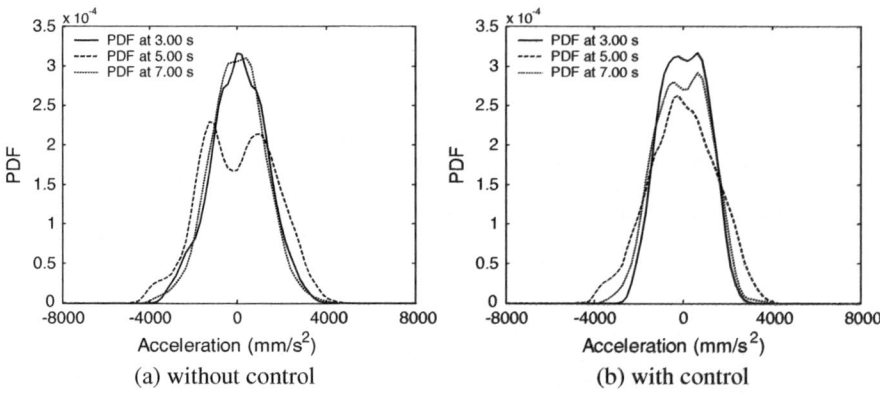

**Fig. 3.23**  PDFs of first story acceleration at typical instants of time

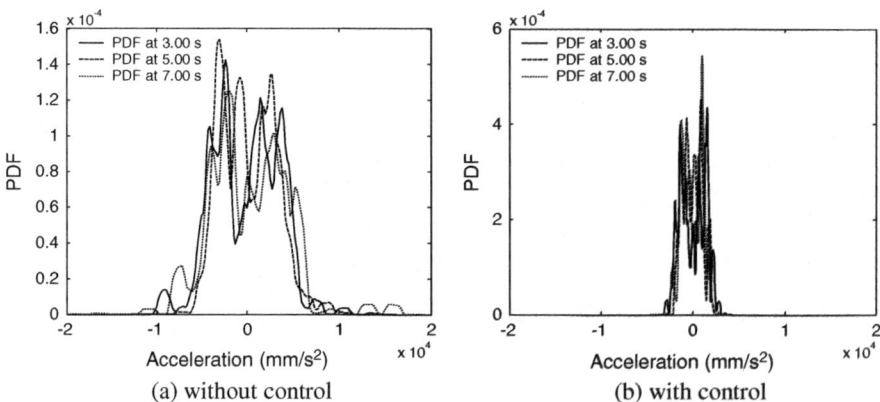

**Fig. 3.24**  PDFs of eighth story acceleration at typical instants of time

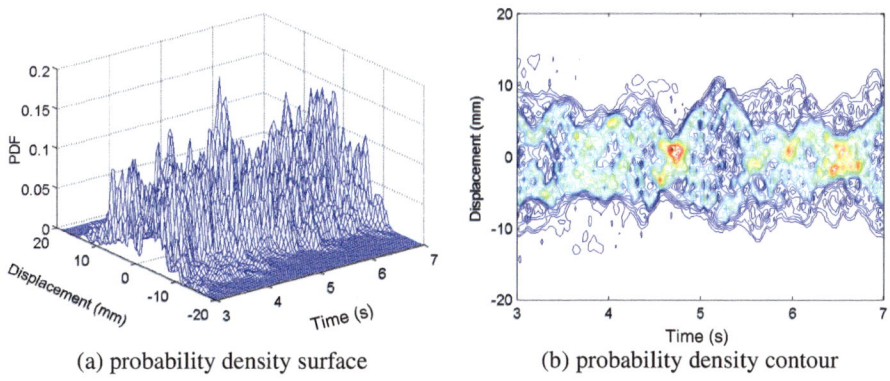

(a) probability density surface                    (b) probability density contour

**Fig. 3.25** Probability density of first interstory drift at typical time interval with control

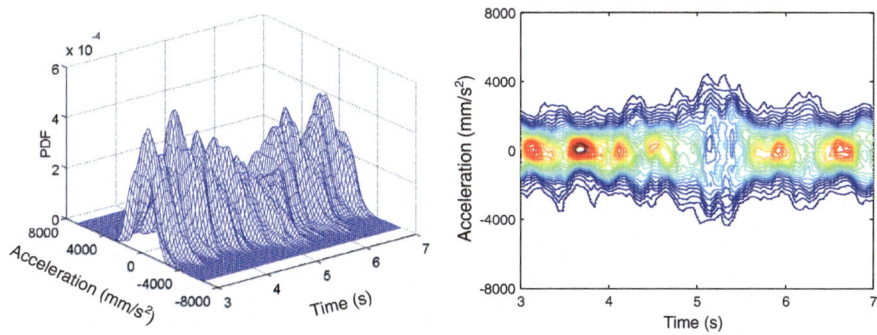

**Fig. 3.26** Probability density of first story acceleration at typical time interval with control

force has certain similarities to the eighth interstory control force except the scale of amplitude, of which the first interstory control force is 10 times larger than the eighth interstory control force. It is also seen that the time histories of the standard deviations of the first and eighth interstory control forces exhibit positive similarities, while the time histories of the means exhibit negative similarities. In view of Figs. 3.19 and 3.20, one might recognize that the responses of the two stories arise to be asynchronous, and accordingly the feedback control forces on the two stories arise to be asynchronous. This phenomenon is also shown in the PDFs of interstory control forces at typical instants of time; see Fig. 3.28.

Control effectiveness of extreme-value responses of the eight-story building structure by active tendon systems is shown in Table 3.1. It is seen that the interstory drift reduces significantly, of which the mean decreases about 70%, the standard deviation decreases about 85%, and the amplitude of the response is reduced nearly the same along story level of the structure. The reduction of acceleration from the fifth story to the eighth story with control is remarkable as well, of which the mean decreases 50% on average and the standard deviation decreases 75% on average. It is noted

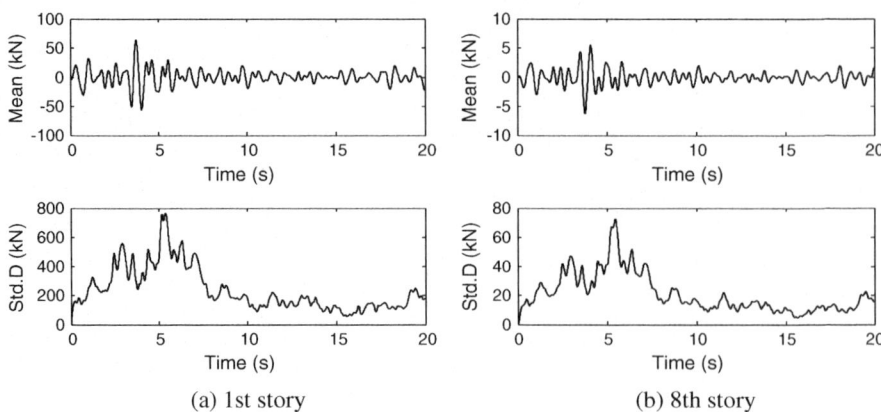

**Fig. 3.27** Time histories of mean and standard deviation of interstory control forces

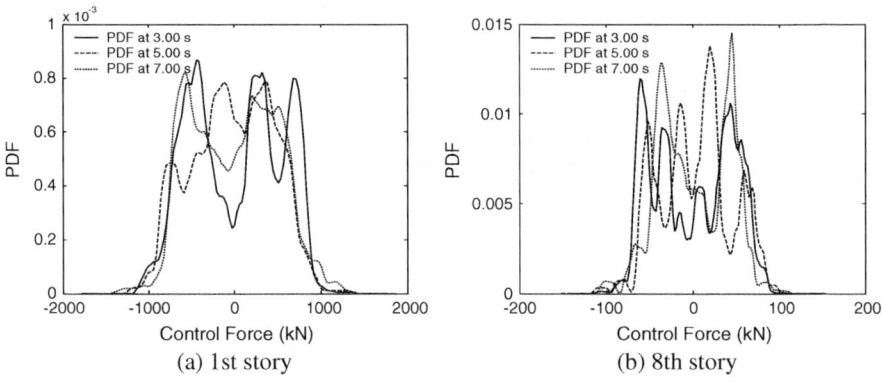

**Fig. 3.28** PDFs of interstory control forces at typical instants of time

that the control effectiveness of acceleration of the low two stories is obviously less than other stories. As it is indicated in Table 3.1, the story acceleration arises to be more uniform along the story level of the structure after the structure is controlled. Moreover, the higher the story level, the smaller the control force arises to be. Therefore, more control cost is required for the low stories in order to attain a uniform story acceleration along the story level. The ratios of extreme-value control forces, besides, between different interstories are almost equal to those of extreme-value interstory drift, which is due to the fact that the control force is a linear function with respect to the interstory drift.

**Table 3.1** Control effectiveness of eight-story shear frame by active tendon systems

| Equ. ext.-value | | | Story number | | | | | | | |
|---|---|---|---|---|---|---|---|---|---|---|
| | | | 1 | 2 | 3 | 4 | 5 | 6 | 7 | 8 |
| Interstory drift (mm) | Mn | Unc. | 29.95 | 28.66 | 26.65 | 24.64 | 21.94 | 18.11 | 13.02 | 6.84 |
| | | Con. | 9.17 | 8.62 | 7.97 | 7.14 | 6.12 | 4.86 | 3.39 | 1.74 |
| | | Eff. | −69.38% | −69.92% | −70.09% | −71.02% | −72.11% | −73.16% | −73.96% | −74.56% |
| | Std. d | Unc. | 14.11 | 13.91 | 13.16 | 11.39 | 9.15 | 6.84 | 4.56 | 2.28 |
| | | Con. | 2.16 | 2.05 | 1.85 | 1.61 | 1.32 | 1.02 | 0.69 | 0.35 |
| | | Eff. | −84.69% | −85.26% | −85.94% | −85.86% | −85.57% | −85.09% | −84.87% | −84.65% |
| Story acceleration (mm/s$^2$) | Mn | Unc. | 3031.9 | 3489.5 | 4140 | 4547.7 | 4645 | 5150.4 | 6118.4 | 6759.5 |
| | | Con. | 2803.3 | 2616.0 | 2435.8 | 2302.3 | 2229.7 | 2230.9 | 2264.0 | 2294.0 |
| | | Eff. | −7.54% | −25.03% | −41.16% | −49.37% | −52.00% | −56.68% | −63.00% | −66.06% |
| | Std. d | Unc. | 918.0 | 1296.2 | 1549.7 | 1671.4 | 1976.6 | 2268.7 | 2249.2 | 2249.5 |
| | | Con. | 862.3 | 693.1 | 526.5 | 405.6 | 376.1 | 410.8 | 455.3 | 483.6 |
| | | Eff. | −6.07% | −46.53% | −66.03% | −75.73% | −80.97% | −81.89% | −79.76% | −78.50% |
| Interstory control force (kN) | Mn | | 1090.25 | 712.66 | 503.30 | 339.16 | 205.52 | 131.26 | 117.39 | 99.49 |
| | Std. d | | 196.63 | 155.79 | 142.61 | 127.91 | 102.72 | 62.04 | 29.67 | 21.58 |

[a]Effectiveness is defined as (Con.-Unc.)/Unc.

## 3.5  Comparative Studies against LQG

In order to validate the physically based stochastic optimal control, comparative studies against the classical LQG and the deterministic control are carried out. For illustrative purposes, the numerical example addressed in Sect. 3.4.1 is employed.

Concerning the controlled structure with active tendon system shown in Fig. 3.4, the equation of motion is given by

$$\ddot{x}(t) + 2\zeta\omega_0\dot{x}(t) + \omega_0^2 x(t) = m^{-1}u(t) - \ddot{x}_g(\Theta, t) \tag{3.5.1}$$

which can be rewritten as a formulation in state space as follows:

$$\dot{\mathbf{Z}}(t) = \mathbf{A}\mathbf{Z}(t) + \mathbf{B}u(t) + \mathbf{D}\ddot{x}_g(\Theta, t) \tag{3.5.2}$$

where

$$\mathbf{Z}(t) = \begin{bmatrix} x(t) \\ \dot{x}(t) \end{bmatrix}, \mathbf{A} = \begin{bmatrix} 0 & 1 \\ -\omega_0^2 & -2\zeta\omega_0 \end{bmatrix}, \mathbf{B} = \begin{bmatrix} 0 \\ m^{-1} \end{bmatrix}, \mathbf{D} = \begin{bmatrix} 0 \\ -1 \end{bmatrix} \tag{3.5.3}$$

The cost function of the LQG is defined as (Chen et al. 1998)

$$J_1(\mathbf{Z}, u) = E\left[ \mathbf{S}(\mathbf{Z}(t_f), t_f) + \frac{1}{2}\int_{t_0}^{t_f} (\mathbf{Z}^{\mathrm{T}}(t)\mathbf{Q}_{\mathbf{Z}}\mathbf{Z}(t) + \mathbf{R}_{\mathrm{U}}u^2(t))\mathrm{d}t \right] \tag{3.5.4}$$

of which the constraint condition is given by

$$\begin{cases} \mathrm{d}\mathbf{Z}(t) = [\mathbf{A}\mathbf{Z}(t) + \mathbf{B}u(t)]\mathrm{d}t + \mathbf{L}\mathrm{d}w(t) \\ \qquad\qquad \mathbf{Z}(t_0) = \mathbf{0} \end{cases} \tag{3.5.5}$$

where $\mathbf{L}$ is the $(2 \times 1)$ matrix denoting the location of external excitation; $w(t)$ denotes the one-dimensional Brownian motion, which is generally modeled as a white Gaussian noise:

$$E[\mathrm{d}w(t)] = 0, E[\mathrm{d}w^2(t)] = 2\beta S_0\mathrm{d}t \tag{3.5.6}$$

where $S_0$ is the spectral intensity factor of random seismic ground motion $\ddot{x}_g(\Theta, t)$, which is estimated by

$$S_0 = \frac{\bar{a}_{\max}^2}{f^2\omega_e} \tag{3.5.7}$$

where $\bar{a}_{\max}$ denotes the mean of the peak ground acceleration of random seismic ground motion; $f$ denotes the peak factor; and $\omega_e$ denotes the spectral area pertain-

**Table 3.2** Relation between spectral intensity factor, peak ground acceleration (PGA), and site class

| Site class/PGA | I | | II | | III | | IV | |
|---|---|---|---|---|---|---|---|---|
| | $0.11g$ | $0.3g$ | $0.11g$ | $0.3g$ | $0.11g$ | $0.3g$ | $0.11g$ | $0.3g$ |
| $f$ | 2.9 | 2.9 | 3.0 | 3.0 | 3.1 | 3.1 | 3.2 | 3.2 |
| $\omega_e$ (rad s$^{-1}$) | 59.50 | 59.50 | 39.71 | 39.71 | 29.93 | 29.93 | 19.95 | 19.95 |
| $S_0$ (m$^2$ s$^{-3}$) | 0.0023 | 0.0173 | 0.0033 | 0.0242 | 0.0040 | 0.0301 | 0.0057 | 0.0423 |

See Chinese Code for Seismic Design of Building Structures (GB50011-2010), $1g = 9.8$ m/s$^2$

ing to unit spectral intensity factor. Table 3.2 shows the spectral intensity factor in the cases of peak ground accelerations $\bar{a}_{max} = 0.1g, 0.3g$ and typical site classes in accordance with the Chinese Code for Seismic Design of Building Structures (GB50011-2010).

It is seen that the mathematical formulation of the equation of motion of the controlled structure with active tendon system, i.e., Eq. (3.5.5) is just the classical Itô stochastic differential equation. As it is mentioned in the previous sections, the measurement noise inherent in system state and control force is out of concern, and is thus ignored in this study. The random seismic ground motion $\ddot{x}_g(\Theta, t)$ is assumed to be a wide-band excitation and mathematically modeled by a nominal white Gaussian noise.

Transferring the constraint extreme-value problem of function Eq. (3.5.4) to an unconstraint extreme-value problem, the solution of the control system can be derived by solving Hamilton–Jacobi–Bellman equation in the context of randomness. Introducing a generalized Hamilton function as follows (Li and Chen 2009):

$$H[\mathbf{Z}^*(t), u(t), t] = \frac{1}{2}\left(\mathbf{Z}^{*T}\mathbf{Q}_Z\mathbf{Z}^* + \mathbf{R}_U u^2\right) + \frac{\partial V}{\partial \mathbf{Z}}\left(\mathbf{AZ}^* + \mathbf{B}u\right) + \pi S_0 \text{Tr}\left(\frac{\partial^2 V}{\partial \mathbf{Z}^2}\mathbf{LL}^T\right) \tag{3.5.8}$$

where $V$ denotes the optimal value function and is assumed to be

$$V(\mathbf{Z}(t), t) = \frac{1}{2}\mathbf{Z}^T(t)\mathbf{P}(t)\mathbf{Z}(t) + v(t) \tag{3.5.9}$$

of which $v(t)$ is a correct term with respect to the randomness associated with the generalized Hamilton function.

Utilizing the dynamic programming method, one could gain the solution

$$u(t) = -\mathbf{R}_U^{-1}\mathbf{B}^T\mathbf{P}(t)\mathbf{Z}(t) \tag{3.5.10}$$

$$v(t) = -\pi S_0 \int_{t_0}^{t_f} \text{Tr}(\mathbf{P}(t)\mathbf{LL}^T)\mathrm{d}t \tag{3.5.11}$$

where $\mathbf{P}(t)$ denotes the Riccati matrix, satisfying with the matrix algebraic Riccati equation; see Eq. (3.3.17).

In view of Eqs. (3.5.10) and (3.3.18), the control law of the LQG has the same formulation to the LQR-based PSO for a closed-loop system in the sense of sample trajectory, and the so-called deterministic equivalence principle is satisfied. It is shown as well that concerning the linear time-invariant system subjected to the white Gaussian noise, the gain matrix can be calculated offline though the Hamilton function includes a random excitation term.

Substituting Eq. (3.5.10) into Eq. (3.5.1) and using the Fourier transform on both sides, one has

$$\{[(\omega_0^2 + m^{-1}\widehat{K}) - \omega^2] + (2\zeta\omega_0 + m^{-1}\widehat{C})(i\omega)\}x(\omega) = -\ddot{x}_g(\boldsymbol{\Theta}, \omega) \qquad (3.5.12)$$

where $\widehat{C}, \widehat{K}$ denote the numerical damping and numerical stiffness provided by the control force $u(t)$, respectively,

$$\widehat{C} = \mathbf{R}_U^{-1}(B_1 P_{12} + B_2 P_{22}), \ \widehat{K} = \mathbf{R}_U^{-1}(B_1 P_{11} + B_2 P_{21}) \qquad (3.5.13)$$

According to the statistical relation between the input and output of linear stochastic systems in frequency domain (Crandall 1958), one has

$$S_X(\omega) = \frac{S_0}{[(\omega_0^2 + m^{-1}\widehat{K}) - \omega^2]^2 + (2\zeta\omega_0 + m^{-1}\widehat{C})^2\omega^2} \qquad (3.5.14)$$

In view of Wiener–Khintchine theorem (Wiener 1964; Chatfield 1989), the mean-square displacement under control is then derived as follows:

$$E[x^2(t)] = \int_{-\infty}^{\infty} \frac{S_0}{[(\omega_0^2 + m^{-1}\widehat{K}) - \omega^2]^2 + (2\zeta\omega_0 + m^{-1}\widehat{C})^2\omega^2} d\omega \qquad (3.5.15)$$

Concerning the integral shown in Eq. (3.5.15), a closed solution can be attained by virtue of a class of specified rules (Roberts and Spanos 1990), which is given by

$$E[x^2(t)] = \frac{\pi S_0}{(2\zeta\omega_0 + m^{-1}\widehat{C})(\omega_0^2 + m^{-1}\widehat{K})} \qquad (3.5.16)$$

Obviously, there is a linear relation between system state and control force in frequency domain, which is given as

$$u(\omega) = [-\widehat{C}(i\omega) - \widehat{K}]x(\omega) \qquad (3.5.17)$$

Then, the mean-square control force can be derived as follows:

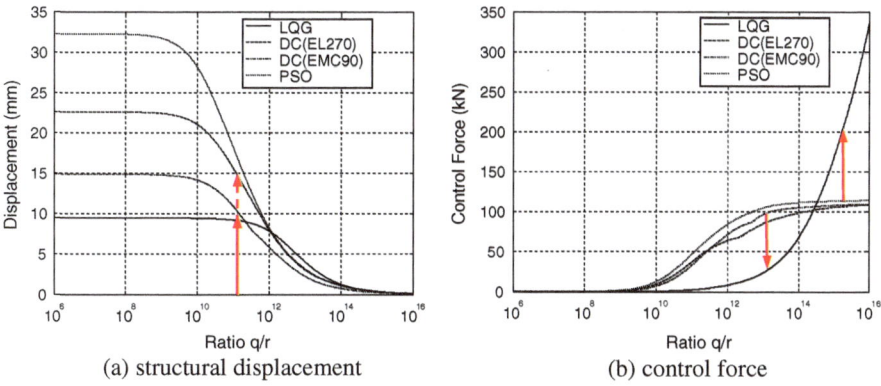

(a) structural displacement          (b) control force

**Fig. 3.29** Relation between root-mean-square quantities and ratio of coefficients of weighting matrices by means of PSO, LQG, and DC

$$E[u^2(t)] = \int_{-\infty}^{\infty} \frac{(\widehat{K}^2 + \widehat{C}^2 \omega^2)S_0}{[(\omega_0^2 + m^{-1}\widehat{K}) - \omega^2]^2 + (2\zeta\omega_0 + m^{-1}\widehat{C})^2\omega^2} \, d\omega \qquad (3.5.18)$$

Using the rule shown in Eq. (3.5.16) once again, one has

$$E[u^2(t)] = \frac{\pi S_0[\widehat{C}^2(\omega_0^2 + m^{-1}\widehat{K}) + \widehat{K}^2]}{(2\zeta\omega_0 + m^{-1}\widehat{C})(\omega_0^2 + m^{-1}\widehat{K})} \qquad (3.5.19)$$

Figure 3.29 shows the relation between root-mean-square quantities and the ratio of coefficients of weighting matrices by means of the physically based stochastic optimal (PSO) control, the LQG control, and the deterministic control (DC) using random seismic ground motion, white Gaussian noise, and recorded seismic ground motions, i.e., EL270 and EMC90, as the external excitation. In this case, the coefficient of state weighting matrix is set as 100. For the LQG, the root-mean-square quantities can be calculated directly from Eqs. (3.5.16) and (3.5.19); while for the PSO and the DC, the root-mean-square quantities are identified as their peaks since the derivations of these quantities are time variant and nonstationary.

It is indicated that (i) as the ratio $10^6 \leq q/r < 10^{12}$, the LQG underestimates the structural displacement, in that the stationary response of the structural system subjected to white Gaussian noise is several times lower than the peaks of the nonstationary response of the structural system subjected to the random and recorded seismic ground motions. The LQG, meanwhile, underestimates the desired control force. It is seen that the control force assessed by the LQG increases exponentially along with the ratio of coefficients of weighting matrices in logarithmic scale. The control force assessed by the PSO, however, increases logarithmically along with the ratio of coefficients of weighting matrices in logarithmic scale; (ii) as ratio

$10^{12} \leq q/r < 4 \times 10^{14}$, the structural displacement controlled by the PSO declines significantly and the peak is close to that controlled by the LQG. The difference of control forces between the two schemes becomes large along with the ratio of coefficients of weighting matrices in logarithmic scale; (iii) as ratio $q/r \geq 4 \times 10^{14}$, the structural responses controlled by the two schemes are almost the same, but the control force of the LQG increases exponentially and surpasses that of the PSO rapidly; and (vi) similar to the structural displacement, the structural velocity and structural acceleration quantified by the LQG change exponentially along with the ratio of coefficients of weighting matrices. In summary, the LQG underestimates the desired control force when the ratio of coefficients of weighting matrices is set at a low level, while it overestimates the desired control force when the ratio of coefficients of weighting matrices is set at a high level. It is thus remarked that employing the LQG with nominal white Gaussian noise as the input cannot attain a reasonable structural control system for civil engineering structures.

It is also seen from Fig. 3.29 that by means of the deterministic control (DC), the structural control system designed as the seismic ground motion EMC90 might be disabled when the structure is subjected to the seismic ground motion EL270. If the ratio of coefficients of weighting matrices is set as $q/r = 2 \times 10^{11}$, for instance, the extreme value of structural displacement is within 10 mm when the structure is subjected to the seismic ground motion EMC90. However, the structural displacement attains 15 mm when the structure is subjected to the seismic ground motion EL270, although the required control forces designed as the two recorded seismic ground motions are almost same. It is thus demonstrated that the deterministic control cannot guarantee a safe structure; while the PSO offers an elegant means for the logical control of structures that can secure a safe structure in the sense of probability.

## 3.6   Discussions and Summaries

The relevant theory and methods for the classical stochastic optimal control such as the LQG still remain open for the control gain design of engineering structures subjected to the nonstationary excitations, e.g., strong earthquakes and high winds. In essence, moreover, the classical stochastic optimal control belongs to a family of moment-based schemes. The physically based stochastic optimal control, however, facilitates the control gain design of engineering structures with strong nonlinearities and subjected to nonstationary excitations, which circumvents the dilemma pertaining to the classical stochastic optimal control. Moreover, the physically based stochastic optimal control can implement the regulation of probability density of structural systems, by virtue of the probabilistic criteria in terms of structural reliability for parameter optimization of control law. The reliability-based probabilistic criteria can be readily applied by the proposed control scheme since the PSO straightforwardly include the solution of probability density of structural responses provided by the probability density evolution method.

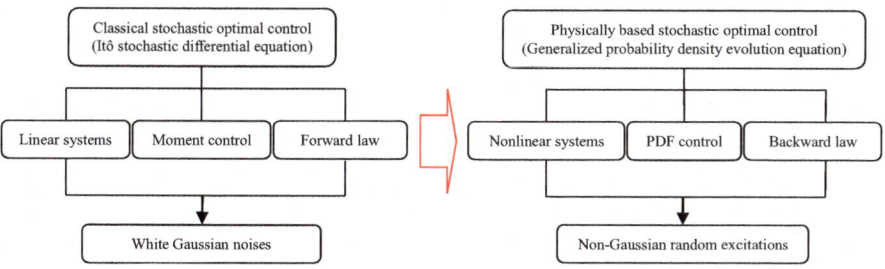

**Fig. 3.30** Schematic diagram of differences between classical stochastic optimal control and PSO

Indeed, a probability-density-based stochastic optimal control can be implemented in conjunction with the classical Fokker–Planck–Kolmogorov equation (FPK equation). However, similar to the situation of the FPK equation in the field of random vibration, the FPK equation with control force term still encounters the challenge of solving the probability density of stochastic systems in practice. Applicability of the FPK equation-based stochastic optimal control is far less than the LQG though the latter merely concerns the second-order moment of structural responses. While the generalized probability density evolution equation involved in the PSO breaks through the dilemma, which forms into the logical basis for the theory and methods of stochastic optimal control of structures. A schematic diagram shows the differences between the classical stochastic optimal control and the PSO; see Fig. 3.30.

It is worth noting that the classical stochastic optimal control is capable of implementing the moment-based control and attaining probability-density-based control in cases of extremely particular situations through defining the FPK equation involving control force terms. However, the physically based stochastic optimal control is readily to implement the probability-density-based control by connecting two families of equations: one is the equation of motion of controlled stochastic systems, e.g., Eq. (3.2.1), which is termed as the physical equation; another is the probability density evolution equation of controlled stochastic systems, e.g., Eqs. (3.2.9) and (3.2.10), which is termed as the evolution equation. The solving of the physical equation is carried out over realizations, which resorts to advanced techniques employed in the deterministic optimal control, such as the linear quadratic regulator (LQR), the optimal polynomial control (OPC), etc. Most of these techniques are optimal control methods based on the Riccati equation and dynamic programming methods based on the Bellman's optimality principle.

# References

Åström KJ (1970) Introduction to stochastic control theory. Academic Press, New York
Athans M, Falb P (1966) Optimal control: an introduction to the theory and its applications. McGraw Hill, New York

Bani-Hani KA, Alawneh MR (2007) Prestressed active post-tensioned tendons control for bridges under moving loads. Struct Control Health Monit 14:357–383

Bucy RS, Kalman RE (1961) New results in linear filtering and prediction theory. ASME Trans J Basic Eng 83:95–108

Chang CC, Yu LO (1998) A simple optimal pole location technique for structural control. Eng Struct 20(9):792–804

Chatfield C (1989) The analysis of time series-an introduction, 4th edn. Chapman and Hall, London

Chen JB, Li J (2008) Strategy for selecting representative points via tangent spheres in the probability density evolution method. Int J Numer Meth Eng 74(13):1988–2014

Chen JB, Liu WQ, Peng YB, Li J (2007) Stochastic seismic response and reliability analysis of base-isolated structures. J Earthq Eng 11(6):903–924.

Chen SP, Li XJ, Zhou XY (1998) Stochastic linear quadratic regulators with indefinite control weight costs. SIAM J Control Optim 36:1685–1702

Chung LL, Reinhorn AM, Soong TT (1988) Experiments on active control of seismic structures. ASCE J Eng Mech 114(2):241–256

Crandall SH (1958) Random vibration. Technology Press of MIT, Wiley, New York

Florentin JJ (1961) Optimal control of continuous time, Markov, stochastic systems. J Electron Control 10:473–488

Ho CC, Ma CK (2007) Active vibration control of structural systems by a combination of the linear quadratic Gaussian and input estimation approaches. J Sound Vib 301:429–449

Josa-Fombellida R, Rincón-Zapatero JP (2007) New approach to stochastic optimal control. J Optimiz Theory and App 135(1):163–177

Kalman RE (1960a) On the general theory of control systems. In: Proceedings of 1st IFAC Moscow congress. Butterworth Scientific Publications

Kalman RE (1960b) A new approach to linear filtering and prediction problems. ASME Trans J Basic Eng 82:35–45

Kohiyama M, Yoshida M (2014) LQG design scheme for multiple vibration controllers in a data center facility. Earthq Struct 6(3):281–300

Kushner HJ (1962) Optimal stochastic control. IRE Trans Autom Control AC-7:120–122

Li J, Ai XQ (2006) Study on random model of earthquake ground motion based on physical process. Earthq Eng Eng Vib 26(5):21–26 (in Chinese)

Li J, Chen JB (2008) The principle of preservation of probability and the generalized density evolution equation. Struct Saf 30:65–77

Li J, Chen JB (2009) Stochastic dynamics of structures. Wiley, Singapore

Li J, Chen JB, Fan WL (2007) The equivalent extreme-value event and evaluation of the structural system reliability. Struct Saf 29(2):112–131

Li J, Liu ZJ (2006) Expansion method of stochastic processes based on normalized orthogonal bases. J Tongji Univ (Nat Sci) 34(10):1279–1283 (in Chinese)

Li J, Peng YB, Chen JB (2010) A physical approach to structural stochastic optimal controls. Probabilistic Eng Mech 25(1):127–141

Lin YK, Cai GQ (1995) Probabilistic structural dynamics: advanced theory and applications. McGraw-Hill, New York

Mathews JH Fink KD (2003) Numerical methods Using Matlab, 4th edn. Prentice-Hall

Øksendal B (2005) Stochastic differential equations: An introduction with applications, 6th edn, Springer-Verlag, Berlin

Roberts JB, Spanos PD (1990) Random vibration and statistical linearization. Wiley, West Sussex

Soong TT (1990) Active structural control: theory and practice. Longman Scientific & Technical, New York

Stengel RF (1986) Stochastic optimal control: theory and application. Wiley, New York

Stengel RF, Ray LR, Marrison CI (1992) Probabilistic evaluation of control system robustness. In: IMA workshop on control systems design for advanced engineering systems: complexity, uncertainty, information and organization, Minneapolis, MN

Sun JQ (2006) Stochastic dynamics and control. Elsevier, Amsterdam

Wiener N (1949) Extrapolation, interpolation and smoothing of stationary time series, with engineering applications. The MIT Press, Cambridge

Wiener N (1964) Time series. The MIT Press, Cambridge

Yang JN (1975) Application of optimal control theory to civil engineering structures. ASCE J Eng Mech Div 101(EM6):819–838

Yang JN, Akbarpour A, Ghaemmaghami P (1987) New optimal control algorithms for structural control. ASCE J Eng Mech 113(9):1369–1386

Yang JN, Li Z, Vongchavalitkul S (1994) Generalization of optimal control theory: linear and nonlinear control. ASCE J Eng Mech 120(2):266–283

Zhang WS, Xu YL (2001) Closed form solution for along-wind response of actively controlled tall buildings with LQG controllers. J Wind Eng Ind Aerodyn 89:785–807

Zhu WQ (2006) Nonlinear stochastic dynamics and control in Hamiltonian formulation. ASME Trans 59:230–248

Zhu WQ, Ying ZG, Soong TT (2001) An optimal nonlinear feedback control strategy for randomly excited structural systems. Nonlinear Dyn 24:31–51

# Chapter 4
# Probabilistic Criteria of Stochastic Optimal Control

## 4.1 Preliminary Remarks

As indicated in Chap. 3, the classical LQG with specified parameters of control law just secures the structural performance in the sense of second-order moment, and cannot be applied to design a logical control system on the basis of nominal white Gaussian noise. The methodology of physically based stochastic optimal control of structures was thus proposed. However, the optimization of parameters of control law in the previous chapter relies upon a trial-and-error procedure. For instance, Eq. 3.3.21 for parameter optimization is a statistical moment-relevant probabilistic criterion that might not ensure the structural safety.

It is revealed in the previous study that the control effectiveness of the LQR-based PSO straightforwardly hinges on the cost function and its deduced control law, in which the parameters of weighting matrices play a critical role. Several strategies for selecting weighting matrices have been developed in recent years. One involves the performance function pertaining to the system stability, including the trial-and-error procedure and the control criterion on Lyapunov asymptotic stability condition. For example, Chen et al. proposed an optimization technique for designing the stabilizer of linear power systems, where the specified weighting matrix was constructed so as to move the dominant eigenvalues of system matrix as far as possible from the imaginary axis until the desired damping ratio was attained (Chen et al. 1992). Yang et al. analyzed a variety of energy-based weighting matrices through solving the Riccati equation, and figured out the applicability of these weighting matrices (Yang et al. 1992a, b). Although this strategy originates from the deterministic optimal control, it is also applicable to the stochastic optimal control since the stability belongs to the intrinsic property of structural systems. An effective algorithm, for instance, for selecting weighting matrices was proposed in the case of optimal control under interval uncertainty (Tsay et al. 1991). Another strategy is via the system second-order statistics evaluation, which has been addressed in details in Chap. 3. It is readily seen that these two strategies both exclude the optimization procedures and algorithms,

© Springer Nature Singapore Pte Ltd. and Shanghai Scientific and Technical Publishers 2019
Y. Peng and J. Li, *Stochastic Optimal Control of Structures*,
https://doi.org/10.1007/978-981-13-6764-9_4

and the derived parameters of weighting matrices hardly guarantee a desired performance of structures. The third strategy includes optimization procedures. In the context of the classical LQG, Stengel et al. developed a family of probabilistically optimal criteria pertaining to system robustness of structures (Stengel et al. 1992). Zhu et al. investigated the weighting matrices of the quadratic cost function based on the stochastic averaging method for quasi-Hamiltonian systems represented by the Hamilton–Jacobi–Bellman (HJB) equation (Zhu et al. 2001).

This chapter attempts to establish a family of probabilistic criteria for the physically based stochastic optimal control. The pertinent procedure of optimizing the cost-function weights is then addressed, with the application of this family of probabilistic criteria to the stochastic optimal control of randomly base-excited structures. The proposed probabilistic criteria include a single-objective one with respect to system state and control force, and a multiple-objective one with respect to the trade-off between system quantities of interest. This treatment is expected to circumvent the adverse situation for designing cost-function weights by trial-and-error procedures.

## 4.2 Gain Matrix of Stochastic Optimal Control

It is indicated in Eq. 3.3.18; the control law of a closed-loop system can be expressed as follows:

$$\mathbf{U}(\mathbf{\Theta}, t) = \mathbf{f}(\mathbf{Q_Z}, \mathbf{R_U})\mathbf{Z}(\mathbf{\Theta}, t) \tag{4.2.1}$$

where $\mathbf{f}(\cdot)$ denotes the gain matrix of stochastic optimal control.

The system state $\mathbf{Z}(t)$ and control force $\mathbf{U}(t)$ are governed by the generalized probability density evolution equations; see Eqs. 3.2.9 and 3.2.10, respectively. A probabilistic criterion based on probability density can be established, whereby the parameters of optimal control law $(\mathbf{Q_Z^*}, \mathbf{R_U^*})$ are then attained. Figure 4.1 shows the PDFs of structural response at the typical instant of time with different parameters of control law.

As it is mentioned in Chap. 3, the physically based stochastic optimal control involves a two-step optimization: the first step is minimizing the cost function so as to build the mapping relation between the set of control parameters and the set of control gains; the second step is minimizing the performance function so as to derive the optimal control parameters. Therefore, utilizing the probabilistic criterion with respect to probability density aims to seek for the potential parameters of optimal control law $(\mathbf{Q_Z^*}, \mathbf{R_U^*})$ so that the performance function pertaining to structural performance is minimum.

It has been noted that the weighting matrices $\mathbf{Q_Z}$ and $\mathbf{R_U}$ in the quadratic cost function are rigorously positive semi-definite and positive definite, respectively, which is necessary for the deterministic control so as to guarantee a convex optimization problem. While this is not true for the classical stochastic optimal control, e.g.,

**Fig. 4.1** PDFs of structural response at typical instant of time with different parameters of control law

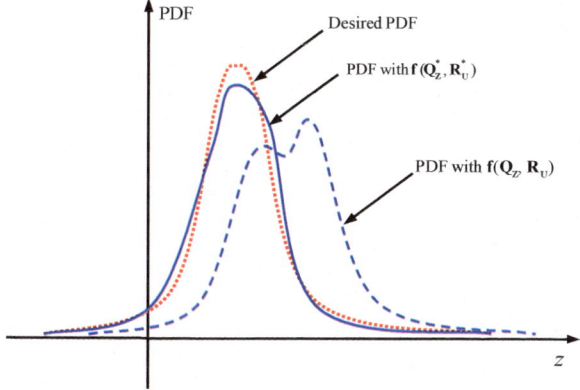

in some stochastic LQR problems, the cost-function weights might be indefinite (specifically, negative definite) but the problem still remains well posed when the diffusion term in the state equation is dependent upon the control action (Chen et al. 1998). Since the proposed physically based stochastic optimal control relies upon the deterministic realizations, the positive definite cost-function weights are preferable which is well suited to search locally optimal solutions. Therefore, $\mathbf{Q_Z}$ and $\mathbf{R_U}$ are rigorously positive semi-definite and positive definite in this study, respectively. Besides, the cost-function weights $\mathbf{Q_Z}$ and $\mathbf{R_U}$ are both theoretically time-dependent, symmetrical, and full matrices (Leondes and Salami 1980), but practically they are often assumed to be time-independent. Some efforts aiming at the optimization of cost-function weights reveal that the diagonal elements are usually far larger than the off-diagonal elements, which indicates that the cross terms between the displacement and velocity can be ignored safely (Chen et al. 1992). It is thus reasonable to assume the cost-function weights as the following formulation (Zhang and Xu 2001):

$$\mathbf{Q_Z} = \begin{bmatrix} \mathbf{Q}_d & \mathbf{0} \\ \mathbf{0} & \mathbf{Q}_v \end{bmatrix}, \quad \mathbf{R_U} = \mathbf{R}_u \tag{4.2.2}$$

## 4.3 Probabilistic Criteria

By virtue of the sample trajectory description, the generalized probability density evolution equation reveals the evolutionary process of stochastic systems. As to stochastic dynamical systems, the extreme value of sample processes in probabilistic space is a random variable. If the extreme value of system quantities of interest is set as the objective of stochastic optimal control, an equivalent extreme-value event including the relevance between the quantities can be readily constructed (Li et al. 2007). For instance, an equivalent extreme-value vector of system state is defined as follows:

**Fig. 4.2** Schematic of mean criterion

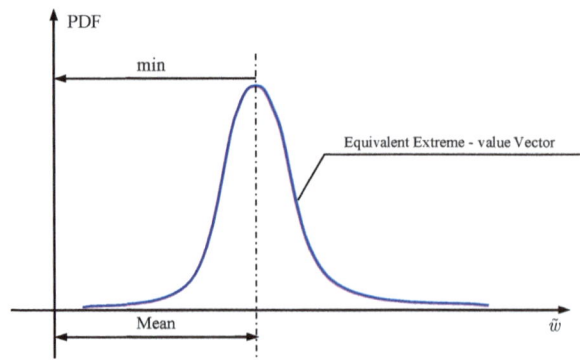

$$\tilde{Z}(\Theta) = \max_{t}\left[\max_{i}|Z_i(\Theta, t)|\right] \tag{4.3.1}$$

where $Z_i(\Theta, t)$ denotes the $i$th component of system state; $t$ denotes the duration of system state.

A variety of probabilistic criteria with physical meanings can be constructed (Li et al. 2011a, b). In view of the statistical moments, one can readily have three probabilistic criteria as follows:

(i)  Mean criterion

$$\min(J_2) = \min\left\{ E[\tilde{\mathbf{W}}^\mathrm{T}\tilde{\mathbf{W}}]\,\Big|\, \bigcup_{k=1}^{M} \{F[\tilde{V}_k] \leq \tilde{V}_{k,\mathrm{con}}\}\right\} \tag{4.3.2}$$

where $\tilde{\mathbf{W}}$ denotes the equivalent extreme-value quantity to be controlled; $\tilde{V}$ denotes the equivalent extreme-value quantity as the constraint. The physical meaning of the mean criterion is that the ensemble average of extreme value of the control objective is minimized. A schematic is shown in Fig. 4.2, where $\tilde{w}$ denotes a component of the equivalent extreme-value quantity $\tilde{\mathbf{W}}$.

(ii)  Mean-standard deviation criterion

$$\min(J_2) = \min\left\{ E[\tilde{\mathbf{W}}^\mathrm{T}\tilde{\mathbf{W}}] + \beta\sigma[\tilde{\mathbf{W}}^\mathrm{T}\tilde{\mathbf{W}}]\,\Big|\, \bigcup_{k=1}^{M} \{F[\tilde{V}_k] \leq \tilde{V}_{k,\mathrm{con}}\}\right\} \tag{4.3.3}$$

where $\beta$ denotes the coefficient of confidence level. The physical meaning of the mean-standard deviation criterion is that the ensemble average plus $\beta$ times standard deviation of extreme value of the control objective is minimized, as shown in Fig. 4.3.

**Fig. 4.3** Schematic of mean-standard deviation criterion

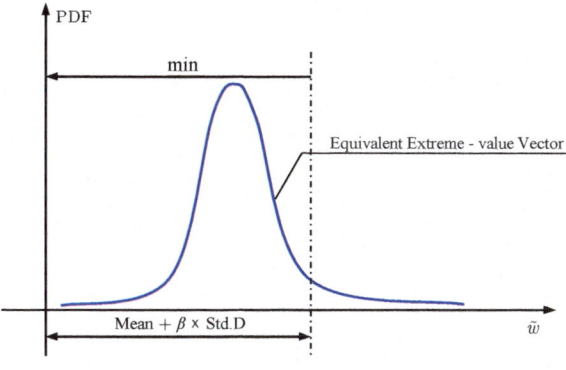

**Fig. 4.4** Schematic of exceedance probabilistic criterion

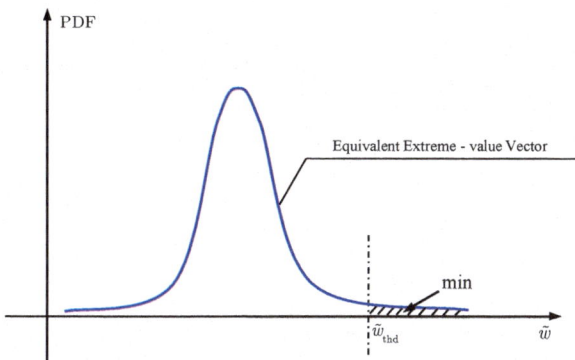

(iii) Exceedance probability criterion

$$\min(J_2) = \min\left\{ \Pr^{\mathrm{T}}(\tilde{\mathbf{W}} - \tilde{\mathbf{W}}_{\mathrm{thd}} > \mathbf{0}) \Pr(\tilde{\mathbf{W}} - \tilde{\mathbf{W}}_{\mathrm{thd}} > \mathbf{0}) \middle| \bigcup_{k=1}^{M} \{F[\tilde{V}_k] \leq \tilde{V}_{k,\mathrm{con}}\} \right\}$$

$$(4.3.4)$$

where $\tilde{\mathbf{W}}_{\mathrm{thd}}$ denotes the threshold of quantity to be controlled; $\Pr\{\cdot\}$ denotes the probability of the random event. The physical meaning of the exceedance probability criterion is that the structural safety can be guaranteed with a minimum exceedance probability; see Fig. 4.4.

It is indicated that from the mean criterion to the exceedance probability criterion, the involved criteria include statistical moments and local information of probability density. Although the exceedance probability criterion does not involve the full details of the probability density, it is widely used in practice due to its relevance directly with the structural reliability.

### 4.3.1  Single-Objective Criteria

#### 4.3.1.1  System State Based Optimization Criteria

There is a practical issue that the remaining quantities of system state might be uncontrollable if the probabilistic criterion is designed merely on a single quantity. A penalty function is often introduced to bypass this challenge. For instance, if the structural displacement pertaining to structural safety is defined as the control objective, and the structural acceleration is considered as a constraint, the probabilistic criteria in Eqs. (4.3.2), (4.3.3), and (4.3.4) can be specified by the following formulations.

(i)  Mean criterion

$$\min\{J_2\} = \min\left\{E[\tilde{D}] + 10^6 \times H(\tilde{A}_{\max} - \tilde{A}_{\mathrm{con}})\right\} \qquad (4.3.5)$$

where $\tilde{D}(\Theta) = \max\limits_{t}\left[\max\limits_{i}|D_i(\Theta, t)|\right]$ denotes the equivalent extreme-value displacement; $H(\cdot)$ denotes a Heaviside step function which exhibits the behaviors as follows:

$$H(\tilde{A}_{\max} - \tilde{A}_{\mathrm{con}}) = \begin{cases} 0, & \tilde{A}_{\max} < \tilde{A}_{\mathrm{con}} \\ 1, & \tilde{A}_{\max} \geq \tilde{A}_{\mathrm{con}} \end{cases} \qquad (4.3.6)$$

where $\tilde{A}_{\max} = \max(\tilde{A}(\Theta))$ denotes the maximum acceleration; $\tilde{A}(\Theta) = \max\limits_{t}\left[\max\limits_{i}|A_i(\Theta, t)|\right]$ denotes the equivalent extreme-value acceleration; $\tilde{A}_{\mathrm{con}}$ denotes the threshold of acceleration constraint.

(ii)  Mean-standard deviation criterion

$$\min\{J_2\} = \min\left\{E[\tilde{D}] + \beta \times \sigma[\tilde{D}] + 10^6 \times H(\tilde{A}_{\max} - \tilde{A}_{\mathrm{con}})\right\} \qquad (4.3.7)$$

(iii)  Exceedance probability criterion

$$\min\{J_2\} = \min\left\{\Pr(\tilde{D} - \tilde{D}_{\mathrm{thd}} > 0) + H(\tilde{A}_{\max} - \tilde{A}_{\mathrm{thd}})\right\} \qquad (4.3.8)$$

#### 4.3.1.2  Control Force Based Optimization Criteria

For a controlled structural system, not only the system state pertaining to system safety is of concern but also the control force pertaining to control cost is of concern in some situations. The probabilistic criteria in function of control force are provided as follows:

(i)  Mean criterion

$$\min(J_2) = \min\{E[\tilde{U}]\} \tag{4.3.9}$$

(ii)  Mean-standard deviation criterion

$$\min(J_2) = \min\{E[\tilde{U}] + \beta \times \sigma[\tilde{U}]\} \tag{4.3.10}$$

(iii)  Exceedance probability criterion

$$\min\{J_2\} = \min\left\{\Pr(\tilde{U} - \tilde{U}_{\text{thd}} > 0)\right\} \tag{4.3.11}$$

where $\tilde{U}(\Theta) = \max\limits_{t}[\max\limits_{i}|U_i(\Theta, t)|]$ denotes the equivalent extreme-value control force; $\tilde{U}_{\text{thd}}$ denotes the threshold of the control force.

### 4.3.1.3  System State and Control Force Based Optimization Criteria

When structural displacement pertaining to the structural safety is set as the control objective, and structural acceleration and control force are set as the constraint, the probabilistic criteria are given by

(i)  Mean criterion

$$\min\{J_2\} = \min\left\{E[\tilde{D}] + 10^6 \times (H(\tilde{A}_{\max} - \tilde{A}_{\text{con}}) + H(\tilde{U}_{\max} - \tilde{U}_{\text{con}}))\right\} \tag{4.3.12}$$

where $\tilde{U}_{\max} = \max(\tilde{U}(\Theta))$ denotes the maximum control force.

(ii)  Mean-standard deviation criterion

$$\min\{J_2\} = \min\left\{E[\tilde{D}] + \beta \times \sigma[\tilde{D}] + 10^6 \times (H(\tilde{A}_{\max} - \tilde{A}_{\text{con}}) + H(\tilde{U}_{\max} - \tilde{U}_{\text{con}}))\right\}$$
$$\tag{4.3.13}$$

(iii)  Exceedance probability criterion

$$\min\{J_2\} = \min\left\{\Pr(\tilde{D} - \tilde{D}_{\text{thd}} > 0) + H(\tilde{A}_{\max} - \tilde{A}_{\text{con}}) + H(\tilde{U}_{\max} - \tilde{U}_{\text{con}})\right\}$$
$$\tag{4.3.14}$$

## 4.3.2  Multiple-Objective Criteria

It is recognized that if the probabilistic criterion merely relies upon a single quantity, the remaining physical quantities pertaining to system performance, e.g., the structural velocity relevant to system serviceability and the structural acceleration relevant to system comfortability, might be neglected. These physical quantities, however, often need to be concerned simultaneously in practice due to their mutual constraints. Therefore, the design principle in trade-off is necessary for a logical control. It thus leads to the proposal of multiple-objective probabilistic criteria.

### 4.3.2.1  Performance Trade-off Based Optimization Criteria

(i)  Mean criterion

In this control criterion, the performance function is defined as the ensemble average of the quadratic combination of system state and control force:

$$J_2 = E\left[\int_{t_0}^{t_f} \frac{1}{2}\{\tilde{Z}_t^{\mathrm{T}}(t)\mathbf{Q}_{\tilde{Z}_t}\tilde{Z}_t(t) + \tilde{U}_t^{\mathrm{T}}(t)\mathbf{R}_{\tilde{U}_t}\tilde{U}_t(t)\}\mathrm{d}t\right] \tag{4.3.15}$$

where $\tilde{Z}_t(\mathbf{\Theta}, t) = \max_i |Z_i(\mathbf{\Theta}, t)|$, $\tilde{U}_t(\mathbf{\Theta}, t) = \max_i |U_i(\mathbf{\Theta}, t)|$ denotes the equivalent processes of system state and control force, respectively.

Introducing the standardized form of matrix

$$\tilde{Z}_t^{\mathrm{T}}\mathbf{Q}_{\tilde{Z}_t}\tilde{Z}_t = \mathrm{Tr}(\tilde{Z}_t^{\mathrm{T}}\mathbf{Q}_{\tilde{Z}_t}\tilde{Z}_t) = \mathrm{Tr}(\mathbf{Q}_{\tilde{Z}_t}\tilde{Z}_t\tilde{Z}_t^{\mathrm{T}}) \tag{4.3.16}$$

where $\mathrm{Tr}(\cdot)$ denotes the trace of matrix, and considering the relation between control force and system state; see Eq. (3.2.18), one has

$$J_2 = \frac{1}{2}\mathrm{Tr}\left\{\int_{t_0}^{t_f} ((\mathbf{Q}_{\tilde{Z}_t} + \mathbf{G}_{\tilde{Z}_t}^{\mathrm{T}}\mathbf{R}_{\tilde{U}_t}\mathbf{G}_{\tilde{Z}_t})E[\tilde{Z}_t(t)\tilde{Z}_t^{\mathrm{T}}(t)])\mathrm{d}t\right\} \tag{4.3.17}$$

where the switch between the ensemble average operator and the integral operator and the switch between ensemble average operator and the trace operator are applied. Besides, the dimensions of weighting matrices $\mathbf{Q}_{\tilde{Z}_t}$, $\mathbf{R}_{\tilde{U}_t}$ and of the gain matrix $\mathbf{G}_{\tilde{Z}_t}$ are both consistent with the equivalent processes of system state and control force $\tilde{Z}_t$, $\tilde{U}_t$, which have the forms as follows:

$$\mathbf{Q}_{\tilde{Z}_t} = \begin{bmatrix} Q_d & 0 \\ 0 & Q_v \end{bmatrix}, \mathbf{R}_{\tilde{U}_t} = R_u, \mathbf{G}_{\tilde{Z}_t} = \mathbf{R}_{\tilde{U}_t}^{-1}\mathbf{B}^{\mathrm{T}}\mathbf{P}_t \tag{4.3.18}$$

where $\mathbf{P}_t$ denotes the Riccati matrix.

The mean criterion with acceleration constraint is thus given by

$$J_2 = \frac{1}{2}\mathrm{Tr}\left\{\int_{t_0}^{t_f}((\mathbf{Q}_{\tilde{Z}_t} + \mathbf{G}_{\tilde{Z}_t}^{\mathrm{T}}\mathbf{R}_{\tilde{U}_t}\mathbf{G}_{\tilde{Z}_t})E[\tilde{Z}_t(t)\tilde{Z}_t^{\mathrm{T}}(t)])dt\right\} + 10^6 \times H(\tilde{A}_{\max} - \tilde{A}_{\mathrm{con}})$$

$$(4.3.19)$$

It is readily seen that the mean criterion based on performance trade-off indicates a minimum mean-square quantity in balance.

(ii)  Exceedance probability criterion

The mean criterion can attain a performance trade-off in the sense of mean. However, it does not meet the requirement of the elaborate design of the control system. Another performance function is defined as the exceedance probability of quadratic combination of system state and control force:

$$J_2 = \int_{L_{\mathrm{thd}}}^{\infty} p(L)dL \tag{4.3.20}$$

where

$$L(\tilde{Z}_t, \tilde{U}_t, \mathbf{\Theta}) = \frac{1}{2}\mathrm{Tr}\left\{\int_{t_0}^{t_f}((\mathbf{Q}_{\tilde{Z}_t} + \mathbf{G}_{\tilde{Z}_t}^{\mathrm{T}}\mathbf{R}_{\tilde{U}_t}\mathbf{G}_{\tilde{Z}_t})[\tilde{Z}_t(t)\tilde{Z}_t^{\mathrm{T}}(t)])dt\right\} \tag{4.3.21}$$

$$L_{\mathrm{thd}} = \frac{1}{2}\left[q_{\mathrm{corr}}[F_{\mathrm{corr}}(\tilde{D})]^2 + q_{\mathrm{corr}}[F_{\mathrm{corr}}(\tilde{V})]^2 + r_{\mathrm{corr}}[F_{\mathrm{corr}}(\tilde{U})]^2\right](t_f - t_0) \tag{4.3.22}$$

where $\tilde{V}(\mathbf{\Theta}) = \max_t\left[\max_i|V_i(\mathbf{\Theta}, t)|\right]$ denotes the equivalent extreme-value velocity; The threshold $L_{\mathrm{thd}}$ is defined as the first-passage failure criterion: if the quantile of any quantity among structural displacement, structural velocity, and control force first attains to its threshold, the remaining quantities are assigned by their present characteristic values, and meanwhile the coefficients of weighting matrices $q_{corr}$, $r_{corr}$ are defined.

The exceedance probability criterion with acceleration constraint is then given by

$$\min(J_2) = \min\left\{\int_{L_{thd}}^{\infty} p(L)dL + H(\tilde{A}_{\max} - \tilde{A}_{\mathrm{con}})\right\} \tag{4.3.23}$$

It is readily seen that the exceedance probability criterion based on performance trade-off indicates a minimum exceedance probability of quantities in balance.

#### 4.3.2.2   Energy Trade-off Based Optimization Criteria

The probabilistic criteria based on performance trade-off refer to a minimum argument consisting of system quantities. This treatment, however, does not straightforwardly meet with the control demand of system quantities of interest. Alternative criteria can be constructed based on a probabilistic measure of system quantities and energy trade-off.

(i)  Mean criterion

The performance function in terms of means of system quantities can be defined as

$$
J_2 = \frac{1}{2} \int_{t_0}^{t_f} \left\{ E^{\mathrm{T}}[\tilde{Z}_t] \mathbf{Q}_{\tilde{Z}_t} E[\tilde{Z}_t] + E^{\mathrm{T}}[\tilde{U}_t] \mathbf{R}_{\tilde{U}_t} E[\tilde{U}_t] \right\} \mathrm{d}t \tag{4.3.24}
$$

which can be deduced into

$$
J_2 = \frac{1}{2} \mathrm{Tr}\{ \int_{t_0}^{t_f} ((\mathbf{Q}_{\tilde{Z}_t} + \mathbf{G}_{\tilde{Z}_t}^{\mathrm{T}} \mathbf{R}_{\tilde{U}_t} \mathbf{G}_{\tilde{Z}_t}) E[\tilde{Z}_t] E^{\mathrm{T}}[\tilde{Z}_t]) \mathrm{d}t \} \tag{4.3.25}
$$

The probabilistic criterion with acceleration constraint is thus given by

$$
J_2 = \frac{1}{2} \mathrm{Tr}\left\{ \int_{t_0}^{t_f} \left( (\mathbf{Q}_{\tilde{Z}_t} + \mathbf{G}_{\tilde{Z}_t}^{\mathrm{T}} \mathbf{R}_{\tilde{U}_t} \mathbf{G}_{\tilde{Z}_t}) E[\tilde{Z}_t] E^{\mathrm{T}}[\tilde{Z}_t] \right) \mathrm{d}t \right\} + 10^6 \times H(\tilde{A}_{\max} - \tilde{A}_{\mathrm{con}}) \tag{4.3.26}
$$

Different from the mean criterion based on performance trade-off, the mean criterion based on energy trade-off indicates a minimum mean energy in balance.

(ii)  Exceedance probability criterion

In order to attain a minimum failure possibility of system quantities, the performance function in terms of exceedance probabilities of system quantities is defined as

$$
J_2 = \frac{1}{2} \int_{t_0}^{t_f} \left[ \mathrm{Pr}_{\tilde{Z}_t}^{\mathrm{T}} (\tilde{Z}_t - \tilde{Z}_{t,\mathrm{thd}} > 0) \mathbf{Q}_{\tilde{Z}} \mathrm{Pr}_{\tilde{Z}_t} (\tilde{Z}_t - \tilde{Z}_{t,\mathrm{thd}} > 0) + \mathrm{Pr}_{\tilde{U}_t}^{\mathrm{T}} (\tilde{U}_t - \tilde{U}_{t,\mathrm{thd}} > 0) \mathbf{R}_{\tilde{U}} \mathrm{Pr}_{\tilde{U}_t} (\tilde{U}_t - \tilde{U}_{t,\mathrm{thd}} > 0) \right] \mathrm{d}t \tag{4.3.27}
$$

where $\tilde{Z}_{t,\mathrm{thd}}$, $\tilde{U}_{t,\mathrm{thd}}$ are thresholds of $\tilde{Z}_t$, $\tilde{U}_t$, respectively.

Obviously, the minimization of performance function Eq. (4.3.27) involves the calculation of time-variant reliability, which dramatically increases the complexity of the optimization problem. According to the extreme-value distribution theorem, only the extreme values of system quantities over the duration interval $[t_0, t_f]$ are concerned. Equation (4.3.27) can thus be simplified to be

$$J_2 = \frac{1}{2}\Big[\Pr{}_{\tilde{Z}}^{\mathrm{T}}(\tilde{Z} - \tilde{Z}_{\mathrm{thd}} > \mathbf{0})\mathbf{Q}_{\tilde{Z}} \Pr{}_{\tilde{Z}}(\tilde{Z} - \tilde{Z}_{\mathrm{thd}} > \mathbf{0}) + \Pr{}_{\tilde{U}}^{\mathrm{T}}(\tilde{U} - \tilde{U}_{\mathrm{thd}} > \mathbf{0})\mathbf{R}_{\tilde{U}} \Pr{}_{\tilde{U}}(\tilde{U} - \tilde{U}_{\mathrm{thd}} > \mathbf{0})\Big] \quad (4.3.28)$$

where $\tilde{Z}_{\mathrm{thd}}$, $\tilde{U}_{\mathrm{thd}}$ are thresholds of $\tilde{Z}$, $\tilde{U}$, respectively. The weighting matrices $\mathbf{Q}_{\tilde{Z}}$, $\mathbf{R}_{\tilde{U}}$ in Eq. (4.3.28) are not the same as the weighting matrices $\mathbf{Q}_{\tilde{Z}_t}$, $\mathbf{R}_{\tilde{U}_t}$ in Eqs. (4.3.15) and (4.3.24), in that they have different dimensions. Since the performance function consists of exceedance probabilities of system quantities, $\mathbf{Q}_{\tilde{Z}}$, $\mathbf{R}_{\tilde{U}}$ are dimensionless arguments and can be defined simply as follows:

$$\mathbf{Q}_{\tilde{Z}} = \begin{bmatrix} 1 & 0 \\ 0 & 1 \end{bmatrix}, \mathbf{R}_{\tilde{U}} = 1 \quad (4.3.29)$$

The exceedance probability criterion with acceleration constraint is thus given by

$$J_2 = \frac{1}{2}[\Pr{}_{\tilde{Z}}^{\mathrm{T}}(\tilde{Z} - \tilde{Z}_{\mathrm{thd}} > \mathbf{0})\mathbf{Q}_{\tilde{Z}} \Pr{}_{\tilde{Z}}(\tilde{Z} - \tilde{Z}_{\mathrm{thd}} > \mathbf{0})$$
$$+ \Pr{}_{\tilde{U}}^{\mathrm{T}}(\tilde{U} - \tilde{U}_{\mathrm{thd}} > \mathbf{0})\mathbf{R}_{\tilde{U}} \Pr{}_{\tilde{U}}(\tilde{U} - \tilde{U}_{\mathrm{thd}} > \mathbf{0})] + H(\tilde{A}_{\max} - \tilde{A}_{\mathrm{con}}) \quad (4.3.30)$$

The exceedance probability criterion based on energy trade-off indicates a minimum probability energy in balance.

### 4.3.3  Comparative Studies

For illustrative purposes, the controlled single-story shear frame shown in Sect. 3.4.1 is investigated. Optimization and design of the active tendon system subjected to random seismic ground motion are carried out using the probabilistic criteria mentioned in the previous section.

#### 4.3.3.1  Single-Objective Criteria

The thresholds of structural displacement, structural velocity, structural acceleration, and control force are denoted by 10 mm, 100 mm/s, 3000 mm/s$^2$, and 200 kN, respectively. Coefficient of confidence level $\beta = 1$. The quantile function for assessing the threshold of the performance function in exceedance probability criterion based on performance trade-off is defined as the mean plus three times of standard deviation. In the cost function of optimal control, the state weighting matrix $\mathbf{Q}_{\mathbf{Z}} = \mathrm{diag}\{Q_d, Q_v\}$, and the control force weighting matrix $\mathbf{R}_{\mathbf{U}} = R_u$. The deterministic dynamic analysis resorts to the transfer function method. The toolkit function of MATLAB, *fmincon*, is used in the optimization procedure, where a sequence quadratic programming (SQR) method solving the subproblem is involved in each iteration. The optimization procedure is essentially a scheme of local nonlinear optimization with constraints based on the solution of the Kuhn–Tucker equation (Mathews and Fink 2003).

The optimization results of control parameters under the single-objective criteria are shown in Table 4.1, Table 4.2, and Table 4.3, respectively. Some remarks can be drawn as follows.

*Remark 1* The system quantities of concern attain to a trade-off step by step from the optimization criterion groups with displacement control and acceleration constraint (S-I) to the optimization criterion group with minimum control force (S-II) and the optimization criterion group with displacement control and acceleration-control force constraint (S-III), which arises to a hierarchical design of probabilistic criteria.

*Remark 2* The optimal parameters of mean criterion (i) and of mean-standard deviation criterion (ii) are nearly identical, either in the case of optimization criterion group S-I, or in the cases of optimization criterion groups S-II and S-III. The discrepancy between mean criterion (i) and mean-standard deviation criterion (ii) is just 3.78%. It is understood that since the first-order moment represents the main part of probabilistic information, mean criterion (i) has thus a similar control effectiveness to mean-standard deviation criterion (ii), indicating that mean-standard deviation criterion, at least for this study, can be substituted by mean criterion.

**Table 4.1** Optimal parameters under single-objective criteria with displacement control and acceleration constraint (S-I)

| Parameters | (i) | | | (ii) ($\beta = 1$) | | | (iii) ($\beta = 1$) | | |
|---|---|---|---|---|---|---|---|---|---|
| | $Q_d$ | $Q_v$ | $R_u$ | $Q_d$ | $Q_v$ | $R_u$ | $Q_d$ | $Q_v$ | $R_u$ |
| Initial value | 100 | 100 | $10^{-10}$ | 100 | 100 | $10^{-10}$ | 100 | 100 | $10^{-10}$ |
| Optimal value | 230.2 | 13983.4 | $10^{-10}$ | 221.5 | 13984.1 | $10^{-10}$ | 101.0 | 195.4 | $10^{-10}$ |
| Objective function | 0.91 mm | | | 1.06 mm | | | 0.0016 | | |

**Table 4.2** Optimal parameters under single-objective criteria with minimum control force (S-II)

| Parameters | (i) | | | (ii) ($\beta = 1$) | | | (iii) ($\beta = 1$) | | |
|---|---|---|---|---|---|---|---|---|---|
| | $Q_d$ | $Q_v$ | $R_u$ | $Q_d$ | $Q_v$ | $R_u$ | $Q_d$ | $Q_v$ | $R_u$ |
| Initial value | 100 | 100 | $10^{-8}$ | 100 | 100 | $10^{-8}$ | 100 | 100 | $10^{-8}$ |
| Optimal value | 99.2 | 0.0 | $10^{-8}$ | 99.2 | 0.0 | $10^{-8}$ | 100.0 | 100.0 | $10^{-8}$ |
| Objective function | 0.11 kN | | | 0.17 kN | | | 0.0000 | | |

**Table 4.3** Optimal parameters under single-objective criteria with displacement control and acceleration–control force constraint (S-III)

| Parameters | (i) | | | (ii) ($\beta = 1$) | | | (iii) ($\beta = 1$) | | |
|---|---|---|---|---|---|---|---|---|---|
| | $Q_d$ | $Q_v$ | $R_u$ | $Q_d$ | $Q_v$ | $R_u$ | $Q_d$ | $Q_v$ | $R_u$ |
| Initial value | 100 | 100 | $10^{-10}$ | 100 | 100 | $10^{-10}$ | 100 | 100 | $10^{-10}$ |
| Optimal value | 102.6 | 383.2 | $10^{-10}$ | 102.5 | 383.2 | $10^{-10}$ | 101.0 | 195.4 | $10^{-10}$ |
| Objective function | 4.72 mm | | | 5.67 mm | | | 0.0016 | | |

*Remark 3*  Exceedance probability criterion (iii) in the optimization criterion groups S-I, S-II, and S-III are significantly different from mean criterion (i) and mean-standard deviation criterion (ii). It owes to the fact the probability density exhibits more information than the statistical moments.

*Remark 4*  The optimal parameters of exceedance probability criterion (iii) in the case of the optimization criterion group S-I and those in the case of the optimization criterion group S-III are identical, as shown in Tables 4.1 and 4.3. It is indicated that in the optimization criterion group S-III, the control force constraint has no influence upon the optimal parameters and objective function of exceedance probability criterion (iii); while the control force constraint has significant influence upon the optimal parameters and objective function of mean criterion (i) and mean-standard deviation criterion (ii), since their optimal parameters and the objective functions are significantly different from the optimization criterion group. It is explained that mean criterion (i) and mean-standard deviation criterion (ii) are devoted to minimizing the first two-order moments of system displacement that needs much more control cost, which strengthens the control force constraint. Exceedance probability criterion (iii), however, aims to minimize the exceedance probability of displacement, which hinges upon the detail of the PDF, especially upon the "tail" of the PDF that does not require a minimum mean and a minimum standard deviation. For illustrative purposes, the PDFs of two alternative cases are shown in Fig. 4.5. It is seen that the area of the "tail" of the PDF with a larger mean or with a larger standard deviation is smaller instead, indicating that a less control cost is needed in exceedance probability criterion. It is thus noted that exceedance probability criterion exhibits more economic and more obvious control effectiveness than mean and mean-standard deviation criteria.

Comparative study of optimal control of the structure using the single-objective criteria is carried out. The numerical results are shown in Table 4.4. It is seen that the control effectiveness seriously relies on the probabilistic criteria relevant to the system quantities of interest. Since the control force serves as the objective quantity, the optimization criterion group S-II attains a minimum control force whatever on

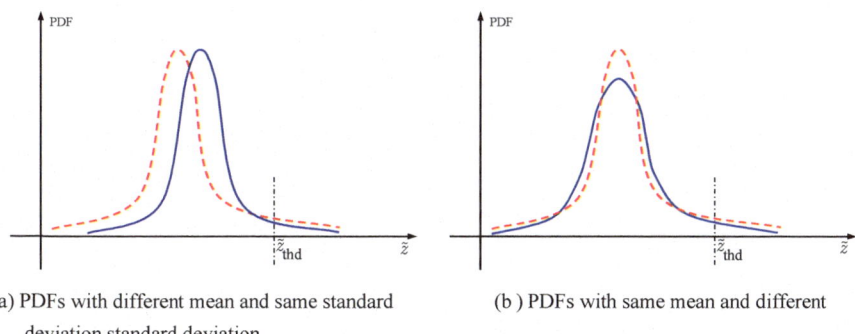

(a) PDFs with different mean and same standard      (b ) PDFs with same mean and different
deviation                                                        standard deviation

**Fig. 4.5**  Comparison between PDFs with different means and standard deviations

the mean or on the standard deviation. The treatment can gain a minimum control cost but might result in a response amplification instead of response reduction. The optimization criterion group S-I attains a minimum displacement under the condition of the constraint, however, it is not mostly economical. In the optimization criterion group S-III, mean criterion (i) has a better displacement and acceleration control than exceedance probability criterion (iii). This is due to the fact that the control force constraint imposed on the optimization criterion group S-III has no influence upon exceedance probability criterion (iii) but has influence upon mean criterion (i). The optimization criterion group S-III is thus more reasonable. In summary, the stochastic optimal control using the probabilistic criterion relevant to exceedance probability with multiple constraints behaves more comprehensively than that using probabilistic criterion relevant to statistical moments with a single constraint. Besides, exceedance probability criterion (iii) in all the optimization criterion groups aims at the maximum system reliability, and thus operates more effective and more economical than mean criterion (i) and mean-standard deviation criterion (ii).

The abovementioned probabilistic criteria are devoted to minimizing the displacement with constraints of the acceleration and control force. It is spontaneously feasible by relocating the objective quantity and constraints.

**Table 4.4** Comparison of control effectiveness using single-objective criteria

| Equ. ext.-value | | | Probabilistic criterion | | | | | |
|---|---|---|---|---|---|---|---|---|
| | | | S-I | | S-II | | S-III | |
| | | | (i) | (iii) | (i) | (iii) | (i) | (iii) |
| Interstory drift (mm) | Mn | Unc. | 28.47 | 28.47 | 28.47 | 28.47 | 28.47 | 28.47 |
| | | Con. | 0.91 | 6.23 | 28.94 | 25.57 | 4.72 | 6.23 |
| | | Eff.[a] | −96.80% | −78.12% | 1.65% | −10.19% | −83.42% | −78.12% |
| | Std.d | Unc. | 13.78 | 13.78 | 13.78 | 13.78 | 13.78 | 13.78 |
| | | Con. | 0.15 | 1.41 | 14.11 | 11.84 | 0.95 | 1.41 |
| | | Eff. | −98.91% | −89.77% | 2.39% | −14.08% | −93.11% | −89.77% |
| Story acceleration (mm/s$^2$) | Mn | Unc. | 3602.66 | 3602.66 | 3602.66 | 3602.66 | 3602.66 | 3602.66 |
| | | Con. | 1075.40 | 1235.60 | 3661.74 | 3245.40 | 1157.05 | 1235.60 |
| | | Eff. | −70.15% | −65.70% | 1.64% | −9.92% | −67.88% | −65.70% |
| | Std.d | Unc. | 1745.59 | 1745.59 | 1745.59 | 1745.59 | 1745.59 | 1745.59 |
| | | Con. | 355.23 | 348.92 | 1786.11 | 1504.00 | 331.04 | 348.92 |
| | | Eff. | −79.65% | −80.01% | 2.32% | −13.84% | −81.04% | −80.01% |
| Interstory control force (kN) | Mn | | 106.33 | 86.55 | 0.11 | 10.78 | 92.78 | 86.55 |
| | Std.d | | 35.68 | 30.71 | 0.06 | 5.44 | 31.78 | 30.71 |

[a]Effectiveness is defined as (Con.-Unc.)/Unc.

### 4.3.3.2 Multiple-Objective Control Criteria

The definition of the threshold of $L_{thd}$ in Eq. (4.3.22) refers to Fig. 3.11. It is seen that if the quantile locates at the mean plus three times of standard deviation, the quantiles of the equivalent extreme-value displacement and equivalent extreme-value control force are 7.41 mm and 188.86 kN, respectively, which are less than their thresholds when the quantile of the equivalent extreme-value velocity is beyond its threshold 100 mm/s. The corresponding ratio of coefficients of weighting matrices is $4 \times 10^{12}$, and the coefficients are set as $q_{corr} = 400$, $r_{corr} = 1 \times 10^{-10}$, respectively. Meanwhile, the threshold $L_{thd}$ is 77.71 in SI units.

The optimal control parameters under multiple-objective criteria are shown in Table 4.5. It is noted again that the control effectiveness of structural control relies on the physical meanings of the probabilistic criteria. The optimized control parameters and objective functions of mean and exceedance probability criteria in optimization criterion group on performance trade-off M-I(i), M-I(ii) and of the mean criterion in optimization criterion group on energy trade-off M-II(i) are nearly identical. This is owing to the fact that these criteria all involve the balance of system quantities in physical dimensions.

Table 4.6 shows the comparison of control effectiveness using the multiple-objective criteria. It is readily seen that the probabilistic criteria in optimization criterion group on performance trade-off M-I aim at minimizing the probabilistic measure of performance function; see the ensemble average and exceedance probability, which does not straightforwardly underline the system quantity control and might result in a disabled control system. Although mean criterion M-I(i) and exceedance probability criterion M-I(ii) exhibit less exceedance probabilities of representative functions shown in Eq. (4.3.21) and attain a better balance between control force and system state, they derive larger exceedance probabilities on both displacement and velocity. Similarly, mean criterion M-II(i) in optimization criterion group on energy trade-off M-II aims at the best balance between control force and system state, and cannot guarantee structural safety as well. However, the exceedance probability criterion M-II(ii) in optimization criterion group on energy trade-off M-II not only guarantees a safe structure but also remains the control cost in a rational range. As shown in Table 4.6, the exceedance probabilities of displacement and velocity using the probabilistic criterion M-II(ii) are significantly less than those using other multiple-objective criteria, and meanwhile, the exceedance probability of control force 0.0002 can be accepted as well. Besides, the acceleration, as constraint to all the multiple-objective criteria, has an exceedance probability less than $5 \times 10^{-5}$, which indicates an effective constraint on the optimization process.

One might recognize that the performance functions included in probabilistic criteria M-I(i) and M-II(i) have certain similarities to the cost function in the classical LQG. It is indicated that the traditional policy controlling the mean-square response of system quantities cannot guarantee the structural safety; while the exceedance probability criterion based control parameter optimization is just the key of stochastic optimal control.

**Table 4.5** Optimal control parameters under multiple-objective criteria

| Parameters | (M-I) | | | | | | (M-II) | | | | | |
| --- | --- | --- | --- | --- | --- | --- | --- | --- | --- | --- | --- | --- |
| | (i) | | | (ii) | | | (i) | | | (ii) | | |
| | $Q_d$ | $Q_v$ | $R_u$ | $Q_d$ | $Q_v$ | $R_u$ | $Q_d$ | $Q_v$ | $R_u$ | $Q_d$ | $Q_v$ | $R_u$ |
| Initial value | 100 | 100 | $10^{-10}$ | 100 | 100 | $10^{-10}$ | 100 | 100 | $10^{-10}$ | 100 | 100 | $10^{-10}$ |
| Optimal value | 0.0 | 80.7 | $10^{-10}$ | 3.6 | 80.7 | $10^{-10}$ | 0.0 | 80.7 | $10^{-10}$ | 1073.6 | 505.0 | $10^{-10}$ |
| Objective function | 35.96 | | | 0.0386 | | | 0.8370 | | | $0.0150 \times 10^{-6}$ | | |

**Table 4.6** Comparison of control effectiveness using multiple-objective criteria

| Exceedance Prob. | Probabilistic criterion | | | | |
| --- | --- | --- | --- | --- | --- |
| | Unc. | M-I | | M-II | |
| | | (i) | (ii) | (i) | (ii) |
| Interstory drift $P_{f,d}$ | 0.9020 | 0.3147 | 0.3146 | 0.3147 | $3.60 \times 10^{-7}$ |
| Interstory velocity $P_{f,v}$ | 0.8941 | 0.5245 | 0.5244 | 0.5245 | $4.88 \times 10^{-5}$ |
| Story acceleration $P_{f,a}$ | 0.5735 | $4.46 \times 10^{-5}$ | $4.46 \times 10^{-5}$ | $4.46 \times 10^{-5}$ | $3.60 \times 10^{-7}$ |
| Interstory control force $P_{f,u}$ | – | $3.60 \times 10^{-7}$ | $3.60 \times 10^{-7}$ | $3.60 \times 10^{-7}$ | $1.66 \times 10^{-4}$ |
| Representative function $P_{f,p}$ | 0.6745 | 0.0339 | 0.0386 | 0.0339 | 0.6981 |

For further investigation of probabilistic criteria, a comparison of structural optimal control using the exceedance probability criterion in optimization criterion group M-II, the exceedance probability criterion in optimization criterion group S-III, the criterion on system second-order statistics evaluation (SSSE), and the criterion on Lyapunov asymptotic stability condition (LASC) is carried out. The numerical results are shown in Table 4.7. It is seen that the criterion on Lyapunov asymptotic stability condition has the best acceleration control on the mean; while it has the worst acceleration control on the standard deviation, and the largest control cost on the standard deviation. The criterion on system second-order statistics evaluation exhibits the best displacement control on the mean and standard deviation; while it exhibits the largest control cost on the mean. The exceedance probability criterion in optimization criterion group S-III has the least control cost on the mean and standard deviation; while it has the worst displacement control on the mean and standard deviation, and the worst acceleration control on the mean. The exceedance probability criterion in optimization criterion group M-II exhibits the best acceleration control on the standard deviation, and it has no worse control of quantities. Moreover, the probabilistic criterion M-II(ii) can provide accurate reliabilities of system quantities of interest simultaneously; while other criteria fail to offer this critical information. It is thus believed that the exceedance probability criterion in optimization criterion group on energy trade-off accommodates the system performance to achieve a better trade-off between response reduction and control cost, which is a preferential criterion in structural control.

## 4.4 Numerical Example

For illustrative purposes, the exceedance probability criterion in optimization criterion group on energy trade-off is applied to the eight-story shear frame attached with fully distributed active tendon systems shown in Sect. 3.4.2. The thresholds of

**Table 4.7** Comparison of stochastic optimal control in terms of different criteria

| Equ. ext.-value | | | Probabilistic criterion | | | |
|---|---|---|---|---|---|---|
| | | | M-II(ii): $\mathbf{Q_Z}$ = diag $\{1073.6,505.0\}$, $\mathbf{R_U} = 10^{-10}$ | S-III(iii): $\mathbf{Q_Z}$ = diag $\{101.0,195.4\}$, $\mathbf{R_U} = 10^{-10}$ | SSSE: $\mathbf{Q_Z}$ = diag $\{80,80\}$, $\mathbf{R_U} = 10^{-11}$ | LASC (Yang et al. 1992a): $\mathbf{Q_Z}$ = diag $\{10^7, 10^9\}$, $\mathbf{R_U} = 8 \times 10^{-6}$ |
| Interstory drift (mm) | Mn | Unc. | 28.47 | 28.47 | 28.47 | 28.47 |
| | | Con. | 4.15 | 6.23 | 3.40 | 5.35 |
| | | Eff.[a] | −85.42% | −78.12%▼ | −88.06%▲ | −81.21% |
| | Std.d | Unc. | 13.78 | 13.78 | 13.78 | 13.78 |
| | | Con. | 0.81 | 1.41 | 0.63 | 1.40 |
| | | Eff. | −94.12% | −89.77%▼ | −95.43%▲ | −89.84% |
| Story acceleration (mm/s$^2$) | Mn | Unc. | 3602.7 | 3602.7 | 3602.7 | 3602.7 |
| | | Con. | 1141.0 | 1235.6 | 1114.7 | 909.9 |
| | | Eff. | −68.33% | −65.70%▼ | −69.06% | −74.74%▲ |
| | Std.d | Unc. | 1745.6 | 1745. 6 | 1745. 6 | 1745. 6 |
| | | Con. | 331.8 | 348.9 | 332.0 | 364.8 |
| | | Eff. | −80.99%▲ | −80.01% | −80.98% | −79.10%▼ |
| Interstory control force (kN) | Mn | | 94.93 | 86.55▲ | 98.02▼ | 88.78 |
| | Std.d | | 32.46 | 30.71▲ | 33.12 | 33.51▼ |

[a]Effectiveness is defined as (Con.-Unc.)/Unc.
[b]▼ indicates a worse control effectiveness against other criteria
▲ indicates a better control effectiveness against other criteria

the interstory drift, interstory velocity, story acceleration, and the interstory control force are 15 mm, 150 mm/s, 8000 mm/s$^2$ and 2000 kN, respectively. Without consideration of the cross terms between state quantities of all stories, and assigning a same state weight to all the controllers on the stories, the cost-function weights are set as follows:

$$\mathbf{Q_Z} = \begin{bmatrix} Q_d\mathbf{I} & \mathbf{0} \\ \mathbf{0} & Q_v\mathbf{I} \end{bmatrix}, \quad \mathbf{R_U} = R_u\mathbf{I} \tag{4.4.1}$$

where $Q_d$, $Q_v$, $R_u$ denote the coefficients of weighting matrices with respect to displacement, velocity and control force, respectively.

Using the same deterministic analysis and optimization technique to the previous section, one can readily attain the optimal values of coefficients of weighting matrices, as shown in Table 4.8. The exceedance probabilities of objective function and system quantities such as interstory drift, interstory velocity, and interstory control force are provided as well. The convergence process of the objective function along the iteration number is shown in Fig. 4.6, which indicates a good trade-off between response reduction and control cost. It is seen the exceedance probabilities of state quantities and control force are in the range of 0.002 and 0.004.

Time histories of the mean and standard deviation of the 1st interstory drift and of the 8th interstory drift with and without controls are shown in Fig. 4.7, respectively. It is seen that the interstory drift decreases significantly and arises to an equally proportional reduction along story level, where the interstory drift is nearly four times smaller than that of uncontrolled structure. Meanwhile, the interval of time histories with significant variation is reduced more seriously than other intervals. Time histories of the mean and standard deviation of the first story acceleration and of the eighth story acceleration with and without controls are shown in Fig. 4.8. It is seen that the first story acceleration with smaller peak is nearly not changed; while the eighth story acceleration with larger peak is reduced significantly, which attains a more smooth story acceleration along structural height. With consideration of interstory drift, the control effectiveness relies on the applied probabilistic criterion that the story exhibiting a larger response attains a better improvement, resulting in the desired structural performance. In comparison with Figs. 3.19 and 3.20, the mean-square responses of the controlled structure are almost identical both using the exceedance probability criterion in optimization criterion group on energy trade-off and the criterion on system second-order statistics evaluation.

A more elaborate representation is the concern with the PDFs of structural responses. Shown in Figs. 4.9 and 4.10 are the PDFs of the first and eighth interstory drifts and story accelerations at typical instants of time with and without controls, respectively. It is seen that by comparison with Figs. 3.21b–3.24b, which are derived from the system second-order statistics evaluation criterion, the application of the exceedance probability criterion in optimization criterion group on energy trade-off attains better displacement and acceleration controls, where the amplitude of the PDFs becomes larger. Meanwhile, the variation of interstory drifts with control

**Table 4.8** Optimization results of cost-function weights

| Parameters | $Q_d$ | $Q_v$ | $R_u$ |
|---|---|---|---|
| Initial value | 100 | 100 | $10^{-12}$ |
| Optimal value | 102.8 | 163.7 | $10^{-12}$ |
| Objective function | $1.122 \times 10^{-5}$ ($P_{f,d} = 0.0023$, $P_{f,v} = 0.0035$, $P_{f,u} = 0.0022$) | | |

**Fig. 4.6** Convergence process of objective function against iteration number

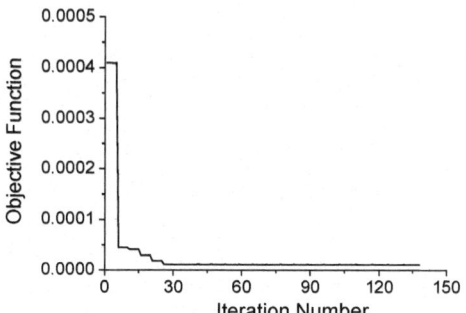

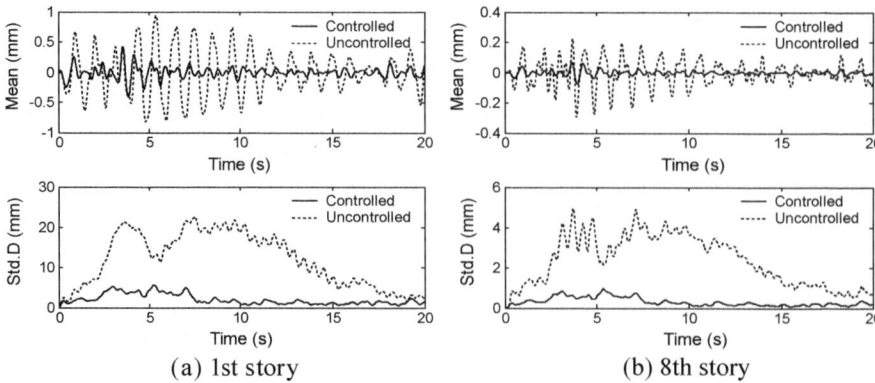

(a) 1st story                                    (b) 8th story

**Fig. 4.7** Time histories of mean and standard deviation of interstory drifts with and without controls

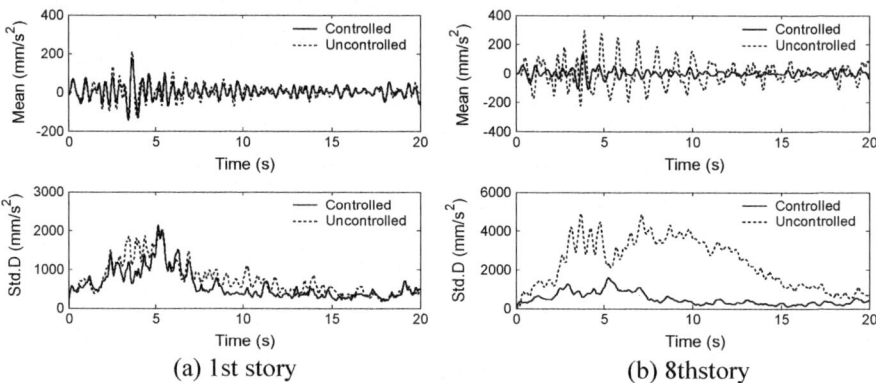

(a) 1st story                                    (b) 8thstory

**Fig. 4.8** Time histories of mean and standard deviation of story accelerations with and without controls

behaves smaller along with the instants of time than that without control, and the PDF tail of interstory drift with control changes very gently along with the time. This tail behavior is also exposed in the PDFs of the first story acceleration at the typical instants of time, although the variation of the story acceleration with control is not seriously improved comparing with that without controls. It is indicated that these results are consistent with the physical meanings of probabilistic criteria. As mentioned in the previous sections, the exceedance probability criterion is not for optimization of parameters of control law on full probability density, and just the tail of probability density function is controlled, no matter of interstory drift or of story acceleration. Figures 4.11 and 4.12 show the surface and contour of probability density at typical time interval of the first interstory drift and story acceleration with control.

The time histories of mean and standard deviation of interstory control forces of the first and eighth stories are shown in Fig. 4.13. It is seen that the time history

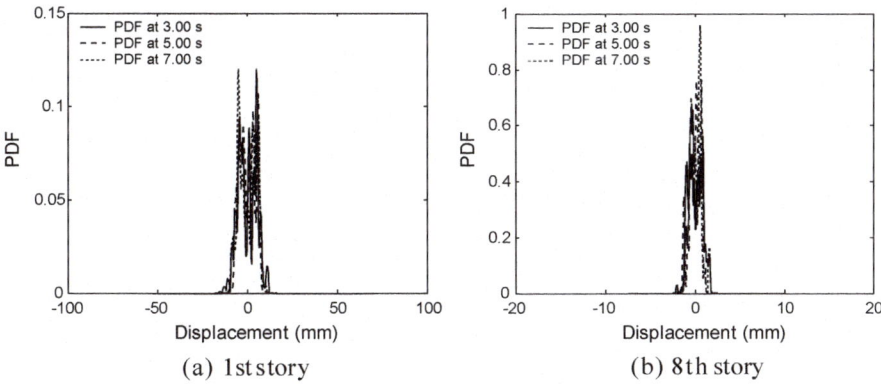

Fig. 4.9   PDFs of interstory drifts with control at typical instants of time

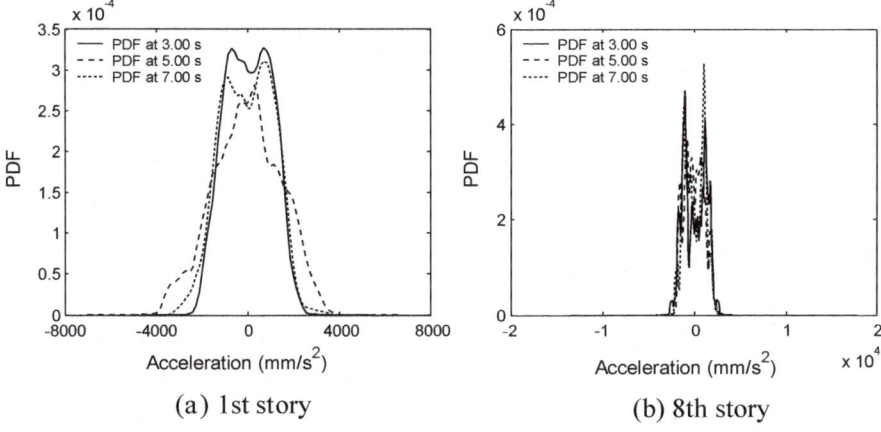

Fig. 4.10   PDFs of story accelerations with control at typical instants of time

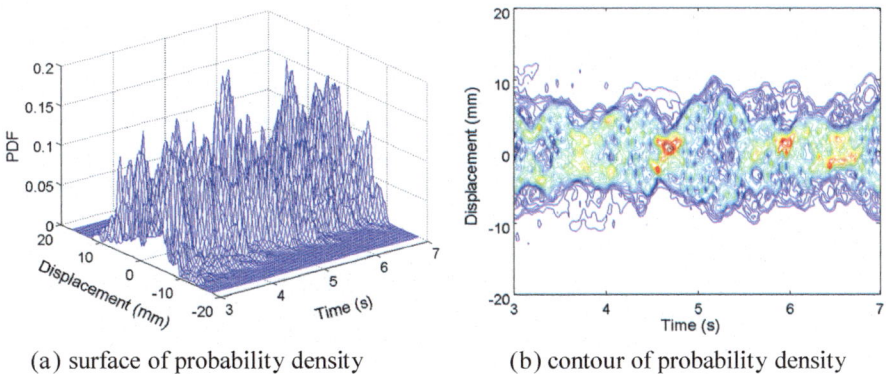

Fig. 4.11   Surface and contour of probability density at typical time interval of 1st interstory drift with control

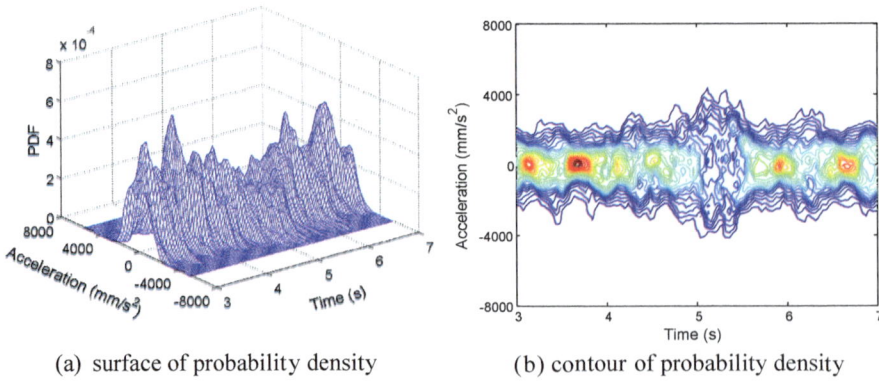

(a) surface of probability density            (b) contour of probability density

**Fig. 4.12** Surface and contour of probability density at typical time interval of 1st story acceleration with control

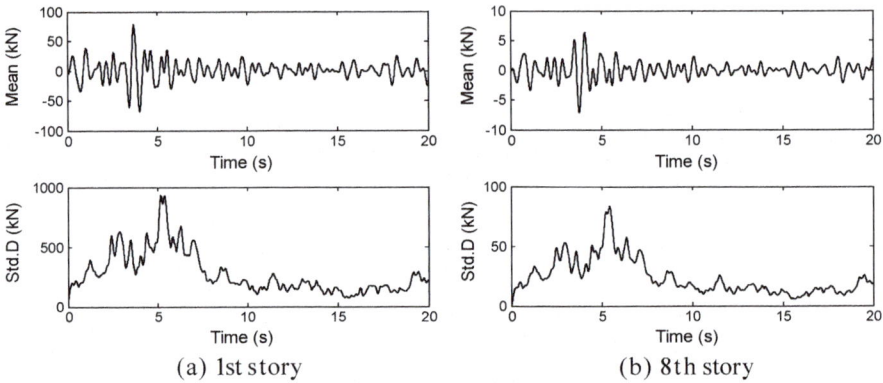

(a) 1st story                              (b) 8th story

**Fig. 4.13** Time histories of mean and standard deviation of interstory control force

curves of control forces have some similarities in details between the two stories: positive similarity inherent in the standard deviation and reverse similarity inherent in the mean, except the amplitudes where the first interstory control force is almost 10 times of the eighth interstory control force, which is similar to the results shown in Fig. 3.27. These similarities represent a coherence of feedback controls between stories where their control forces exhibit reverse similarity, which is also indicated in the PDFs of interstory control force between the first and eighth stories; see Fig. 4.14. It is understood that the control force is a linear combination of the displacement and velocity multiplying their corresponding elements of gain matrix.

The control effectiveness of extreme-value responses of the eight-story shear frame is shown in Table 4.9. It is seen that the interstory drifts along structural height are all reduced significantly, where the mean of extreme values decreases almost 75%, and the standard deviation of extreme values decreases almost 88%. The story acceleration with control gains a significant improvement as well. Except the first

**Table 4.9** Control effectiveness of eight-story shear frame

| Equ. ext.-value | | | Story number | | | | | | | |
|---|---|---|---|---|---|---|---|---|---|---|
| | | | 1 | 2 | 3 | 4 | 5 | 6 | 7 | 8 |
| Interstory drift (mm) | Mn | Unc. | 29.95 | 28.66 | 26.65 | 24.64 | 21.94 | 18.11 | 13.02 | 6.84 |
| | | Con. | 8.24 | 7.58 | 6.80 | 5.92 | 4.93 | 3.83 | 2.62 | 1.33 |
| | | Eff.[a] | −72.49% | −73.55% | −74.48% | −75.97% | −77.53% | −78.85% | −79.88% | −80.56% |
| | Std.d | Unc. | 14.11 | 13.91 | 13.16 | 11.39 | 9.15 | 6.84 | 4.56 | 2.28 |
| | | Con. | 1.75 | 1.64 | 1.49 | 1.30 | 1.08 | 0.83 | 0.57 | 0.29 |
| | | Eff. | −87.60% | −88.21% | −88.68% | −88.59% | −88.20% | −87.87% | −87.50% | −87.28% |
| Story acceleration (mm/s²) | Mn | Unc. | 3031.9 | 3489.5 | 4140.0 | 4547.7 | 4645.0 | 5150.4 | 6118.4 | 6759.5 |
| | | Con. | 2818.5 | 2651.5 | 2490.9 | 2361.0 | 2273.2 | 2228.2 | 2212.3 | 2209.8 |
| | | Eff. | −7.04% | −24.01% | −39.83% | −48.08% | −51.06% | −56.74% | −63.84% | −67.31% |
| | Std.d | Unc. | 918.0 | 1296.2 | 1549.7 | 1671.4 | 1976.6 | 2268.7 | 2249.2 | 2249.5 |
| | | Con. | 862.6 | 704.6 | 562.9 | 462.4 | 414.3 | 405.7 | 415.0 | 424.4 |
| | | Eff. | −6.03% | −45.64% | −63.68% | −72.33% | −79.04% | −82.12% | −81.55% | −81.13% |
| Interstory control force (kN) | Mn | | 1306.59 | 788.13 | 525.04 | 339.28 | 207.32 | 149.21 | 139.37 | 114.12 |
| | Std.d | | 269.04 | 186.52 | 154.24 | 129.19 | 98.45 | 58.73 | 34.38 | 26.02 |

[a]Effectiveness is defined as (Con.-Unc.)/Unc.

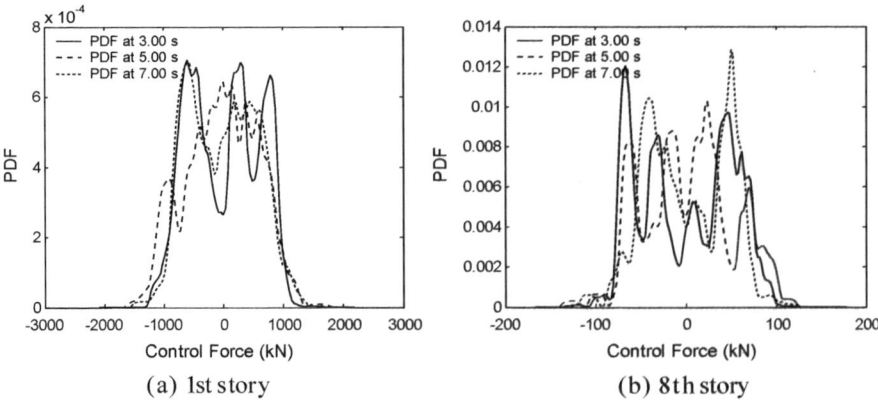

**Fig. 4.14**   PDFs of interstory control force at typical instants of time

story acceleration with mean and standard deviation both less than 10%, the remaining story accelerations gain 25–65% reduction on mean and 45–80% reduction on standard deviation. It is revealed that although the mitigation ratio of story acceleration is not homogeneous along structural height, the story acceleration arises to be more uniform after the active tendons are deployed. These characteristics have similarities with the application of the criterion on system second-order statistics evaluation addressed in Sect. 3.4.2: significant reduction attained on interstory drift and better distribution attained on story acceleration. The difference, however, between the two probabilistic criteria lies in that the former attains a better displacement and acceleration control than the latter, and thus guarantees the structural safety and structural comfortability to a greater extent. Of course, the former needs larger control forces acting on the stories. It is readily seen in Fig. 4.7 that the reason causing the difference of control effectiveness between two probabilistic criteria is still the physical meanings behind the probabilistic criteria. The quantiles of displacement constraint and evaluation quantities in the criterion on system second-order statistics evaluation are all defined as the mean plus one time of standard deviation, which is really a probabilistic criterion with less failure probability. It is further revealed in the exceedance probabilities of system quantities shown in Table 4.9 that the application of the exceedance probability criterion in optimization criterion group on energy trade-off gains a more reasonable structural performance.

## 4.5   Discussions and Summaries

Similar to the classical stochastic optimal control, a critical step in the physically based stochastic optimal control is the design and optimization of parameters of control law. Bypassing the trial-and-error procedure for optimization of control parameters, a family of probabilistic criteria for stochastic optimal control is proposed according to the probabilistic distribution and statistical moments of equivalent extreme-value quantities of structural systems. This treatment can implement the ready definition of the optimal parameters of control law for the physically based stochastic optimal control.

As to the single-objective criteria, the second-order moment and the tail of probability density of equivalent extreme-value quantities are investigated. As to the multiple-objective criteria, the ensemble average and exceedance probability of equivalent extreme-value quantities on performance trade-off and on energy trade-off are addressed, respectively. It is revealed that the exceedance probability criterion on energy trade-off can attain a rational balance between response reduction and control cost, which is thus a preferential criterion for designing the optimal control law of stochastic dynamical systems.

It is noted that the probabilistic criteria proposed in this chapter are not the criteria for regulating the full probability density. In fact, the present criteria underlie the control design of structural systems involving the first-passage problem, which exhibits a broad sense since the first-passage problem is of concern for a large family of engineering structures. As to the durability challenge, however, of structures on the basis of cumulative damage criterion, the accurate assessment on structural performance and the full regulation on probability density might be expected. For instance, assuming the maximum information entropy as the most rational configuration of physical quantities, a control logic tracing the desired probability density was designed (Sun 2006). Structural control on the optimal probability density can thus be implemented by defining the probability density with maximum entropy as an objective.

# References

Chen GP, Malik OP, Qin YH, Xu GY (1992) Optimization technique for the design of a linear optimal power system stabilizer. IEEE Trans Energy Convers 7(3):453–459

Chen SP, Li XJ, Zhou XY (1998) Stochastic linear quadratic regulators with indefinite control weight costs. SIAM J Control Optim 36:1685–1702

Leondes CT, Salami MA (1980) Algorithms for the weighting matrices in sampled-data linear time-invariant optimal regulator problems. Comput Electr Eng 7:11–23

Li J, Chen JB, Fan WL (2007) The equivalent extreme-value event and evaluation of the structural system reliability. Struct Saf 29(2):112–131

Li J, Peng YB, Chen JB (2011a) Probabilistic criteria of structural stochastic optimal controls. Probab Eng Mech 26(2):240–253

Li J, Peng YB, Chen JB (2011b) Nonlinear stochastic optimal control strategy of hysteretic structures. Struct Eng Mech 38(1):39–63

Mathews JH, Fink KD (2003) Numerical methods using Matlab, 4th edn. Prentice-Hall

Stengel RF, Ray LR, Marrison CI (1992) Probabilistic evaluation of control system robustness. In: IMA workshop on control systems design for advanced engineering systems: complexity, uncertainty, information and organization, Minneapolis, MN

Sun JQ (2006) Stochastic Dynamics and Control. Elsevier, Amsterdam

Tsay SC, Fong IK, Kuo TS (1991) Robust linear quadratic optimal–control for systems with linear uncertainties. Int J Control 53(1):81–96

Yang JN, Li Z, Liu SC (1992a) Stable controllers for instantaneous optimal control. ASCE J Eng Mech 118(8):1612–1630

Yang JN, Li Z, Liu SC (1992b) Control of hysteretic system using velocity and acceleration feedbacks. ASCE J Eng Mech 118(11):2227–2245

Zhang WS, Xu YL (2001) Closed form solution for along-wind response of actively controlled tall buildings with LQG controllers. J Wind Eng Ind Aerodyn 89:785–807

Zhu WQ, Ying ZG, Soong TT (2001) An optimal nonlinear feedback control strategy for randomly excited structural systems. Nonlinear Dyn 24:31–51

# Chapter 5
# Generalized Optimal Control Policy

## 5.1 Preliminary Remarks

In Chaps. 3 and 4, we proposed the methodology of stochastic optimal control in the context of the probability density evolution method, and developed the pertinent probabilistic criteria of stochastic optimal control, which form into a complete theoretical framework of the physically based stochastic optimal control. The control effectiveness, however, relies not only upon the probabilistic criteria for optimization of parameters of control law but also upon the constraints of the finiteness of available structural space for control device placement. There involves a broad topic on the stochastic optimal control of structures: how to define the control modality, how to optimize the control parameters pertaining to the devices, and how to determine the optimal number of control devices and their placements in the structural system. This topic is referred to as the generalized optimal control policy. The first issue of defining control modality is critical but often relies upon practical experiences. For instance, the passive modality with tuned mass dampers and viscous dampers is preferable for wind-induced vibration control of high-rise buildings; the semiactive modality with magnetorheological dampers is preferable for wind–rain-induced vibration control of cable-stayed bridges. The second issue has been addressed in preceding chapters, in which the optimization of control parameters pertaining to the active modality is carried out taking into account the randomness inherent in external excitations. The third exhibits a multifold perspective across the former two issues, which can be explored from two aspects. One aims at maximizing control effectiveness using a number of available control devices through control law design, controller (control device) parameter optimization, and control device placement allocation. The other aims at minimizing control cost to attain a similar structural performance as the objective.

Some freedoms towards alleviating the burden of these constraints have been gained by allowing a finite number of control devices to be arbitrarily located in space, and considering the problem of optimal control device placement. However,

© Springer Nature Singapore Pte Ltd. and Shanghai Scientific and Technical Publishers 2019
Y. Peng and J. Li, *Stochastic Optimal Control of Structures*,
https://doi.org/10.1007/978-981-13-6764-9_5

this subject has not been investigated as extensively as the problem of control force optimization in practical engineering (Amini and Tavassoli 2005). Criteria for optimizing control device placement were proposed that included the minimization of a modal control index (Chang and Soong 1980), the minimization of an energy index (Chen et al. 1991), the minimization of a system failure index (Vander and Carignan 1984; Ibidapo-Obe 1985), and the minimization of a performance index (Kim et al. 2003; Park et al. 2004). Pioneering work also contributed to the control device placement using the concept of degree of controllability (Laskin 1982; Lindberg and Longman 1984; Cheng and Pantelies 1988; Zhang and Soong 1992).

It is commonly acknowledged that a system is either controllable or not depending on whether there is a potential control device transferring the system state to an objective, which is often related to the stability domain of system state. However, this knowledge cannot reveal to what extent the structural system is controlled. The concept of degree of controllability was thus provided as an alternative index for evaluating the controlled system, and it was primarily used to determine the optimal number and placement of control devices. A controllability index for evaluating the response reduction of seismic structures was proposed by Cheng and Pantelides (Cheng and Pantelides 1988). In their findings, the optimal control device placement shall be the structural story exhibiting a maximum interstory drift. It is readily recognized that this scheme of designing control systems excludes the interaction between the control device placement and the influence of randomness inherent in seismic ground motions. In view of the limitation of the traditional controllability index, a sequential procedure for optimal damper placement was developed, using a matrix transfer method to solve the random seismic response of structures (Zhang and Soong 1992).

It is indicated that the two aspects of structural control, i.e., control force design and control device placement, can be arranged into an unified framework. For this reason, this chapter is first devoted to addressing the unified formula of optimal control law in the classical passive modality, active modality, semiactive modality, and hybrid modality. A generalized optimal control law, which serves as the means implementing the generalized optimal control policy, for the physically based stochastic optimal control is then proposed. In order to efficiently attain the optimal placement of control devices, an exceedance probability based probabilistic controllability index and its gradient are introduced. A diagram of the generalized optimal control policy (GOCP) and the generalized optimal control law (GOCL) in structural control logic is shown in Fig. 5.1.

## 5.2 Unified Formula of Optimal Control Law

Consider an $n$-degree-of-freedom system of controlled structures subjected to random excitations and governed by the equation of motion as follows:

$$\mathbf{M\ddot{X}}(t) + \mathbf{C\dot{X}}(t) + \mathbf{KX}(t) = \mathbf{B}_s\mathbf{U}(t) + \mathbf{D}_s\mathbf{F}(\mathbf{\Theta}, t), \qquad (5.2.1)$$

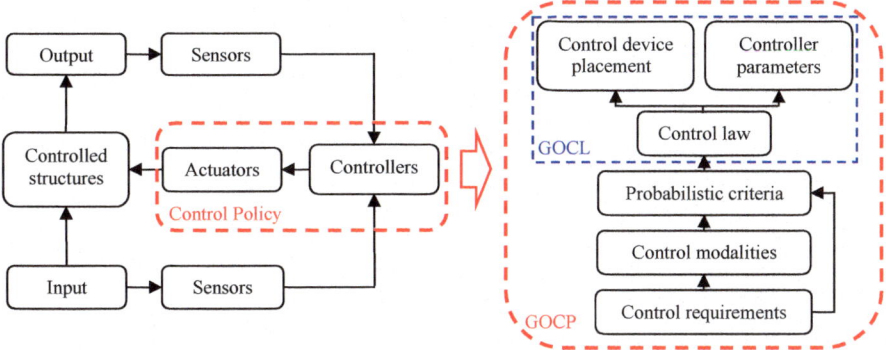

**Fig. 5.1**   Diagram of generalized formulation of control policy in structural control logic

where $\mathbf{X}(t)$ is the $n$-dimensional column vector denoting structural displacement; $\mathbf{F}(\Theta, t)$ is the $p$-dimensional column vector denoting random excitation. $\mathbf{M}$, $\mathbf{C}$ and $\mathbf{K}$ are $n \times n$ mass, damping, and stiffness matrices, respectively; $\mathbf{D}_s$ is the $n \times p$ matrix denoting the location of external excitation; $\mathbf{B}_s$ is the $n \times r$ matrix denoting the location of control devices; $\mathbf{U}(t)$ is the $r$-dimensional column vector denoting control force.

### 5.2.1   Passive Control Modality

Suppose that the $n$-degree-of-freedom system is controlled by passive control devices, the equation of motion of controlled structures is written as

$$\mathbf{M}\ddot{\mathbf{X}}(t) + \mathbf{C}\dot{\mathbf{X}}(t) + \mathbf{K}\mathbf{X}(t) = \mathbf{B}_{sp}\mathbf{U}_p(t) + \mathbf{D}_s\mathbf{F}(\Theta, t) \tag{5.2.2}$$

where $\mathbf{B}_{sp}$ is the $n \times r$ matrix denoting the location of passive control devices; $\mathbf{U}_p(t)$ is the $r$-dimensional vector denoting passive control force.

When the passive control force is modeled as a linear function of structural displacement, velocity, and acceleration, the pertinent control law can be written as

$$\mathbf{U}_p(t) = -\bar{\mathbf{M}}\ddot{\mathbf{X}}(t) - \bar{\mathbf{C}}\dot{\mathbf{X}}(t) - \bar{\mathbf{K}}\mathbf{X}(t) \tag{5.2.3}$$

where $\bar{\mathbf{M}}$, $\bar{\mathbf{C}}$ and $\bar{\mathbf{K}}$ are mass, damping, and stiffness matrices pertaining to the control system, respectively.

The equation of motion of controlled structures is then given by

$$(\mathbf{M} + \mathbf{B}_{sp}\bar{\mathbf{M}})\ddot{\mathbf{X}}(t) + (\mathbf{C} + \mathbf{B}_{sp}\bar{\mathbf{C}})\dot{\mathbf{X}}(t) + (\mathbf{K} + \mathbf{B}_{sp}\bar{\mathbf{K}})\mathbf{X}(t) = \mathbf{D}_s\mathbf{F}(\Theta, t) \tag{5.2.4}$$

It is readily seen that optimal control law of the passive control modality just relies upon the optimal design of additional mass (physical mass) $\mathbf{B}_{sp}\bar{\mathbf{M}}$, additional damping (physical damping) $\mathbf{B}_{sp}\bar{\mathbf{C}}$, and additional stiffness (physical stiffness) $\mathbf{B}_{sp}\bar{\mathbf{K}}$. Two aspects, i.e., the optimization of matrix denoting location of control devices $\mathbf{B}_{sp}$, and the optimization of parameters of control devices $\bar{\mathbf{M}}$, $\bar{\mathbf{C}}$ and $\bar{\mathbf{K}}$, ought to be included in the control law design.

### 5.2.2   Active Control Modality

As to the active control, the equation of motion of controlled structures can be written as

$$\mathbf{M}\ddot{\mathbf{X}}(t) + \mathbf{C}\dot{\mathbf{X}}(t) + \mathbf{K}\mathbf{X}(t) = \mathbf{B}_{sa}\mathbf{U}_a(t) + \mathbf{D}_s\mathbf{F}(\boldsymbol{\Theta}, t) \tag{5.2.5}$$

where $\mathbf{B}_{sa}$ is the $n \times r$ matrix denoting the location of active control devices; $\mathbf{U}_a(t)$ is the $r$-dimensional vector denoting active control force.

The formulation of optimal control law of controlled structures is given by

$$\mathbf{U}_a(t) = -\mathbf{f}_{\mathbf{M}}(\mathbf{Q}_{\mathbf{Z}}, \mathbf{R}_{\mathbf{U}})\ddot{\mathbf{X}}(t) - \mathbf{f}_{\mathbf{C}}(\mathbf{Q}_{\mathbf{Z}}, \mathbf{R}_{\mathbf{U}})\dot{\mathbf{X}}(t) - \mathbf{f}_{\mathbf{K}}(\mathbf{Q}_{\mathbf{Z}}, \mathbf{R}_{\mathbf{U}})\mathbf{X}(t) \tag{5.2.6}$$

where $\mathbf{Q}_{\mathbf{Z}}$ denotes the $3n \times 3n$ semi-positive weighting matrix with respect to system state; $\mathbf{R}_{\mathbf{U}}$ denotes the $r \times r$ positive weighting matrix with respect to control force; $\mathbf{f}_{\mathbf{M}}(\cdot)$, $\mathbf{f}_{\mathbf{C}}(\cdot)$, $\mathbf{f}_{\mathbf{K}}(\cdot)$ denote the components of control law pertaining to structural acceleration, velocity, and displacement. The equation of motion of controlled structures is then written as

$$[\mathbf{M} + \mathbf{B}_{sa}\mathbf{f}_{\mathbf{M}}(\mathbf{Q}_{\mathbf{Z}}, \mathbf{R}_{\mathbf{U}})]\ddot{\mathbf{X}}(t) + [\mathbf{C} + \mathbf{B}_{sa}\mathbf{f}_{\mathbf{C}}(\mathbf{Q}_{\mathbf{Z}}, \mathbf{R}_{\mathbf{U}})]\dot{\mathbf{X}}(t) + [\mathbf{K} + \mathbf{B}_{sa}\mathbf{f}_{\mathbf{K}}(\mathbf{Q}_{\mathbf{Z}}, \mathbf{R}_{\mathbf{U}})]\mathbf{X}(t) = \mathbf{D}_s\mathbf{F}(\boldsymbol{\Theta}, t) \tag{5.2.7}$$

It is seen that the optimal control law of active control modality relies upon the optimal design of artificial mass (numerical mass) $\mathbf{B}_{sa}\mathbf{f}_{\mathbf{M}}(\mathbf{Q}_{\mathbf{Z}}, \mathbf{R}_{\mathbf{U}})$, artificial damping (numerical damping) $\mathbf{B}_{sa}\mathbf{f}_{\mathbf{C}}(\mathbf{Q}_{\mathbf{Z}}, \mathbf{R}_{\mathbf{U}})$ and artificial stiffness (numerical stiffness) $\mathbf{B}_{sa}\mathbf{f}_{\mathbf{K}}(\mathbf{Q}_{\mathbf{Z}}, \mathbf{R}_{\mathbf{U}})$, which involves the optimization of matrix denoting the location of control devices $\mathbf{B}_{sa}$ and the optimization of parameters of control devices $\mathbf{Q}_{\mathbf{Z}}$, $\mathbf{R}_{\mathbf{U}}$.

### 5.2.3   Semiactive Control Modality

The equation of motion of semiactively controlled structural systems can be written as

$$\mathbf{M}\ddot{\mathbf{X}}(t) + \mathbf{C}\dot{\mathbf{X}}(t) + \mathbf{K}\mathbf{X}(t) = \mathbf{B}_{ss}\mathbf{U}_s(t) + \mathbf{D}_s\mathbf{F}(\Theta, t) \tag{5.2.8}$$

where $\mathbf{B}_{ss}$ is the $n \times r$ matrix denoting the location of semiactive control devices; $\mathbf{U}_s(t)$ is the $r$-dimensional column vector denoting semiactive control force.

Semiactive control is typically categorized into the active variable stiffness modality and the active variable damping modality. The optimal control law of controlled structures is then given by

$$\mathbf{U}_s(t) = -\mathbf{f}_{\mathbf{C}}(\widehat{\mathbf{C}}, \bar{\mathbf{C}})\dot{\mathbf{X}}(t) - \mathbf{f}_{\mathbf{K}}(\widehat{\mathbf{K}}, \bar{\mathbf{K}})\mathbf{X}(t) \tag{5.2.9}$$

where $\widehat{\mathbf{C}}, \bar{\mathbf{C}}$ denote the tuned and non-tuned damping matrices, respectively; $\widehat{\mathbf{K}}, \bar{\mathbf{K}}$ denote the tuned and non-tuned stiffness matrices. The equation of motion of controlled structures is then rewritten as

$$\mathbf{M}\ddot{\mathbf{X}}(t) + [\mathbf{C} + \mathbf{B}_{ss}\mathbf{f}_{\mathbf{C}}(\widehat{\mathbf{C}}, \bar{\mathbf{C}})]\dot{\mathbf{X}}(t) + [\mathbf{K} + \mathbf{B}_{ss}\mathbf{f}_{\mathbf{K}}(\widehat{\mathbf{K}}, \bar{\mathbf{K}})]\mathbf{X}(t) = \mathbf{D}_s\mathbf{F}(\Theta, t) \tag{5.2.10}$$

It is seen that the optimal control law of semiactive control modality relies upon the optimal design of the cross-term between additional damping (non-tuned) and artificial damping (tuned)$\mathbf{B}_{ss}\mathbf{f}_{\mathbf{C}}(\widehat{\mathbf{C}}, \bar{\mathbf{C}})$, and the cross-term between additional stiffness (non-tuned) and artificial stiffness (tuned) $\mathbf{B}_{ss}\mathbf{f}_{\mathbf{K}}(\widehat{\mathbf{K}}, \bar{\mathbf{K}})$. The optimization of matrix denoting the location of control devices $\mathbf{B}_{ss}$ and the optimization of parameters of control devices $\widehat{\mathbf{C}}, \bar{\mathbf{C}}, \widehat{\mathbf{K}}, \bar{\mathbf{K}}$ are included.

### 5.2.4 Hybrid Control Modality

Hybrid control is typically a combination modality of passive control and active control (or semiactive control). The equation of motion of controlled structures is written as

$$\mathbf{M}\ddot{\mathbf{X}}(t) + \mathbf{C}\dot{\mathbf{X}}(t) + \mathbf{K}\mathbf{X}(t) = \mathbf{B}_{sh}\mathbf{U}_h(t) + \mathbf{D}_s\mathbf{F}(\Theta, t) \tag{5.2.11}$$

The formulation of optimal control law of controlled structures is given by

$$\mathbf{U}_h(t) = -[\mathbf{f}_{\mathbf{M}}(\mathbf{Q}_{\mathbf{Z}}, \mathbf{R}_{\mathbf{U}}) + \bar{\mathbf{M}}]\ddot{\mathbf{X}}(t) - [\mathbf{f}_{\mathbf{C}}(\mathbf{Q}_{\mathbf{Z}}, \mathbf{R}_{\mathbf{U}}) + \bar{\mathbf{C}}]\dot{\mathbf{X}}(t) - [\mathbf{f}_{\mathbf{K}}(\mathbf{Q}_{\mathbf{Z}}, \mathbf{R}_{\mathbf{U}}) + \bar{\mathbf{K}}]\mathbf{X}(t) \tag{5.2.12}$$

or

$$\mathbf{U}_h(t) = -\bar{\mathbf{M}}\ddot{\mathbf{X}}(t) - [\mathbf{f}_{\mathbf{C}}(\widehat{\mathbf{C}}, \bar{\mathbf{C}}) + \bar{\mathbf{C}}]\dot{\mathbf{X}}(t) - [\mathbf{f}_{\mathbf{K}}(\widehat{\mathbf{K}}, \bar{\mathbf{K}}) + \bar{\mathbf{K}}]\mathbf{X}(t) \tag{5.2.13}$$

It is readily seen that similar to the semiactive control, the hybrid control provides the artificial mass, the artificial damping, and the artificial stiffness as the active control, and also provides the additional mass, the additional damping, and the additional stiffness as the passive control. The parameters of control law optimization include the matrix denoting the location of control devices $\mathbf{B}_{sh}$ and the parameters of control devices $\bar{\mathbf{M}}, \bar{\mathbf{C}}, \bar{\mathbf{K}}, \mathbf{Q}_Z, \mathbf{R}_U, \hat{\mathbf{C}}, \hat{\mathbf{K}}$.

One might recognize from Eqs. 5.2.3, 5.2.6, 5.2.9, and 5.2.12 that the optimal control law has a unified formula as follows:

$$\mathbf{U}(t) = -\mathbf{f}(\tilde{\mathbf{M}}, \tilde{\mathbf{C}}, \tilde{\mathbf{K}})[\,\ddot{\mathbf{X}}(t)\ \dot{\mathbf{X}}(t)\ \mathbf{X}(t)\,]^{\mathrm{T}} \qquad (5.2.14)$$

where $\tilde{\mathbf{M}}, \tilde{\mathbf{C}}, \tilde{\mathbf{K}}$ denote generalized mass, generalized damping, and generalized stiffness, respectively; $\mathbf{f}(\cdot)$ denotes gain matrix of state-feedback control.

It is indicated as well that the optimal control law not only relies upon the output actualized by the control device but also upon the deployment of the control device. A generalized formula of optimal control law is then given by

$$\mathbf{U}(t) = \mathbf{f}(\tilde{\mathbf{M}}, \tilde{\mathbf{C}}, \tilde{\mathbf{K}}, \mathbf{B}_s)[\,\ddot{\mathbf{X}}(t)\ \dot{\mathbf{X}}(t)\ \mathbf{X}(t)\,]^{\mathrm{T}} \qquad (5.2.15)$$

Equations (5.2.15) is the so-called generalized optimal control law; $\mathbf{f}(\tilde{\mathbf{M}}, \tilde{\mathbf{C}}, \tilde{\mathbf{K}}, \mathbf{B}_s)$ denotes the gain matrix of generalized optimal control law, which has a general formulation $\mathbf{f}(\mathbf{I}^*, \mathbf{L}^*)$ where $\mathbf{I}^* = [I_{\tilde{\mathbf{M}}}^*, I_{\tilde{\mathbf{C}}}^*, I_{\tilde{\mathbf{K}}}^*]$ denotes the optimal parameters describing the generalized mass, generalized damping, and generalized stiffness; $\mathbf{L}^* = [L_x^*, L_y^*, L_z^*]$ denotes the optimal placement vector describing the deployment of control devices in the three-dimensional space $(x, y, z)$ of structures. For instance, as to the control devices deployed in a two-dimensional (2D) structure shown in Fig. 5.2, the matrix of optimal placement can be represented by the number of column-beam lattices of structures as follows:

$$\mathbf{L}_{xz}^* = \begin{bmatrix} 2 & 0 & 0 \\ 0 & 0 & 0 \\ 0 & 0 & 0 \\ 0 & 0 & 1 \\ 0 & 0 & 0 \end{bmatrix}_{5\times 3}^{\mathrm{T}} \qquad (5.2.16)$$

While as to a three-dimensional (3D) structure, the axis $y$ can be viewed as the unfold of the axis $x$, and the matrix of optimal placement is given by

$$\mathbf{L}_{xyz}^* = \begin{bmatrix} 1 & 0 & 4 & 0 & 0 & 0 & 0 & 0 & 0 & 0 & 0 & 0 & 0 & 0 & 0 & 0 & 0 \\ 0 & 0 & 0 & 0 & 0 & 0 & 0 & 0 & 0 & 0 & 3 & 0 & 0 & 0 & 0 & 0 \\ 0 & 0 & 0 & 2 & 0 & 0 & 0 & 0 & 0 & 0 & 0 & 0 & 0 & 0 & 0 & 0 & 0 \end{bmatrix}_{3\times 17}^{\mathrm{T}} \qquad (5.2.17)$$

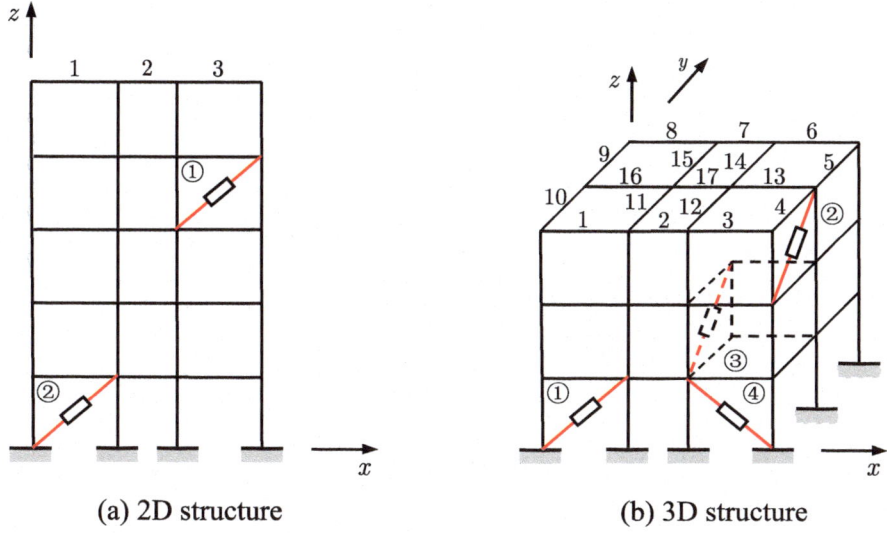

**(a) 2D structure**    **(b) 3D structure**

**Fig. 5.2** Schematic diagram of control devices deployed in structures

where the element zero indicates that there are no control devices in the column-beam lattice; the element nonzero indicates that there is a control device with the associated sequence of deployment in the column-beam lattice.

It has been revealed in Chaps. 3 and 4 that the critical step in the stochastic optimal control is the derivation of optimal control law and the optimization of control parameters. The essence of solving the generalized optimal control law is thus the optimization of the parameters $(\mathbf{I}^*, \mathbf{L}^*)$. In fact, the generalized optimal control law of the physically based stochastic optimal control places the three-level design objectives into an integral way: the control law design, the controller (control device) parameter optimization and the optimal control device placement. These design objectives correspond to the pertinent principles and probabilistic criteria, respectively, as shown in Fig. 5.3.

For illustrative purposes, the three-level design of stochastic optimal control by means of active modality is summarized in steps as follows:

*Step 1*: Using Pontryagin's maximum principle or Bellman's optimality principle, the cost function is minimized so as to derive the formulation of optimal control law; see Eq. (3.3.18).

*Step 2*: According to the exceedance probability criterion on energy trade-off; see Eq. (4.3.30), the performance function pertaining to parameter optimization is minimized so as to gain the optimal controller parameters.

*Step 3*: In view of the criterion on minimum story controllability index gradient (Min-SCIG), the performance function pertaining to placement optimization is minimized so as to define the optimal control device deployment.

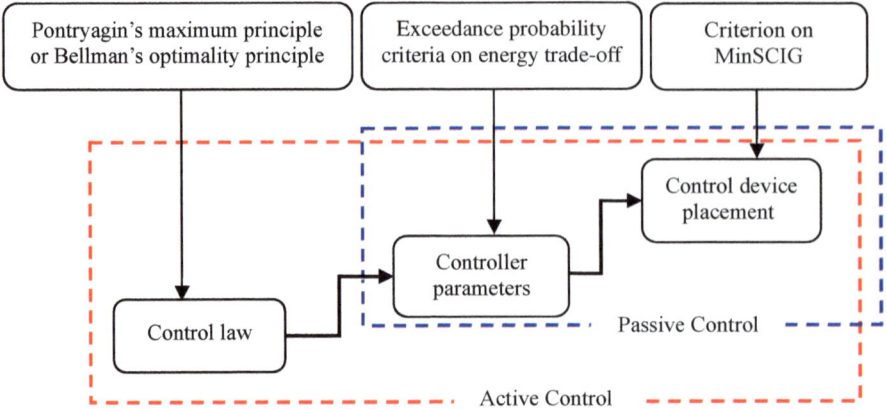

**Fig. 5.3** Three-level design for generalized optimal control law

*Steps 1* and *2* have been addressed in details in Chaps. 3 and 4. The optimization in *Step 3* will be illustrated in the following sessions. Besides, as to the stochastic optimal control by means of passive modality, just *Step 2* and *3* are included since there is no feedback logic involved.

## 5.3  Probabilistic Controllability Index

It is revealed from the previous sections that the optimization of parameters of the optimal control law resorts to a certain probabilistic criterion with the specified performance function. In order to identify the optimal placement of control devices in structural space, a probabilistic controllability index in function of exceedance probability of quantities is defined as follows:

$$\rho_i = \tfrac{1}{2}[\Pr^{\mathrm{T}}_{\tilde{Z}_i}(\tilde{Z}_i - \tilde{Z}_{i,\mathrm{thd}} > \mathbf{0})\mathbf{Q}_{\tilde{Z}_i}\Pr_{\tilde{Z}_i}(\tilde{Z}_i - \tilde{Z}_{i,\mathrm{thd}} > \mathbf{0}) \\ + \Pr^{\mathrm{T}}_{\tilde{U}_i}(\tilde{U}_i - \tilde{U}_{i,\mathrm{thd}} > \mathbf{0})\mathbf{R}_{\tilde{U}_i}\Pr_{\tilde{U}_i}(\tilde{U}_i - \tilde{U}_{i,\mathrm{thd}} > \mathbf{0})] , \quad i = 1, 2, \ldots, n$$

$$(5.3.1)$$

where $\tilde{Z}_i = [\,\max_t|X_i(\mathbf{\Theta}, t)|\ \max_t|\dot{X}_i(\mathbf{\Theta}, t)|\ \max_t|\ddot{X}_i(\mathbf{\Theta}, t)|\,]^{\mathrm{T}}$, $\tilde{U}_i = \max_t|U_i(\mathbf{\Theta}, t)|$ denote the extreme-value state vector and the extreme-value control force of the $i$th element of structural systems in the time interval $[t_0, t_f]$, respectively; $\Pr(\cdot)$ operates component-wise on its vector argument with denotation of exceedance probability; $\tilde{Z}_{i,\mathrm{thd}}, \tilde{U}_{i,\mathrm{thd}}$ are the thresholds of $\tilde{Z}_i, \tilde{U}_i$; the weighting matrices $\mathbf{Q}_{\tilde{Z}_i}, \mathbf{R}_{\tilde{U}_i}$ denote the relative importance of system state and control force, which are defined for simplicity in formulation with uniform weights on the quantities as follows:

$$\mathbf{Q}_{\tilde{Z}_i} = \begin{bmatrix} 1 & 0 & 0 \\ 0 & 1 & 0 \\ 0 & 0 & 1 \end{bmatrix}, \mathbf{R}_{\tilde{U}_i} = 1 \tag{5.3.2}$$

It is indicated in Eq. 5.3.1 that the exceedance probability based probabilistic controllability index characterizes system safety represented by the interstory drift, system serviceability represented by the interstory velocity, system comfortability represented by the story acceleration, and control device workability represented by the control force and their trade-off. This definition is more comprehensive than the previous definition of controllability index with single and deterministic quantity (Zhang and Soong 1992).

Moreover, a controllability index gradient is defined as

$$J_3 = \Delta\rho_i^j = \frac{\rho_i^{j-1} - \rho_i^j}{\rho_i^{j-1}}, \quad i = 1, 2, \ldots, n; \, j = 1, 2, \ldots, r \tag{5.3.3}$$

where $\rho_i^0$ denotes the probabilistic controllability index of uncontrolled structures.

In order to determine the optimal control device placement in each sequential case, a criterion on the minimum controllability index gradient is constructed, i.e., the design of sequentially optimal placement of control devices involves the following optimization problem:

$$L^* = \{x_i^{j,*}, y_i^{j,*}, z_i^{j,*}\} = \arg\min_{x,y,z}\{J_3\}, \quad i = 1, 2, \ldots, n; \quad j = 1, 2 \ldots, r \tag{5.3.4}$$

Therefore, the $j$th control device will be deployed next at the $i$th element with the minimum controllability index gradient, i.e., $\Delta\rho_i^{j-1} = \min\{\Delta\rho_1^{j-1}, \Delta\rho_2^{j-1}, \ldots, \Delta\rho_n^{j-1}\}$. The corresponding placement vector $\{x_i^{j,*}, y_i^{j,*}, z_i^{j,*}\}$ is updated until the predetermined structural performance objective is achieved. Since the controllability index gradient of uncontrolled structures makes no sense, the first control device is considered to be deployed at the $i$th element with the maximum controllability index, i.e., $\rho_i^0 = \max\{\rho_1^0, \rho_2^0, \ldots, \rho_n^0\}$. It is noted that the control device is optimally deployed in the column-beam lattice at the current stage, and the remaining control devices will be placed in other column-beam lattices even if the occupied lattice needs more control devices at following stages. Besides, the controllability index and the controllability index gradient are termed as the story controllability index and the story controllability index gradient for easy-to-understand cases of building structures.

It is worth noting that the criterion on maximum story controllability index (MaxSCI) developed in previous studies (Zhang and Soong 1992) is different from the criterion on minimum story controllability index gradient (MinSCIG) proposed in the present study. Using the former criterion, the deployment of the $j$th control device to be carried out next relies upon the maximum story controllability index at this stage with $(j-1)$ control devices. The former criterion achieves the performance objective more slowly than that of the latter criterion, which will be illustrated in details in the following numerical examples.

Obviously, the exceedance probability in the controllability index can be readily solved since system state $\tilde{Z}_i$ and control force, $\tilde{U}_i$ are separately governed by the GDEEs; see Eqs. (3.2.9) and (3.2.10).

## 5.4  Solution Procedure

### 5.4.1  Probabilistic Criteria

Details are now provided in implementing the generalized optimal control law. As was mentioned previously, the solving of the gain matrix $\mathbf{f}(\mathbf{I}^*, \mathbf{L}^*)$ of the generalized optimal control law involves optimization procedures, including the minimization of controllability index gradient in sequential cases to identify the optimal placement of control devices in structural space and the minimization of performance function to identify the optimal parameters of control devices or controllers. In view of the discussions on probabilistic criteria in Chap. 4, an exceedance probability criterion on energy trade-off is applied here; see Eq. (4.3.30), i.e., the definition of performance function pertaining to parameter optimization without acceleration constraint as follows:

$$J_2 = \frac{1}{2}[\Pr_{\tilde{Z}}^{\mathrm{T}}(\tilde{Z} - \tilde{Z}_{\mathrm{thd}} > \mathbf{0})\mathbf{Q}_{\tilde{Z}}\Pr_{\tilde{Z}}(\tilde{Z} - \tilde{Z}_{\mathrm{thd}} > \mathbf{0}) + \Pr_{\tilde{U}}^{\mathrm{T}}(\tilde{U} - \tilde{U}_{\mathrm{thd}} > \mathbf{0})\mathbf{R}_{\tilde{U}}\Pr_{\tilde{U}}(\tilde{U} - \tilde{U}_{\mathrm{thd}} > \mathbf{0})]$$

(5.4.1)

where $\tilde{U} = \max_t(\max_i|U_i(\boldsymbol{\Theta}, t)|)$ denotes the equivalent extreme-value control force; $\tilde{Z} = [\max_t(\max_i|X_i(\boldsymbol{\Theta}, t)|)\ \max_t(\max_i|\dot{X}_i(\boldsymbol{\Theta}, t)|)\ \max_t(\max_i|\ddot{X}_i(\boldsymbol{\Theta}, t)|)]^{\mathrm{T}}$ denotes the equivalent extreme-value system state; $\tilde{Z}_{\mathrm{thd}}, \tilde{U}_{\mathrm{thd}}$ denote the thresholds of equivalent extreme-value $\tilde{Z}, \tilde{U}$, respectively. The weighting matrices $\mathbf{Q}_{\tilde{Z}}, \mathbf{R}_{\tilde{U}}$ are set as the same as $\mathbf{Q}_{\tilde{Z}_i}, \mathbf{R}_{\tilde{U}_i}$ in Eq. (5.3.1) owing to the dimension similarities between the performance function and the probabilistic controllability index pertaining to parameter and placement optimizations, respectively.

### 5.4.2  Flowchart of Solution Procedure

The solution procedure of the generalized optimal control law is included in the following steps:

*Step 1*: Computation of controllability index of uncontrolled structural systems. The numerical solution process involves:

(i)  Probability-assigned space partition to determine the representative point set $\sigma_{\text{res}} \triangleq \{\boldsymbol{\theta}_q = (\theta_{1,q}, \theta_{2,q}, \cdots, \theta_{s,q}) | q = 1, 2, \ldots, n_{\text{res}}\}$ and the corresponding assigned probabilities $P_q$ (Li and Chen 2007; Chen and Li 2008).

(ii)  Deterministic dynamic analysis of controlled structural systems with respect to the representative points using the transfer function method to attain the system state $Z(\boldsymbol{\theta}_q, t)$ and its derivative process $\dot{Z}(\boldsymbol{\theta}_q, t)$, the control force $U(\boldsymbol{\theta}_q, t)$, and its derivative process $\dot{U}(\boldsymbol{\theta}_q, t)$.

(iii)  Using the finite difference method to solve the GDEEs and to attain the numerical solutions of $p_{Z\Theta}(z, \boldsymbol{\theta}_q, t)$, $p_{U\Theta}(u, \boldsymbol{\theta}_q, t)$, where the modified Lax–Wendroff difference scheme with TVD nature is applied (Li and Chen 2004; Chen and Li 2005).

(iv)  Repeating (ii) and (iii) and running over all the representative points $q = 1, 2, \ldots, n_{\text{res}}$, the probability density solution can be readily obtained by integral:

$$p_Z(z, t) = \sum_{q=1}^{n_{\text{res}}} p_{Z\Theta}(z, \boldsymbol{\theta}_q, t) S_q, \quad p_U(u, t) = \sum_{q=1}^{n_{\text{res}}} p_{U\Theta}(u, \boldsymbol{\theta}_q, t) S_q \quad (5.4.2)$$

where $S_q$ denotes the area measure of the subdomain represented by the representative point $\boldsymbol{\theta}_q$, which is relevant to the partition strategy of probability-assigned space $\Omega_\Theta$.

Having the numerical solutions of PDFs, the statistical moments and reliabilities of system quantities can be readily solved. However, in the canonical scheme of reliability-based structural optimal control (Spencer et al. 1994; May and Beck 1998), the joint probability density function for evaluating the expected level-crossing rate in the Rice formulae must be provided first, inevitably resulting in unguaranteed errors. It has been proved, however, that the structural reliability can be efficiently assessed by means of the GDEEs, as well by introducing the extreme-value distribution theorem with pseudo random process (Chen and Li 2007).

(v)  Computing the story controllability index (gradient) by Eqs. (5.3.1) and (5.3.3).

*Step 2*: Placement and parameter optimization of a new control device. The numerical procedure involves:

(i)  Deploying the control device into the structural space in accordance with the criterion of the MinSCIG.

(ii)  Intializing the values of design parameters based on their physical meanings; as to the passive control modality, for example, the design parameters are $\bar{\mathbf{C}}, \bar{\mathbf{K}}$; while as to the active control modality, the design parameters are $\mathbf{Q_Z}, \mathbf{R_U}$.

(iii)  Searching for the optimal parameters by minimizing the performance function, which involves an iterative process of performing (ii) (iii) and (iv) in *Step 1*, where the toolbox function of DAKOTA, quasi-Newton based OPT ++, is used to invoke the numerical routine (Eldred et al. 2007).

*Step 3*: Calculation of story controllability index of controlled structural systems, running (ii), (iii), (iv), and (v) in *Step 1* on the controlled structural system with the newly added control device, and determining the optimal place deploying a next control device.

Repeating *Step 2* and *Step 3* until the structural performance objective is achieved. The flowchart of these steps is shown in Fig. 5.4. The hybrid programming integrating the DAKOTA and MATLAB can be used to readily solve the generalized optimal control law.

The proposed algorithm for sequential placement of control devices can be used to gradually increase the reliability of structural systems until a target of reliability level is attained. It is clear that the traditional one-stage placement of control devices might

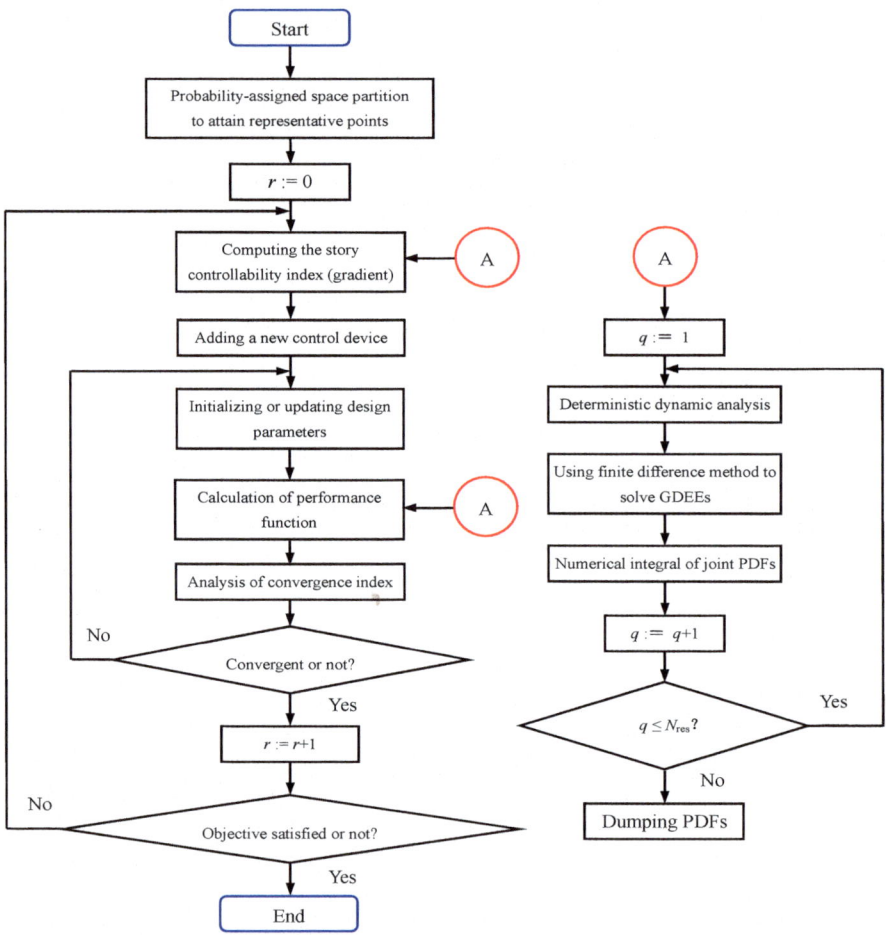

**Fig. 5.4** Flowchart of solving generalized optimal control law

be optimal for a prespecified number of control devices, which does not, however, provide the present flexibility in terminating the control device deployment prior to placing all the control devices.

## 5.5 Numerical Examples

For illustrative purposes, two randomly base-excited structures with control systems designed as the generalized optimal control law are investigated, including a ten-story shear frame attached with viscoelastic dampers and an eight-story shear frame attached with active tendon systems.

### 5.5.1 Viscoelasticity Damped Structure

A ten-story shear frame controlled by $r (1 \leq r \leq 10)$ viscoelastic dampers is investigated. The story mass and interstory stiffness of the uncontrolled structure are: $m_1 = m_2 = 2.4 \times 10^4$ kg, $m_3 = m_4 = 2.0 \times 10^4$ kg, $m_5 = m_6 = 1.8 \times 10^4$ kg, $m_7 = m_8 = 1.6 \times 10^7$ kg, $m_9 = m_{10} = 1.2 \times 10^4$ kg; $k_1 = k_2 = 18$ kN/mm, $k_3 = k_4 = 14$ kN/mm, $k_5 = k_6 = 12$ kN/mm, $k_7 = k_8 = 10$ kN/mm, $k_9 = k_{10} = 9.6$ kN/mm, respectively. The damping ratios of the first two vibrational modes are both 0.02. The Raleigh damping matrix $\mathbf{C} = a\mathbf{M} + b\mathbf{K}$ is employed to represent the model damping. The computed natural frequencies of the model are 4.53, 11.92, 19.19, 25.92, 31.94, 38.82, 43.44, 47.40, 50.08, and 50.92 rad/s. The relation between the output of the viscoelastic damper and the associated interstory drift and interstory velocity is denoted by (Soong and Dargush 1997):

$$U_p(t) = c_D \dot{\bar{X}}(t) + k_D \bar{X}(t) \tag{5.5.1}$$

where $\dot{\bar{X}}(t)$, $\bar{X}(t)$ denote the interstory velocity and the interstory drift, respectively; $c_D, k_D$ denote the damping coefficient and stiffness coefficient of the interstory viscoelastic damper, respectively. The thresholds of interstory drift, interstory velocity, story acceleration, and damper force are assumed to 10 mm, 100 mm/s, and 3000 mm/s$^2$, 200 kN, respectively. The physically motivated random seismic ground motion model shown in Sect. 3.4.1 is employed to simulate the seismic ground motions. The peak ground acceleration is set as 0.11 $g$. The predetermined objective of structural optimal control is determining the parameters and placement of the least number of viscoelastic dampers so as to attain the control effectiveness of viscoelastic dampers fully distributed in the structure. As an assessment objective of structural performance, the fully distributed viscoelastic dampers are placed in the structure simultaneously, and their parameters are designed as the same through optimization procedure.

**Table 5.1** Optimal placements and parameters of the newly added viscoelastic dampers in sequences

| Sequence no. | Placement vector | Parameters of newly added viscoelastic dampers[a] | |
|---|---|---|---|
| | | $c_D$(kNs/mm) | $k_D$(kN/mm) |
| 0 | $[0\,0\,0\,0\,0\,0\,0\,0\,0\,0]^T$ | – | – |
| 1 | $[0\,0\,0\,0\,0\,0\,1\,0\,0\,0]^T$ | 0.253 | 0.111 |
| 2 | $[0\,0\,0\,0\,0\,0\,1\,0\,0\,2]^T$ | 0.100 | 0.100 |
| 3 | $[0\,0\,0\,0\,3\,0\,1\,0\,0\,2]^T$ | 0.155 | 0.098 |
| Fully distributed | $[1\,1\,1\,1\,1\,1\,1\,1\,1\,1]^T$ | 0.374 | 0.127 |

[a]Initial values of parameters are $c_D = 0.1$ kNs/mm, $k_D = 0.1$ kN/mm

The optimal placement and parameters of the newly added viscoelastic dampers in sequences are shown in Table 5.1. It is seen that with only three optimally deployed viscoelastic dampers, a similar performance is achieved to the structural system with fully distributed viscoelastic dampers. The three viscoelastic dampers are deployed in the seventh interstory, the tenth interstory, and the fifth interstory in turn. Furthermore, the parameters of generalized optimal control law are given by

$$(c_D^*, k_D^*, L^*) = \begin{bmatrix} 0 & 0 & 0 & 0 & 0.155 & 0 & 0.252 & 0 & 0 & 0.100 \\ 0 & 0 & 0 & 0 & 0.098 & 0 & 0.111 & 0 & 0 & 0.100 \\ 0 & 0 & 0 & 0 & 3 & 0 & 1 & 0 & 0 & 2 \end{bmatrix}^T \qquad (5.5.2)$$

where the damping and stiffness parameters of viscoelastic dampers have been embedded into the first and second rows of the parameter matrix. The placement matrix of viscoelastic dampers are folded into a vector since the shear frame structure can be viewed as a one-dimensional model with lumped masses along $z$ axis, where zero denotes no viscoelastic dampers in the interstory and nonzero denotes the viscoelastic damper and its deployment sequence.

Figure 5.5 shows the relation between the added viscoelastic dampers and the controllability index. It is readily seen that the controllability index decreases very rapidly after the viscoelastic damper with present optimal parameters is deployed on its optimal place. Besides, the controllability index of the seventh interstory is always maximum among those of the interstories in all the sequences; while the optimal damper placement is not close to the interstories near to the seventh interstory. This owes to the fact that it is the design criterion of the MinSCIG other than of the MaxSCI employed in the present study. A comparison of control effectiveness between the two design criteria is shown in Fig. 5.6. It is seen that just using three viscoelastic dampers, the objective of structural control is attained when the MinSCIG is employed; while four viscoelastic dampers are required to achieve the same control effectiveness if the MaxSCI is employed. Figure 5.7 shows the viscoelastic damper deployments as

**Fig. 5.5** Relation between
added viscoelastic dampers
and controllability index

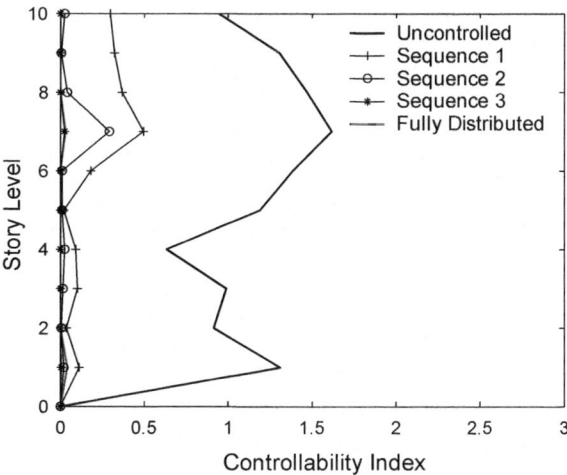

**Fig. 5.6** Comparison
between design criteria of
MinSCIG and MaxSCI in the
case of viscoelasticity
damped structures

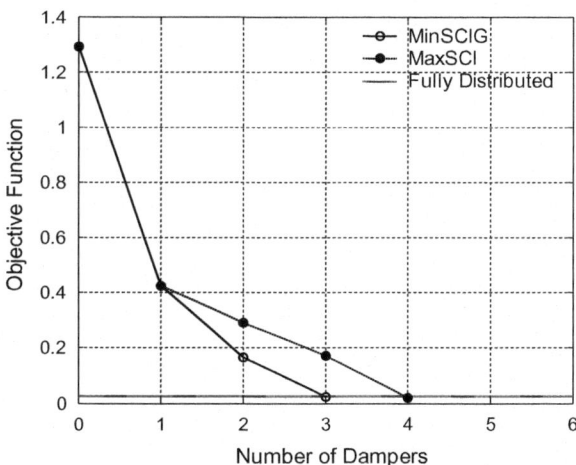

design criteria of the MinSCIG and the MaxSCI, respectively, in which the number
denotes the sequences of viscoelastic damper placements.

The exceedance probabilities of interstory drift, interstory velocity, story acceleration, and interstory control force in sequences of viscoelastic damper deployments
are listed in Table 5.2. It is seen that the exceedance probabilities of the system
quantities of concern gradually decrease along with the deployment of viscoelastic
dampers, till the objective function attains the same level with the case with fully
distributed viscoelastic dampers; say, their objective functions have the same magnitude. It is revealed that the structural control has attained the system safety, system
serviceability, system comfortability, and system workability. In comparison with
the case with fully distributed viscoelastic dampers, the case of Sequence 3 exhibits
a better interstory drift control but a worse story acceleration control. This difference

**Fig. 5.7** Viscoelastic damper deployments as design criteria of MinSCIG and MaxSCI

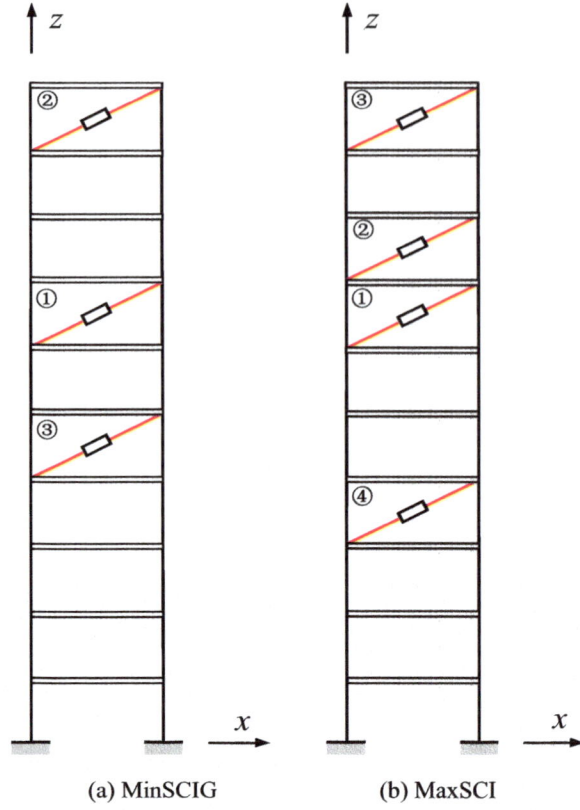

(a) MinSCIG          (b) MaxSCI

can be seen as well from the probability density distributions of equivalent extreme values of interstory drift and story acceleration; see Fig. 5.8. Meanwhile, it is seen from the objective function that the control effectiveness of the case of Sequence 3 is better than the case with fully distributed viscoelastic dampers. It is thus remarked that the structural control employing the generalized optimal control law attains a better trade-off between system quantities than that employing traditional control schemes.

In addition, an index relevant to mean-square control force is defined to evaluate the control cost:

$$\epsilon_u = \int_{t_0}^{t_f} E[\mathbf{U}(t)\mathbf{U}^{\mathrm{T}}(t)]\mathrm{d}t \tag{5.5.3}$$

The control cost along with the viscoelastic damper deployment in sequences is shown in Table 5.2 as well. It is readily seen that the control cost of the case

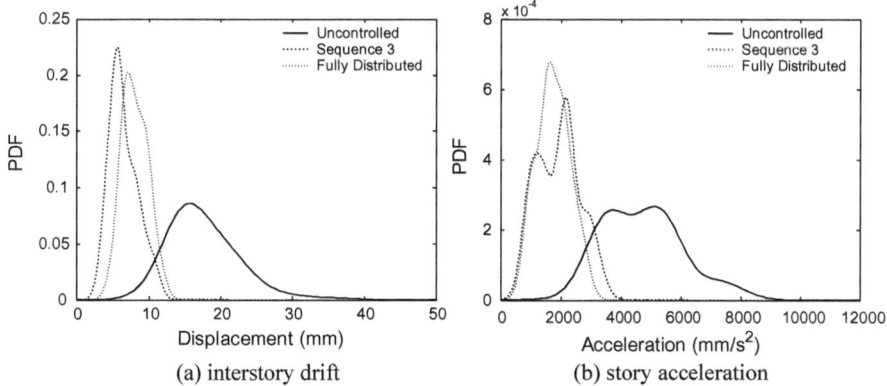

**Fig. 5.8**  PDFs of equivalent extreme values of interstory drift and story acceleration in the case of viscoelasticity damped structures

of Sequence 3 is minimum, and that of the case with fully distributed viscoelastic dampers is maximum. It is explained that the control cost is related both to the number of control devices and to the interstory drift and interstory velocity. One might recognize that the case of Sequence 3 exhibits a better displacement control and an almost same velocity control as the case with fully distributed viscoelastic dampers. Therefore, the control cost of the former is nearly five times less than that of the latter, which proves once again that the optimal control employing the generalized optimal control law attains a better trade-off between system quantities.

The second-order moment of equivalent extreme-value interstory drift in sequences is shown in Fig. 5.9. It is seen that the mean of interstory drift becomes smaller along with the viscoelastic damper deployment; whereas the standard deviation of interstory drift above the seventh story increases when the first viscoelastic damper is deployed in the seventh interstory. It reduces rapidly when the second

**Table 5.2**  Seismic mitigation by viscoelastic dampers in sequences

| Sequence no. | Placement vector | Exceedance probabilities | | | | Objective function $J_2$ | Mean-square control force $\epsilon_u$ (kN$^2$) |
|---|---|---|---|---|---|---|---|
| | | $P_{f,d}$ | $P_{f,v}$ | $P_{f,a}$ | $P_{f,u}$ | | |
| 0 | $[0\,0\,0\,0\,0\ 0\,0\,0\,0\,0]^T$ | 0.9952 | 0.8188 | 0.9599 | – | 1.2911 | – |
| 1 | $[0\,0\,0\,0\,0\ 0\,1\,0\,0\,0]^T$ | 0.4195 | 0.6085 | 0.5483 | $3.60 \times 10^{-7}$ | 0.4235 | 69824.2 |
| 2 | $[0\,0\,0\,0\,0\ 0\,1\,0\,0\,2]^T$ | 0.3226 | 0.4373 | 0.1888 | 0.0000 | 0.1655 | 31682.6 |
| 3 | $[0\,0\,0\,0\,3\ 0\,1\,0\,0\,2]^T$ | 0.0545 | 0.1853 | 0.0671 | 0.0000 | 0.0209 | 20856.2 |
| Fully distributed | $[1\,1\,1\,1\,1\ 1\,1\,1\,1\,1]^T$ | 0.1217 | 0.1886 | 0.0052 | 0.0000 | 0.0252 | 106749.0 |

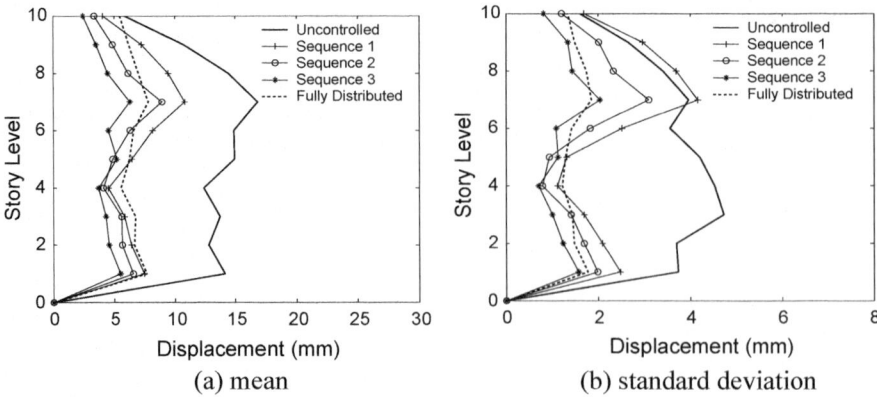

**Fig. 5.9**  Second-order moment of equivalent extreme-value interstory drift in the case of viscoelasticity damped structures

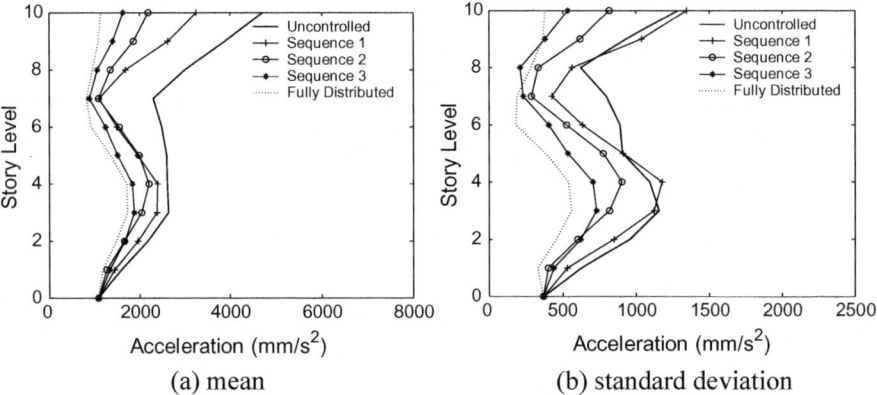

**Fig. 5.10**  Second-order moment of equivalent extreme-value story acceleration in the case of viscoelasticity damped structures

viscoelastic damper is deployed. One might realize that too few control devices may result in a response amplification on the local stories of the structure. This is also indicated in the second-order moment of equivalent extreme-value story acceleration in sequences, as shown in Fig. 5.10. It is seen that the standard deviations of story accelerations of the fourth story, the ninth story, and the tenth story are larger than those stories of the uncontrolled structure when the first viscoelastic damper is deployed. They gradually decrease as the deployment of the next viscoelastic dampers. It is also seen that the mean of story acceleration becomes smaller when the viscoelatic dampers are deployed. In brief, the control effectiveness of interstory drift by the case with the third viscoelastic damper deployment is larger than the case with fully distributed viscoelastic dampers. The reverse is true on the control effectiveness of story acceleration.

## 5.5.2  Active Tendon Exerted Structure

The eight-story shear frame shown in Sect. 3.4.2 controlled by $r(1 \leq r \leq 8)$ active tendons is investigated as another numerical example. The thresholds of interstory drift, interstory velocity, story acceleration, and interstory control force are 15 mm, 150 mm/s, 8000 mm/s$^2$, and 2000 kN, respectively. The structural input is represented by the physically motivated random seismic ground motion model addressed in Sect. 3.4.1. The peak ground acceleration is set as 0.3 $g$. The predetermined objective of structural control is optimizing the parameters and placements of the least number of active tendons to attain the control effectiveness of the case with fully distributed tendons. Similar to Sect. 5.5.1, the fully distributed tendons are deployed in the structural space simultaneously, and their parameters are optimized as the same.

The form of cost-function weights shown in Eq. (4.2.2) is employed in this example, where the cross terms between system states are ignored:

$$\mathbf{Q_Z} = \text{diag}\{Q_{d_1}, \ldots, Q_{d_n}, Q_{v_1}, \ldots, Q_{v_n}\}, \quad \mathbf{R_U} = \text{diag}\{R_{u_1}, \ldots, R_{u_r}\}. \tag{5.5.4}$$

The optimal placement and parameters of the newly added active tendons in sequences are shown in Table 5.3. It is seen that five active tendons with optimal parameters and optimal placements are required to attain the predetermined objective of structural control. The five control devices are deployed in the first interstory, the second interstory, the sixth interstory, the seventh interstory, and the fourth interstory in turn. Thus, the parameters of generalized optimal control law are given by

$$(Q_d^*, Q_v^*, R_u^*, L^*) = \begin{bmatrix} 155.4 & 14360.0 & 0 & 99.8 & 0 & 0.0 & 11.6 & 0 \\ 240.0 & 9.6 & 0 & 89.5 & 0 & 0.0 & 0.0 & 0 \\ 10^{-12} & 10^{-12} & 0 & 10^{-12} & 0 & 10^{-12} & 10^{-12} & 0 \\ 1 & 2 & 0 & 5 & 0 & 3 & 4 & 0 \end{bmatrix}^T \tag{5.5.5}$$

where the parameters and placement of active tendons have been folded into the first, second, and third rows of the parameter matrix. In the placement vector $L^*$, zero denotes no active tendons in the interstory and nonzero denotes the active tendon and its deployment sequence.

The relation between the added active tendons and the controllability index is shown in Fig. 5.11. It is clear that the controllability index decreases rapidly after the active tendon with the optimal parameters is deployed in the present optimal interstory. Comparison between the design criteria of the minimum story controllability index gradient (MinSCIG) and of the maximum story controllability index (MaxSCI) is shown in Fig. 5.12. It is readily seen that the case using the five designed active tendons can attain the almost control effectiveness as the case with fully distributed active tendons when the criterion of MinSCIG is employed. In regard to the criterion of MaxSCI, however, six active tendons are needed to achieve a similar control effectiveness. It is indicated, moreover, in Fig. 5.12 that the structural

**Table 5.3** Optimal placement and parameters of newly added active tendons in sequences

| Sequence no. | Placement vector | Parameters of newly added active tendons[a] | | |
|---|---|---|---|---|
| | | $Q_d$ | $Q_v$ | $R_u$ |
| 0 | $[0\,0\,0\,0\,0\,0\,0\,0]^T$ | – | – | – |
| 1 | $[1\,0\,0\,0\,0\,0\,0\,0]^T$ | 155.4 | 240.0 | $10^{-12}$ |
| 2 | $[1\,2\,0\,0\,0\,0\,0\,0]^T$ | 14360.0 | 9.6 | $10^{-12}$ |
| 3 | $[1\,2\,0\,0\,0\,3\,0\,0]^T$ | 0.0 | 0.0 | $10^{-12}$ |
| 4 | $[1\,2\,0\,0\,0\,3\,4\,0]^T$ | 11.6 | 0.0 | $10^{-12}$ |
| 5 | $[1\,2\,0\,5\,0\,3\,4\,0]^T$ | 99.8 | 89.5 | $10^{-12}$ |
| Fully distributed | $[1\,1\,1\,1\,1\,1\,1\,1]^T$ | 118.2 | 163.5 | $10^{-12}$ |

[a]Initial values of parameters are $Q_d = 100$, $Q_v = 100$, $R_u = 10^{-12}$

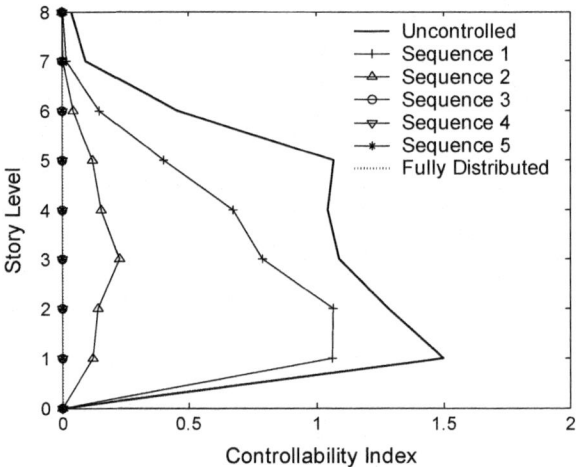

**Fig. 5.11** Relation between the added active tendons and controllability index

performance approaches the predetermined objective after the third active tendon is placed employing the criterion of MinSCIG; while the structural performance with the same number of active tendons employing the criterion of MaxSCI is still far from the predetermined objective. As mentioned in Sect. 5.5.1, the proposed criterion of minimum story controllability index gradient can attain the performance objective more efficiently than the criterion of maximum story controllability index. The active

**Fig. 5.12** Comparison between design criteria of MinSCIG and MaxSCI in the case of active tendon exerted structures

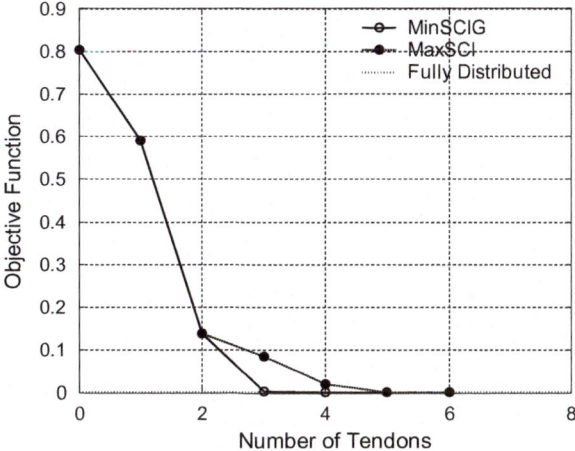

**Fig. 5.13** Active tendon deployments as design criteria of MinSCIG and MaxSCI

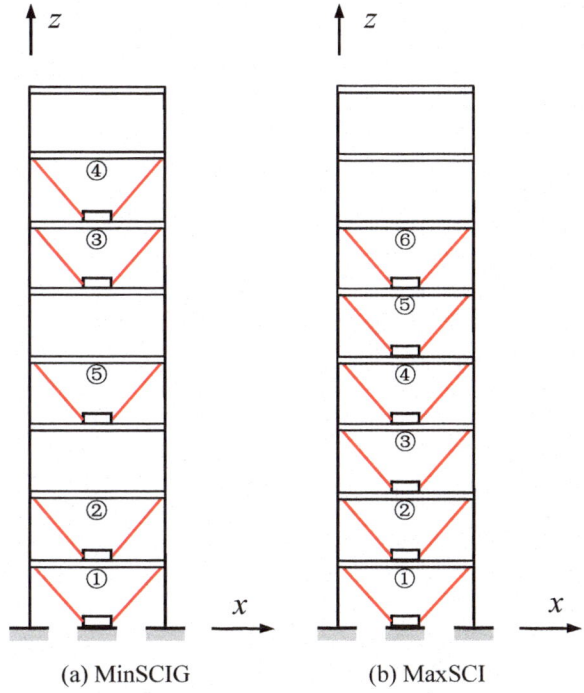

(a) MinSCIG  (b) MaxSCI

tendon deployments as the design criteria of MinSCIG and MaxSCI are shown in Fig. 5.13.

The exceedance probabilities of system quantities of concern, the objective function, and the mean-square control force are presented in Table 5.4. It is seen that the exceedance probabilities of system quantities gradually decrease towards approach-

**Table 5.4** Seismic mitigation by active tendons in sequences

| Sequence no. | Placement vector | Exceedance probabilities | | | | Objective Function $J_2$ | Mean-square control force $\epsilon_u(\text{kN}^2)$ |
|---|---|---|---|---|---|---|---|
| | | $P_{f,d}$ | $P_{f,v}$ | $P_{f,a}$ | $P_{f,u}$ | | |
| 0 | $[0\ 0\ 0\ 0$ $0\ 0\ 0\ 0]^T$ | 0.9963 | 0.7582 | 0.1992 | – | 0.8036 | – |
| 1 | $[1\ 0\ 0\ 0$ $0\ 0\ 0\ 0]^T$ | 0.9620 | 0.4519 | 0.0764 | 0.2098 | 0.5898 | $4.710 \times 10^8$ |
| 2 | $[1\ 2\ 0\ 0$ $0\ 0\ 0\ 0]^T$ | 0.3976 | 0.3267 | 0.0181 | 0.1088 | 0.1385 | $3.219 \times 10^8$ |
| 3 | $[1\ 2\ 0\ 0$ $0\ 3\ 0\ 0]^T$ | 0.0141 | 0.0626 | 0.0009 | 0.0002 | 0.0021 | $8.708 \times 10^7$ |
| 4 | $[1\ 2\ 0\ 0$ $0\ 3\ 4\ 0]^T$ | 0.0134 | 0.0570 | 0.0002 | $3.61 \times 10^{-7}$ | 0.0017 | $8.196 \times 10^7$ |
| 5 | $[1\ 2\ 0\ 5$ $0\ 3\ 4\ 0]^T$ | 0.0001 | 0.0130 | 0.0032 | 0.0004 | $8.95 \times 10^{-5}$ | $8.722 \times 10^7$ |
| Fully dis-tributed | $[1\ 1\ 1\ 1$ $1\ 1\ 1\ 1]^T$ | 0.0022 | 0.0035 | $3.60 \times 10^{-7}$ | 0.0022 | $1.11 \times 10^{-5}$ | $1.737 \times 10^8$ |

ing the predetermined performance objective of the structure, and the same control level as the case with fully distributed active tendons is attained when the fifth active tendon is deployed. Meanwhile, by comparison with the case with the fully distributed tendons, Sequence 5 exhibits a better displacement control and a worse velocity and acceleration control, which is included as well in the comparison between the PDFs of equivalent extreme values of interstory drift and story acceleration; see Fig. 5.14. Besides, the objective function of Sequence 5 is larger than that of the case with fully distributed tendons, indicating that the latter attains a better trade-off between system quantities than the former. However, the control cost of Sequence 5 is two times less than that of the case with fully distributed tendons, which indicates that the former is more energy-saving than the latter. In summary, the structural control employing the generalized optimal control law attains a better trade-off between system quantities than the traditional control design.

The second-order moment of equivalent extreme values of interstory drift and story acceleration varying along the height of the structure with the deployment of active tendons are shown in Figs. 5.15 and 5.16, respectively. It is seen that the means and standard deviations of system quantities of concern become smaller with the active tendon deployment, and they become smoother along the height of the structure. The response profile of the controlled structure along story level really connects to a desired structural performance.

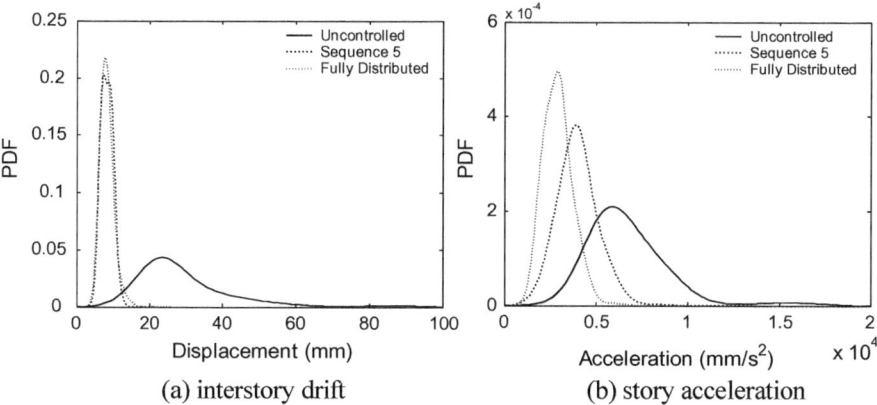

**Fig. 5.14** PDFs of equivalent extreme values of interstory drift and story acceleration in the case of active tendon exerted structures

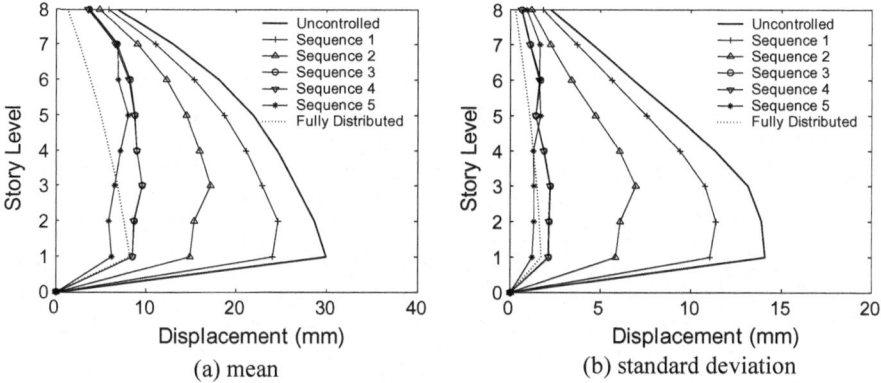

**Fig. 5.15** Second-order moment of equivalent extreme-value interstory drift in the case of active tendon exerted structures

## 5.6 Discussions and Summaries

Modern building designs are not only in demand of accommodating safe residences but also in demand of providing available spaces as possible. The optimal parameter design and optimal placement design of control devices thus ought to be of the same practical significance. Utilizing a less number of control devices, the cost of investment can be reduced, and the space occupied by the structural components can be saved to a certain extent as well. On this basis, this chapter is devoted to addressing the optimal deployment and parameter optimization of control devices in the cases of active and passive control modalities. The control laws on both aspects are investigated and summarized into the generalized optimal control law for implementing

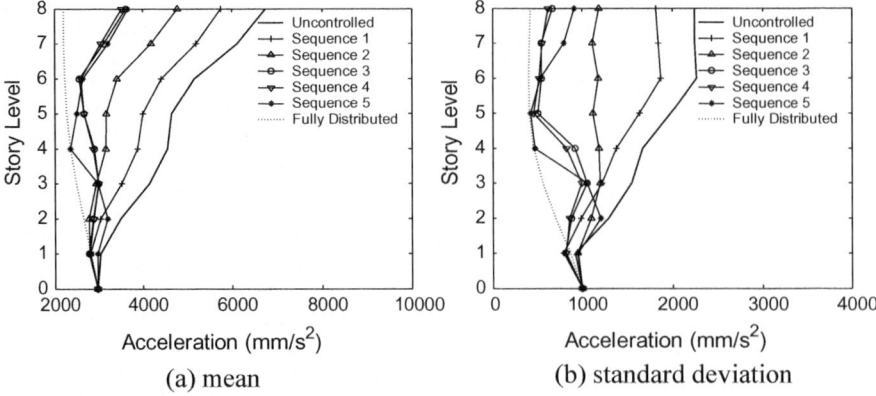

(a) mean                                    (b) standard deviation

**Fig. 5.16** Second-order moment of equivalent extreme-value story acceleration in the case of active tendon exerted structures

the generalized optimal control policy in the context of physically based stochastic optimal (PSO) control.

By comparison with the previously defined degree of controllability in function of single system quantity, the exceedance probability based probabilistic controllability index exhibits more benefits, which characterizes system safety denoted by the interstory drift, system serviceability denoted by the interstory velocity, system comfortability denoted by the story acceleration, and control device workability denoted by the control force and their trade-off.

By virtue of the criterion of minimum story controllability index gradient (MinSCIG), the optimal placement of control devices in each sequence can be attained efficiently. In the solution procedure of the generalized optimal control law, the criterion for parameter optimization can refer to as the exceedance probability criterion on energy trade-off, which has the same physical meanings with the criterion of the MinSCIG. Numerical examples prove that utilizing the generalized optimal control policy can attain the objective of structural control, i.e., gaining the maximum control effectiveness with the minimum control cost; meanwhile, the criterion of the MinSCIG has a better optimization capacity, which achieves the performance objective more efficiently than the criterion of the MaxSCI currently in use.

Besides, system reliability associated with structural responses is used as the kernel of generalized optimal control policy to aid in sequentially locating control devices. Following the deployment of each control device, decision regarding the policy can be updated, in a multistage manner, before deciding on whether to proceed with deploying additional control devices.

# References

Amini F, Tavassoli MR (2005) Optimal structural active control force, number and placement of controllers. Eng Struct 27:1306–1316

Chang Min IJ, Soong TT (1980) Optimal controller placement in modal control of complex systems. J Math Anal Appl 75(2):340–358

Chen JB, Li J (2005) Dynamic response and reliability analysis of non-linear stochastic structures. Probab Eng Mech 20:33–44

Chen JB, Li J (2007) The extreme value distribution and dynamic reliability analysis of nonlinear structures with uncertain parameters. Struct Saf 29(2):77–93

Chen JB, Li J (2008) Strategy for selecting representative points via tangent spheres in the probability density evolution method. Int J Numer Meth Eng 74(13):1988–2014

Chen GS, Robin JB, Salama M (1991) Optimal placement of active/passive members in truss structures using simulated annealing. AIAA J 29(8):1327–1334

Cheng FY, Pantelides CP (1988) Optimal placement of actuators for structural control. Technical Report NCEER-88-0037, National Centre for Earthquake Engineering Research, State University of New York, Buffalo, New York

Eldred MS, Adams BM, Gay DM, et al. (2007) DAKOTA, a multilevel parallel object-oriented framework for design optimization, parameter estimation, uncertainty quantification, and sensitivity analysis (Version 4.1+ User's Manual). Sandia National Laboratories, SAND 2006-6337

Ibidapo-Obe O (1985) Optimal actuators placement for the active control of flexible structures. J Math Anal Appl 105(1):12–25

Kim J, Choi H, Min KW (2003) Performance-based design of added viscous dampers using capacity spectrum method. J Earthquake Eng 7(1):1–24

Laskin RA (1982) Aspects of the dynamics and controllability of large flexible structures. PhD Dissertation, Columbia University

Li J, Chen JB (2004) Probability density evolution method for dynamic reliability analysis of stochastic structures. J Vib Eng 17(2):121–125 (in Chinese)

Li J, Chen JB (2007) The number theoretical method in response analysis of nonlinear stochastic structures. Comput Mech 39(6):693–708

Lindberg RE Jr, Longman RW (1984) On the number and placement of actuator for independent modal space control. J Guid 7(2):215–221

May BS, Beck JL (1998) Probabilistic control for the active mass driver benchmark structural model. Earthquake Eng Struct Dynam 27:1331–1346

Park KS, Koh HM, Hahm D (2004) Integrated optimum design of viscoelastically damped structural systems. Eng Struct 26:581–591

Soong TT, Dargush GF (1997) Passive energy dissipation systems in structural engineering. John Wiley & Sons, New York

Spencer Jr BF, Kaspari Jr DC, Sain MK (1994) Structural control design: a reliability-based approach. Proc Am Control Conf Baltimore Maryland 1062–1066

Vander Velde WE, Carignan CR (1984) Number and placement of control system components considering possible failures. J Guid 7(6):703–709

Zhang RH, Soong TT (1992) Seismic design of viscoelastic dampers for structural applications. ASCE J Struct Eng 118(5):1375–1391

# Chapter 6
# Stochastic Optimal Control of Nonlinear Structures

## 6.1 Preliminary Remarks

The previous chapters are devoted to the stochastic optimal control of linear structural systems. However, since the external excitations such as strong earthquakes, high winds, and huge waves often arise significant randomness inherent in their occurrence time, space, and amplitude. The response of structural systems inevitably excurses inelastic range back and forth when subjected to these severe hazardous actions. Therefore, it is necessary to further investigate the stochastic optimal control of nonlinear structural systems.

The stochastic optimal control of nonlinear systems has been a challenging issue (Housner et al. 1997). For instance, Shefer and Breakwell proposed an optimal control scheme with digital feedback logic for a family of nonlinear systems, by taking into account the non-Gaussian characteristics of the state conditional distribution of systems, where the state estimation involved third-order and higher order moments (Shefer and Breakwell 1987). Yang et al. investigated a stochastic hybrid control of a base-isolated structure under seismic ground motions using the statistical linearization technique for random vibration. In their work, the seismic ground motion is modeled as a filtered shot noise (Yang et al. 1994). Zhu et al. proposed a nonlinear stochastic optimal control strategy for hysteretic systems under random excitations utilizing the stochastic averaging method and the stochastic dynamic programming, and the Itô equation describing the total energy of the system was defined as a one-dimensional controlled diffusion process (Zhu et al. 2000).

In practice, a family of nonlinear structural systems with large deformation is usually modeled as hardening or softening Duffing systems, e.g., the free vibration of gyroscopes, the vibration of bridge plate with hinge joints, and the vibration of fluid–solid coupling dynamical systems (Sekar and Narayanan 1994). The vibration of another family of nonlinear systems gives rise to hysteretic behaviors such as the engineering structures under serious dynamic actions. Many mathematical models have been developed to efficiently describe the hysteretic behaviors of structures,

© Springer Nature Singapore Pte Ltd. and Shanghai Scientific and Technical Publishers 2019
Y. Peng and J. Li, *Stochastic Optimal Control of Structures*,
https://doi.org/10.1007/978-981-13-6764-9_6

of which two classes are widely used in structural engineering and materials, i.e., the bilinear elastoplastic model (Iwan 1961; Clough and Johnson 1966) and the Bouc–Wen hysteretic model (Bouc 1967; Wen 1976; Baber and Wen 1981; Baber and Noori 1985).

In conjunction with the optimal polynomial control, this chapter attempts to extend the theory of physically based stochastic optimal control into the stochastic control of nonlinear and hysteretic structural systems. For illustrative purposes, the stochastic optimal control of typical nonlinear dynamical systems subjected to random seismic ground motions has been carried out, including a family of hardening Duffing oscillators, multi-degree-of-freedom structural systems with hysteretic behaviors represented by Clough hysteretic model and by Bouc–Wen hysteretic model, respectively.

## 6.2  Stochastic Optimal Polynomial Control

The scheme of optimal polynomial control is derived from the Bellman's optimality principle and Hamilton–Jacobi theoretical framework (Suhardjo et al. 1992), which is an extended formulation of the LQR control in essence.

Without loss of generality, an $n$-degree-of-freedom nonlinear system subjected to random excitation is investigated. The vector equation of motion is given by

$$\mathbf{M}\ddot{\mathbf{X}}(t) + \mathbf{f}[\mathbf{X}(t), \dot{\mathbf{X}}(t)] = \mathbf{B}_s\mathbf{U}(t) + \mathbf{D}_s\mathbf{F}(\mathbf{\Theta}, t), \mathbf{X}(t_0) = \mathbf{x}_0; \dot{\mathbf{X}}(t_0) = \dot{\mathbf{x}}_0, \quad (6.2.1)$$

where $\mathbf{X}(t)$ is the $n$-dimensional column vector denoting system displacement; $\mathbf{U}(t)$ is the $r$-dimensional column vector denoting control force; $\mathbf{M}$ is the $n \times n$ mass matrix; $\mathbf{f}[\cdot]$ is the $n$-dimensional column vector denoting nonlinear internal force, including the nonlinear damping force and nonlinear restoring force; $\mathbf{F}(\cdot)$ is the $p$-dimensional column vector denoting random excitation.

In order to transform Eq. (6.2.1) to the equation of motion as Eq. (3.3.1) and to the state equation as Eq. (3.3.2), an expansion on the nonlinear internal force $\mathbf{f}[\cdot]$ needs to be carried out so as to extract the system matrix. Usually, the nonlinear internal force can be expanded into the following Maclaurin series:

$$\mathbf{f}[\mathbf{X}(t), \dot{\mathbf{X}}(t)] = \mathbf{f}[\mathbf{0}, \mathbf{0}] + \left(\frac{\partial\mathbf{f}[\mathbf{0}, \mathbf{0}]}{\partial\mathbf{X}} \cdot \mathbf{X} + \frac{\partial\mathbf{f}[\mathbf{0}, \mathbf{0}]}{\partial\dot{\mathbf{X}}} \cdot \dot{\mathbf{X}}\right)$$

$$+ \frac{1}{2!}\left(\frac{\partial^2\mathbf{f}[\mathbf{0}, \mathbf{0}]}{\partial\mathbf{X}^2}{:}\mathbf{X}^2 + 2\frac{\partial^2\mathbf{f}[\mathbf{0}, \mathbf{0}]}{\partial\mathbf{X}\partial\dot{\mathbf{X}}}{:}\mathbf{X}\dot{\mathbf{X}} + \frac{\partial^2\mathbf{f}[\mathbf{0}, \mathbf{0}]}{\partial\dot{\mathbf{X}}^2}{:}\dot{\mathbf{X}}^2\right)$$

$$+ \cdots + \frac{1}{m!}\left(\frac{\partial^m\mathbf{f}[\mathbf{0}, \mathbf{0}]}{\partial\mathbf{X}^m} \cdot^m \mathbf{X}^m + \sum_{k=1}^{m-1}\frac{m!}{(m-k)!\, k!}\frac{\partial^m\mathbf{f}[\mathbf{0}, \mathbf{0}]}{\partial\mathbf{X}^{m-k}\partial\dot{\mathbf{X}}^k} \cdot^m\mathbf{X}^{m-k}\dot{\mathbf{X}}^k\right.$$

$$\left. + \frac{\partial^m\mathbf{f}[\mathbf{0}, \mathbf{0}]}{\partial\dot{\mathbf{X}}^m} \cdot^m \dot{\mathbf{X}}^m\right) \quad (6.2.2)$$

where $\overset{m}{\cdot}$ denotes $m$ times of dot product for tensor contraction; $\mathbf{X}^m = \mathbf{X} \underbrace{\otimes \cdots \otimes}_{m} \mathbf{X}$ denotes $m$ times of union product for tensor extension, and the union product sign $\otimes$ is omitted in Eq. (6.2.2).

As to a general system of nonlinear structures, the cross terms between $\mathbf{X}^i$ and $\dot{\mathbf{X}}^i$ is far less than other terms on the contribution to the nonlinear internal force, which can thus be ignored safely. Moreover, the nonlinear internal force is typically zero when the state vector is zero, and the first term of the Maclaurin series can thus be ignored, i.e.,

$$\mathbf{f}[\mathbf{X}(t), \dot{\mathbf{X}}(t)] = \left( \frac{\partial \mathbf{f}[0,0]}{\partial \mathbf{X}} \cdot \mathbf{X}^0 + \frac{1}{2!} \frac{\partial^2 \mathbf{f}[0,0]}{\partial \mathbf{X}^2} : \mathbf{X} + \cdots + \frac{1}{m!} \frac{\partial^m \mathbf{f}[0,0]}{\partial \mathbf{X}^m} \overset{m}{\cdot} \mathbf{X}^{m-1} \right) \mathbf{X}$$
$$+ \left( \frac{\partial \mathbf{f}[0,0]}{\partial \dot{\mathbf{X}}} \cdot \dot{\mathbf{X}}^0 + \frac{1}{2!} \frac{\partial^2 \mathbf{f}[0,0]}{\partial \dot{\mathbf{X}}^2} : \dot{\mathbf{X}} + \cdots + \frac{1}{m!} \frac{\partial^m \mathbf{f}[0,0]}{\partial \dot{\mathbf{X}}^m} \overset{m}{\cdot} \dot{\mathbf{X}}^{m-1} \right) \dot{\mathbf{X}}$$

$$(6.2.3)$$

In the state space, Eq. (6.2.1) can be written as

$$\dot{\mathbf{Z}}(t) = \mathbf{\Lambda}(\mathbf{Z})\mathbf{Z}(t) + \mathbf{B}\mathbf{U}(t) + \mathbf{D}\mathbf{F}(\boldsymbol{\Theta}, t) \qquad (6.2.4)$$

with the initial condition $\mathbf{Z}(t_0) = \mathbf{z}_0$, where $\mathbf{Z}(t)$ is the $2n$-dimensional column vector denoting system state; $\mathbf{\Lambda}(\mathbf{Z})$ is the $2n \times 2n$ gradient matrix (system matrix); $\mathbf{B}$ is the $2n \times r$ matrix denoting the location of control devices, and $\mathbf{D}$ is the $2n \times p$ matrix denoting the location of external excitation, respectively:

$$\mathbf{Z}(t) = \begin{bmatrix} \mathbf{X}(t) \\ \dot{\mathbf{X}}(t) \end{bmatrix}, \mathbf{B} = \begin{bmatrix} \mathbf{0} \\ \mathbf{M}^{-1}\mathbf{B}_s \end{bmatrix}, \mathbf{D} = \begin{bmatrix} \mathbf{0} \\ \mathbf{M}^{-1}\mathbf{D}_s \end{bmatrix}$$

$$\mathbf{\Lambda}(\mathbf{Z}) = \begin{bmatrix} \mathbf{0} & \mathbf{I} \\ -\mathbf{M}^{-1} \sum_{i=1}^{m} \frac{1}{i!} \frac{\partial^i \mathbf{f}[0,0]}{\partial \mathbf{X}^i} \overset{i}{\cdot} \mathbf{X}^{i-1} & -\mathbf{M}^{-1} \sum_{i=1}^{m} \frac{1}{i!} \frac{\partial^i \mathbf{f}[0,0]}{\partial \dot{\mathbf{X}}^i} \overset{i}{\cdot} \dot{\mathbf{X}}^{i-1} \end{bmatrix} \qquad (6.2.5)$$

where $m$ denotes the highest order of the Maclaurin's series, which is equal to the highest order of the nonlinear internal force. The terms of series with $(m+1)$ or higher orders are all zeros.

A polynomial cost function with stochastic vector $\boldsymbol{\Theta}$ is given by (Yang et al. 1996)

$$J_1(\mathbf{Z}, \mathbf{U}, \boldsymbol{\Theta}) = \phi(\mathbf{Z}(t_f), t_f) + \frac{1}{2} \int_{t_0}^{t_f} [\mathbf{Z}^T(t)\mathbf{Q}_\mathbf{Z}\mathbf{Z}(t) + \mathbf{U}^T(t)\mathbf{R}_\mathbf{U}\mathbf{U}(t) + h(\mathbf{Z}, t)]dt$$

$$(6.2.6)$$

where $\phi(\mathbf{Z}(t_f), t_f)$ denotes the terminal cost; $\mathbf{Z}(t_f)$ denotes the terminal state; $t_0, t_f$ denote the initial and terminal times, respectively; $\mathbf{Q}_\mathbf{Z}$ is the $2n \times 2n$ positive semi-definite weighting matrix with respect to system state; $\mathbf{R}_\mathbf{U}$ is the $r \times r$ positive definite weighting matrix with respect to control force; $\mathbf{h}(\mathbf{Z}, t)$ denotes the high-order term of the cost function of which the orders are higher than the quadratic term. It is readily seen that the terminal cost, together with the first two terms of the integrand in Eq. (6.2.6) conducts the canonical LQR control; see Eq. (3.3.5).

For a given realization $\boldsymbol{\theta}$ of the stochastic vector $\boldsymbol{\Theta}$, the minimization of the polynomial cost function Eq. (6.2.6) can refer to the Euler–Lagrange equation according to the Pontryagin's maximum principle, also to the celebrated Hamilton–Jacobi—Bellman equation according to the Bellman's optimality principle. Using the Hamilton–Jacobi–Bellman equation (Anderson and Moore 1990), one has

$$\frac{\partial V(\mathbf{Z}, t)}{\partial t} = -\min_{\mathbf{U}}[H(\mathbf{Z}, \mathbf{U}, V'(\mathbf{Z}, t), \boldsymbol{\Theta}, t)] \qquad (6.2.7)$$

where the prime denotes the differentiation with respect to $\mathbf{Z}$; $V(\mathbf{Z}, t)$ is the optimal cost function which satisfies all the properties of a Lyapunov function (Bernstein 1993), and can be considered as

$$V(\mathbf{Z}, t) = \frac{1}{2}\mathbf{Z}^{\mathrm{T}}(t)\mathbf{P}(t)\mathbf{Z}(t) + g(\mathbf{Z}, t) \qquad (6.2.8)$$

where $\mathbf{P}(t)$ denotes the $2n \times 2n$ Riccati matrix; $g(\mathbf{Z}, t)$ denotes a positive definite multinomial in function of $\mathbf{Z}(t)$. The necessary condition for the minimization of the right-hand side of Eq. (6.2.7) is

$$\frac{\partial H(\mathbf{Z}, \mathbf{U}, V'(\mathbf{Z}, t), \boldsymbol{\Theta}, t)}{\partial \mathbf{U}} = \mathbf{0} \qquad (6.2.9)$$

The probabilistic criterion used in the stochastic optimal control of structures relies on the structural performance, which actually includes the influence of external excitations. Therefore, the excitation-relevant term in the expression of feedback logic can be safely ignored, resulting in a closed-loop control with the state feedback. The extended Hamiltonian function is then defined as (Yang et al. 1996)

$$H(\mathbf{Z}, \mathbf{U}, V'(\mathbf{Z}, t), \boldsymbol{\Theta}, t) = \frac{1}{2}[\mathbf{Z}^{\mathrm{T}}(t)\mathbf{Q}_{\mathbf{Z}}\mathbf{Z}(t) + \mathbf{U}^{\mathrm{T}}(t)\mathbf{R}_{\mathbf{U}}\mathbf{U}(t) + h(\mathbf{Z}, t)]$$
$$+ [V'(\mathbf{Z}, t)]^{\mathrm{T}}(\mathbf{\Lambda}(\mathbf{Z})\mathbf{Z}(t) + \mathbf{B}\mathbf{U}(t)) \qquad (6.2.10)$$

Substituting Eq. (6.2.10) into Eq. (6.2.8), we have

$$H(\mathbf{Z}, \mathbf{U}, V'(\mathbf{Z}, t), \boldsymbol{\Theta}, t) = \frac{1}{2}[\mathbf{Z}^{\mathrm{T}}(t)\mathbf{Q}_{\mathbf{Z}}\mathbf{Z}(t) + \mathbf{U}^{\mathrm{T}}(t)\mathbf{R}_{\mathbf{U}}\mathbf{U}(t) + h(\mathbf{Z}, t)]$$
$$+ [\mathbf{Z}^{\mathrm{T}}(t)\mathbf{P}(t) + (g'(\mathbf{Z}, t))^{\mathrm{T}}](\mathbf{\Lambda}(\mathbf{Z})\mathbf{Z}(t) + \mathbf{B}\mathbf{U}(t)) \quad (6.2.11)$$

Substituting Eq. (6.2.11) into Eq. (6.2.9) leads to

$$\mathbf{R}_{\mathbf{U}}\mathbf{U}(t) + \mathbf{B}^{\mathrm{T}}\mathbf{P}(t)\mathbf{Z}(t) + \mathbf{B}^{\mathrm{T}}g'(\mathbf{Z}, t) = \mathbf{0} \qquad (6.2.12)$$

The optimal nonlinear controller is then given by

$$\mathbf{U}(t) = -\mathbf{R}_\mathbf{U}^{-1}\mathbf{B}^\mathrm{T}\mathbf{P}(t)\mathbf{Z}(t) - \mathbf{R}_\mathbf{U}^{-1}\mathbf{B}^\mathrm{T}g'(\mathbf{Z}, t) \qquad (6.2.13)$$

It is also readily seen from Eq. (6.2.13) that

$$\frac{\partial^2 H(\mathbf{Z}, \mathbf{U}, V'(\mathbf{Z}, t), \mathbf{\Theta}, t)}{\partial \mathbf{U}^2} = \mathbf{R}_\mathbf{U} > \mathbf{0} \qquad (6.2.14)$$

Therefore, the minimization of polynomial cost function definitely exists.

Substituting Eqs. (6.2.8), (6.2.11) and (6.2.13) into Eq. (6.2.7), and separating the terms relevant to $\mathbf{Z}(t)$ and the terms relevant to $g(\mathbf{Z}, t)$, one obtains

$$-\dot{\mathbf{P}}(t) = \mathbf{P}(t)\mathbf{\Lambda}(\mathbf{Z}) + \mathbf{\Lambda}^\mathrm{T}(\mathbf{Z})\mathbf{P}(t) - \mathbf{P}(t)\mathbf{B}\mathbf{R}_\mathbf{U}^{-1}\mathbf{B}^\mathrm{T}\mathbf{P}(t) + \mathbf{Q}_\mathbf{Z} \qquad (6.2.15)$$

$$-\dot{g}(\mathbf{Z}, t) = \frac{1}{2}h(\mathbf{Z}, t) - \frac{1}{2}(g'(\mathbf{Z}, t))^\mathrm{T}\mathbf{B}\mathbf{R}_\mathbf{U}^{-1}\mathbf{B}^\mathrm{T}g'(\mathbf{Z}, t)$$
$$+ (g'(\mathbf{Z}, t))^\mathrm{T}[\mathbf{\Lambda}(\mathbf{Z}) - \mathbf{B}\mathbf{R}_\mathbf{U}^{-1}\mathbf{B}^\mathrm{T}\mathbf{P}(t)]\mathbf{Z}(t) \qquad (6.2.16)$$

The following equivalent formulation of matrices has been utilized in the deduction of Eq. (6.2.15):

$$2\mathbf{P}(t)\mathbf{\Lambda}(\mathbf{Z}) = \mathbf{P}(t)\mathbf{\Lambda}(\mathbf{Z}) + \mathbf{\Lambda}^\mathrm{T}(\mathbf{Z})\mathbf{P}(t) \qquad (6.2.17)$$

In order to gain the explicit expression of nonlinear controller as Eq. (6.2.13), a positive polynomial function of $g(\mathbf{Z}, t)$ is chosen as follows (Yang et al. 1996):

$$g(\mathbf{Z}, t) = \sum_{i=2}^{k} \frac{1}{i}[\mathbf{Z}^\mathrm{T}(t)\mathbf{M}_i(t)\mathbf{Z}(t)]^i \qquad (6.2.18)$$

where $\mathbf{M}_i(t)$, $i = 2, 3, \ldots, k$ denote $2n \times 2n$ Lyapunov matrices.

Substituting Eq. (6.2.18) into Eq. (6.2.16), we then have

$$-2\sum_{i=2}^{k}[\mathbf{Z}^\mathrm{T}(t)\mathbf{M}_i(t)\mathbf{Z}(t)]^{i-1}\mathbf{Z}^\mathrm{T}(t)\dot{\mathbf{M}}_i(t)\mathbf{Z}(t) = h(\mathbf{Z}, t)$$

$$-4\left\{\sum_{i=2}^{k}[\mathbf{Z}^\mathrm{T}(t)\mathbf{M}_i(t)\mathbf{Z}(t)]^{i-1}\mathbf{M}_i(t)\mathbf{Z}(t)\right\}^\mathrm{T}\mathbf{B}\mathbf{R}_\mathbf{U}^{-1}\mathbf{B}^\mathrm{T}\left\{\sum_{i=2}^{k}[\mathbf{Z}^\mathrm{T}(t)\mathbf{M}_i(t)\mathbf{Z}(t)]^{i-1}\mathbf{M}_i(t)\mathbf{Z}(t)\right\}$$

$$+4\left\{\sum_{i=2}^{k}[\mathbf{Z}^\mathrm{T}(t)\mathbf{M}_i(t)\mathbf{Z}(t)]^{i-1}\mathbf{M}_i(t)\mathbf{Z}(t)\right\}^\mathrm{T}[\mathbf{\Lambda}(\mathbf{Z}) - \mathbf{B}\mathbf{R}_\mathbf{U}^{-1}\mathbf{B}^\mathrm{T}\mathbf{P}(t)]\mathbf{Z}(t) \qquad (6.2.19)$$

If the high-order term of cost function has the formulation as follows:

$$h(\mathbf{Z}, t) = 2 \sum_{i=2}^{k} [\mathbf{Z}^{\mathrm{T}}(t)\mathbf{M}_i(t)\mathbf{Z}(t)]^{i-1} \mathbf{Z}^{\mathrm{T}}(t)\mathbf{Q}_{\mathbf{Z},i}(t)\mathbf{Z}(t)$$

$$+ 4 \left\{ \sum_{i=2}^{k} [\mathbf{Z}^{\mathrm{T}}(t)\mathbf{M}_i(t)\mathbf{Z}(t)]^{i-1} \mathbf{M}_i(t)\mathbf{Z}(t) \right\}^{\mathrm{T}}$$

$$\mathbf{BR}_{\mathbf{U}}^{-1}\mathbf{B}^{\mathrm{T}} \left\{ \sum_{i=2}^{k} [\mathbf{Z}^{\mathrm{T}}(t)\mathbf{M}_i(t)\mathbf{Z}(t)]^{i-1} \mathbf{M}_i(t)\mathbf{Z}(t) \right\} \quad (6.2.20)$$

a simple analytical solution of nonlinear controllers can be obtained. In Eq. (6.2.20), $\mathbf{Q}_{\mathbf{Z},i}, i = 2, 3, \ldots, k$ denote $2n \times 2n$ positive semi-definite weighting matrices with respect to system state. One thus has the following formulation to solve the Lyapunov matrix:

$$-\dot{\mathbf{M}}_i(t) = \mathbf{M}_i(t)[\mathbf{\Lambda}(\mathbf{Z}) - \mathbf{BR}_{\mathbf{U}}^{-1}\mathbf{B}^{\mathrm{T}}\mathbf{P}(t)] + [\mathbf{\Lambda}(\mathbf{Z}) - \mathbf{BR}_{\mathbf{U}}^{-1}\mathbf{B}^{\mathrm{T}}\mathbf{\Lambda}^{\mathrm{T}}(\mathbf{Z})\mathbf{P}(t)]^{\mathrm{T}}\mathbf{M}_i(t) + \mathbf{Q}_{\mathbf{Z},i}$$
$$(6.2.21)$$

where the equivalent formulation of matrices shown in Eq. (6.2.17) has been utilized here again.

It is noted that Eqs. (6.2.15) and (6.2.21) are the matrix Riccati and Lyapunov equations, respectively.

One might recognize that the Riccati matrix and the Lyapunov matrix are both in function of $\mathbf{\Lambda}(\mathbf{Z})$, which indicates a fact that the control law parameters of the polynomial controller cannot be optimized in an offline manner. Hence, the analytical solution of optimal polynomial controller can be obtained analytically as the following formulation:

$$\mathbf{U}(t) = -\mathbf{R}_{\mathbf{U}}^{-1}\mathbf{B}^{\mathrm{T}}\mathbf{P}(t)\mathbf{Z}(t) - \mathbf{R}_{\mathbf{U}}^{-1}\mathbf{B}^{\mathrm{T}} \sum_{i=2}^{k} [\mathbf{Z}^{\mathrm{T}}(t)\mathbf{M}_i(t)\mathbf{Z}(t)]^{i-1} \mathbf{M}_i(t)\mathbf{Z}(t) \quad (6.2.22)$$

It is readily seen that the polynomial controller consists of linear and nonlinear terms, where the former is the first-order function of state quantity, and the latter is the odd high-order functions of state quantity, e.g. cubic, quintic, etc.

Substituting the solutions of system state $\mathbf{Z}(\mathbf{\Theta}, t)$ and control force $\mathbf{U}(\mathbf{\Theta}, t)$ into the generalized probability density evolution equations as follows:

$$\frac{\partial p_{Z\Theta}(z, \boldsymbol{\theta}, t)}{\partial t} + \dot{Z}(\boldsymbol{\theta}, t)\frac{\partial p_{Z\Theta}(z, \boldsymbol{\theta}, t)}{\partial z} = 0 \quad (6.2.23)$$

$$\frac{\partial p_{U\Theta}(u, \boldsymbol{\theta}, t)}{\partial t} + \dot{U}(\boldsymbol{\theta}, t)\frac{\partial p_{U\Theta}(u, \boldsymbol{\theta}, t)}{\partial u} = 0 \quad (6.2.24)$$

the probability density of system quantities of concern can be readily attained. In Eqs. (6.2.23) and (6.2.24), $Z(t), U(t)$ denote the component formulation of $\mathbf{Z}(t), \mathbf{U}(t)$, respectively.

In regard to a family of optimal control systems with infinite time, if the gradient matrix is not related to the time, the Riccati matrix and Lyapunov matrix are the steady solutions of Eqs. (6.2.15) and (6.2.21), respectively, i.e. involving the solutions of the matrix algebraic Riccati and Lyapunov equations as follows:

$$\mathbf{P}\mathbf{\Lambda}_0 + \mathbf{\Lambda}_0^{\mathrm{T}}\mathbf{P} - \mathbf{P}\mathbf{B}\mathbf{R}_{\mathrm{U}}^{-1}\mathbf{B}^{\mathrm{T}}\mathbf{P} + \mathbf{Q}_{\mathbf{Z}} = \mathbf{0}, \tag{6.2.25}$$

$$\mathbf{M}_i(\mathbf{\Lambda}_0 - \mathbf{B}\mathbf{R}_{\mathrm{U}}^{-1}\mathbf{B}^{\mathrm{T}}\mathbf{P}) + (\mathbf{\Lambda}_0 - \mathbf{B}\mathbf{R}_{\mathrm{U}}^{-1}\mathbf{B}^{\mathrm{T}}\mathbf{P})^{\mathrm{T}}\mathbf{M}_i$$
$$+ \mathbf{Q}_{\mathbf{Z},i} = \mathbf{0}, \quad i = 2, 3, \ldots, k \tag{6.2.26}$$

where $\mathbf{\Lambda}_0 = \mathbf{\Lambda}(\mathbf{Z})|_{\mathbf{z}_0}$ denotes the value of the gradient matrix $\mathbf{\Lambda}(\mathbf{Z})$ at the initial state $\mathbf{z}_0$ (Yang et al. 1996). The solving of matrices $\mathbf{P}$, $\mathbf{M}_i$ can refer to the traditional numerical schemes or to the toolbox functions available in MATLAB.

However, as to the issue addressed in the present study: optimal control system with finite time and time-variant gradient matrix, the system matrix at a small time scale of numerical integral steps can be viewed as time invariant. Therefore, the Riccati matrix and Lyapunov matrix at the $j$th integral step are the approximate solutions of the equations as follows:

$$\mathbf{P}(t_j)\mathbf{\Lambda}(\mathbf{Z}) + \mathbf{\Lambda}^{\mathrm{T}}(\mathbf{Z})\mathbf{P}(t_j) - \mathbf{P}(t_j)\mathbf{B}\mathbf{R}_{\mathrm{U}}^{-1}\mathbf{B}^{\mathrm{T}}\mathbf{P}(t_j) + \mathbf{Q}_{\mathbf{Z}} = \mathbf{0} \tag{6.2.27}$$

$$\mathbf{M}_i(t_j)[\mathbf{\Lambda}(\mathbf{Z}) - \mathbf{B}\mathbf{R}_{\mathrm{U}}^{-1}\mathbf{B}^{\mathrm{T}}\mathbf{P}(t_j)] + [\mathbf{\Lambda}(\mathbf{Z}) - \mathbf{B}\mathbf{R}_{\mathrm{U}}^{-1}\mathbf{B}^{\mathrm{T}}\mathbf{P}(t_j)]^{\mathrm{T}}\mathbf{M}_i(t_j) + \mathbf{Q}_{\mathbf{Z},i} = \mathbf{0}$$
$$\tag{6.2.28}$$

## 6.3 Stochastic Optimal Control of Nonlinear Oscillators

For illustrative purposes, the stochastic optimal control of a family of hardening Duffing oscillators subjected to the random seismic ground motion is investigated. The equation of motion of the nonlinear oscillator is given by

$$\ddot{x}(t) + 2\zeta\omega_0\dot{x}(t) + \omega_0^2[x(t) + \mu x^3(t)] = u(t) + F(\mathbf{\Theta}, t), \quad x(t_0) = \dot{x}(t_0) = 0 \tag{6.3.1}$$

where $x(t)$ denotes the oscillator displacement; $\zeta$, $\omega_0$ denote the damping ratio and the vibrational frequency of the oscillator system; $\mu$ denotes the coefficient describing the nonlinearity level of the oscillator system; $u(t)$ denotes the control force in acceleration unit; $F(\cdot)$ denotes the random excitation in acceleration unit as well.

Introducing the Maclaurin series and ignoring the cross terms between displacement and velocity in the nonlinear internal force, Eq. (6.3.1) can be rewritten to state equation as follows:

$$\dot{\mathbf{Z}}(t) = \mathbf{\Lambda}(\mathbf{Z})\mathbf{Z}(t) + \mathbf{B}u(t) + \mathbf{D}F(\mathbf{\Theta}, t) \qquad (6.3.2)$$

where

$$\mathbf{Z}(t) = \begin{bmatrix} x(t) \\ \dot{x}(t) \end{bmatrix}, \mathbf{\Lambda}(\mathbf{Z}) = \begin{bmatrix} 0 & 1 \\ -[1 + 6\mu x^2(t)]\omega_0^2 & -2\zeta\omega_0 \end{bmatrix}, \mathbf{B} = \begin{bmatrix} 0 \\ 1 \end{bmatrix}, \mathbf{D} = \begin{bmatrix} 0 \\ 1 \end{bmatrix}$$
$$(6.3.3)$$

### 6.3.1   Performance of Active Tendon Control

A practical problem is the randomly base-excited vibration of a bridge plate with hinge joints and its stochastic optimal control using active tendon systems, as shown in Fig. 6.1. This structural system can be modeled as the controlled hardening Duffing oscillator, as mentioned in the previous section (Peng and Li 2011). The damping ratio $\zeta$ and the natural frequency $\omega_0$ are assumed to be 0.02 and 2 rad/s, respectively; the coefficient $\mu$ employs 200 and 10 to represent strong and weak nonlinearities, respectively.

The physically motivated random seismic ground motion model is employed as the input, of which the peak ground acceleration is set as $0.3g$. Using the adaptive step size Runge–Kutta method, the nonlinear dynamic analysis of the Duffing oscillator is carried out. According to the criterion of minimizing the performance function in exceedance probability, i.e., Eq. (5.4.1), the optimal parameters of control law are defined. The thresholds of oscillator displacement, velocity, acceleration, and control force are 50 mm, 300 mm/s, 2000 mm/s², 5000 mm/s², respectively. The weighting matrices employ a diagonal form as follows:

$$\mathbf{Q}_{\mathbf{Z}} = \begin{bmatrix} Q_d & 0 \\ 0 & Q_v \end{bmatrix}, \mathbf{Q}_{\mathbf{Z},i} = \begin{bmatrix} Q_{d,i} & 0 \\ 0 & Q_{v,i} \end{bmatrix}, \mathbf{R}_{\mathbf{U}} = R_u, \quad i = 2, 3, \ldots, k \qquad (6.3.4)$$

The optimal parameters of the polynomial controllers acting on the Duffing oscillators in two nonlinearity levels are shown in Table 6.1. It is seen that the optimal parameters in high-order terms of nonlinear controllers (third-order and fifth-order terms) are numerically almost zeros, and the objective function of different-order controllers are almost the same for the Duffing oscillator with strong and weak nonlinearities. It is revealed that utilizing the criterion of minimizing the performance

**Fig. 6.1** Hinge jointed bridge plate with active tendon systems

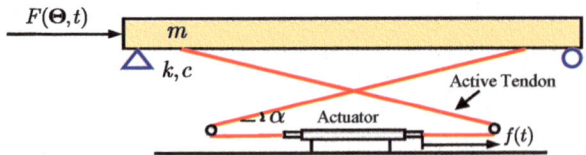

**Table 6.1**  Optimal parameters of polynomial controllers

| Nonlinearity level | Order of controller | Weighting matrices[a] | | | | | | | Objective function $J_2$ |
|---|---|---|---|---|---|---|---|---|---|
| | | $Q_d$ | $Q_v$ | $Q_{d,2}$ | $Q_{v,2}$ | $Q_{d,3}$ | $Q_{v,3}$ | $R_u$ | |
| $\mu = 200$ | First | 0.0 | 39.4 | – | – | – | – | 1.0 | 0.0161 |
| | Third | 0.0 | 39.4 | 0.0 | 0.0 | – | – | 1.0 | 0.0161 |
| | Fifth | 0.0 | 39.4 | 0.0 | 0.0 | 0.0 | 16.8 | 1.0 | 0.0161 |
| $\mu = 10$ | First | 0.0 | 40.4 | – | – | – | – | 1.0 | 0.0153 |
| | Third | 0.0 | 40.4 | 0.0 | 0.0 | – | – | 1.0 | 0.0153 |
| | Fifth | 0.0 | 40.4 | 0.0 | 0.0 | 0.0 | 0.0 | 1.0 | 0.0153 |

[a]Initial values of parameters are $Q_d = Q_v = 100$, $Q_{d,2} = Q_{v,2} = Q_{d,3} = Q_{v,3} = 20$, $R_u = 1$

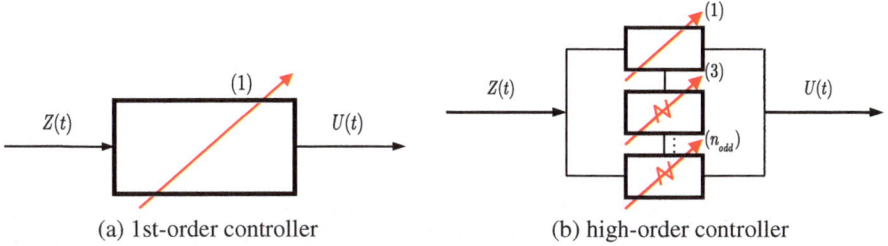

(a) 1st-order controller  (b) high-order controller

**Fig. 6.2**  State-feedback model of first-order and high-order controllers

function in exceedance probability, the linear control by means of the first-order controller can implement the control effectiveness of the nonlinear control by means of
high-order controllers. One might recognize that the linear controller is a promising
control law in that the nonlinear controller may cause the dynamical system unstable
since the nonlinear controller would amplify the measurement noise and weaken the
control effectiveness.

Figure 6.2 shows the state-feedback model of the first-order and high-order controllers. It is seen that the first-order controller is a linear feedback with respect to
system state; while the high-order controller is a nonlinear feedback with respect
to system state, involving the online calculation of high-order terms of oscillator
responses, which tends to result in a serious deviation from the desired control force
and an unstable structural system once the time delay occurs. The applications of linear controllers or the low-order nonlinear controllers are thus preferable in practice.
It is revealed that using the polynomial controller designed as the criterion of minimizing the performance function in exceedance probability, a first-order controller
with linear feedback can attain the control effectiveness of high-order controllers
with nonlinear feedback. This finding features practical significance.

Besides, the optimal weighting matrices of the oscillator system on different nonlinearity levels are very close. It is indicated that the optimal polynomial control, at

least to the family of nonlinear oscillators, exhibits high robustness to the nonlinearities of systems.

Time histories of mean and standard deviation of displacement and acceleration of the oscillator systems with and without the first-order controllers are shown in Fig. 6.3. It is readily seen that the displacements of oscillators both with strong and weak nonlinearities are reduced significantly after the controller is placed on the system. The acceleration of the oscillator with the strong nonlinearity is also improved greatly though a slight enlargement occurs in the initial period due to the intervention of control force; while the acceleration of the oscillator with the weak nonlinearity has an increment other than a reduction. It is explained that the uncontrolled nonlinear system serves as a filter, thereby the acceleration response is filtered significantly. The filtering performance, however, becomes weaker when the control action is posed upon the system. It is seen, moreover, that the time histories of mean and standard deviation of the controlled oscillator with strong and weak nonlinearities are nearly identical since the polynomial controllers, designed for the oscillator in the two nonlinearity levels, are almost the same. It is proved again

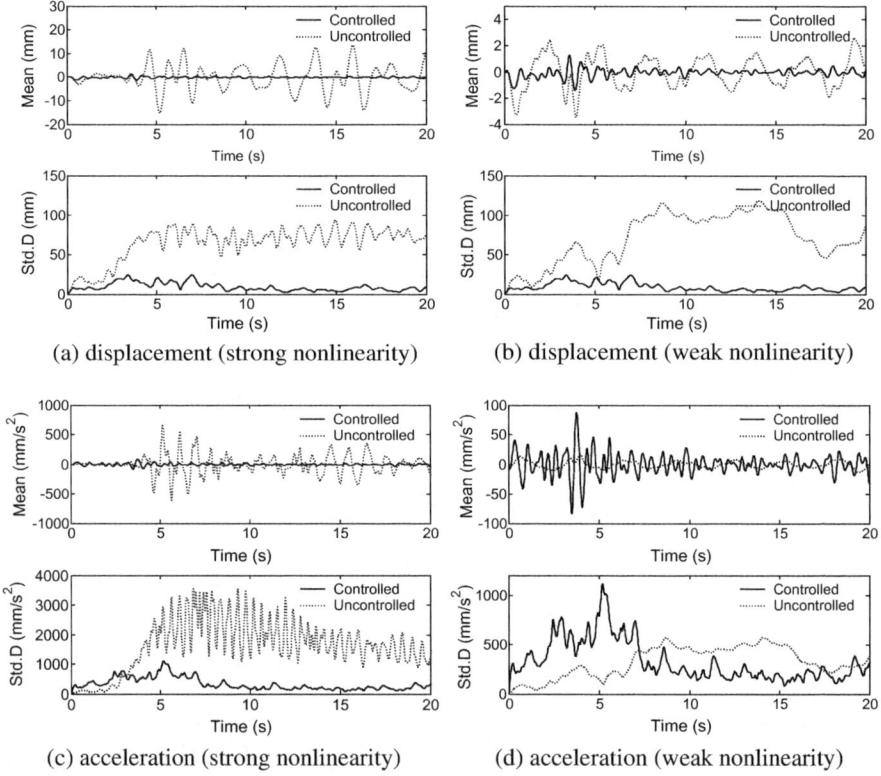

(a) displacement (strong nonlinearity)    (b) displacement (weak nonlinearity)

(c) acceleration (strong nonlinearity)    (d) acceleration (weak nonlinearity)

**Fig. 6.3** Time histories of mean and standard deviation of responses of Duffing oscillators with and without controls

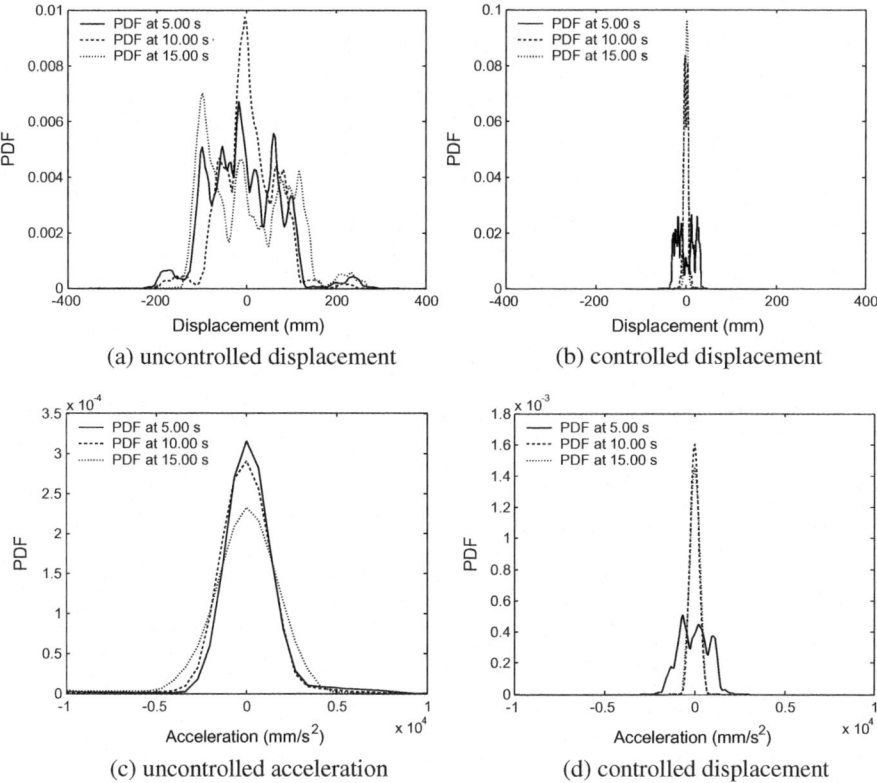

(a) uncontrolled displacement          (b) controlled displacement

(c) uncontrolled acceleration          (d) controlled displacement

**Fig. 6.4**   PDFs of responses of Duffing oscillator at typical instants of time with and without controls

that the polynomial controller is insensitive to the nonlinearity level of the Duffing oscillators.

Figure 6.4 shows the PDFs of displacement and acceleration of the oscillator system with strong nonlinearity at typical instants of time with and without controls. It is seen that the distribution range of PDFs of oscillator responses becomes smaller after the oscillator is controlled, which exhibits a consistency with Fig. 6.3a, c. It is revealed that the variation of oscillator responses is reduced significantly, especially at the instant 15 s, the PDF no matter of the displacement or of the acceleration approximates to be a pulse. Figures 6.5 and 6.6 further show the surface and contour of probability density of the oscillator displacement and acceleration at the typical time interval.

The probabilistic characteristics of optimal control force of the oscillator system with strong nonlinearity are shown in Fig. 6.7, involving time histories of mean and standard deviation, and the PDFs at typical instants of time. It is seen that there are some similarities in the time histories and probability density functions between the

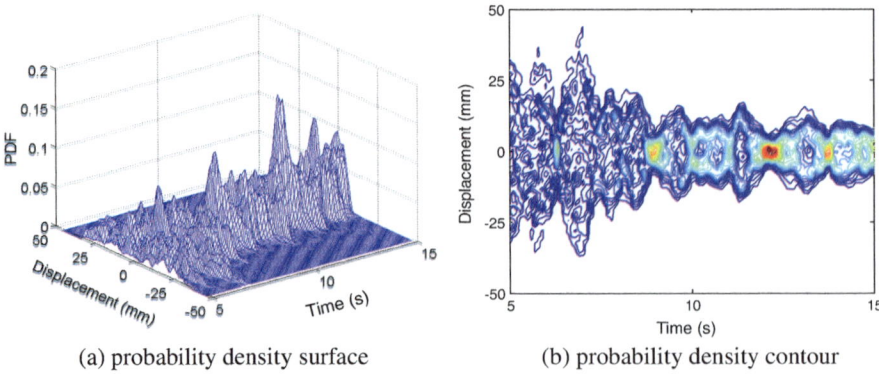

(a) probability density surface                    (b) probability density contour

**Fig. 6.5** Probability density surface and contour of displacement of controlled oscillator system with strong nonlinearity at typical time interval

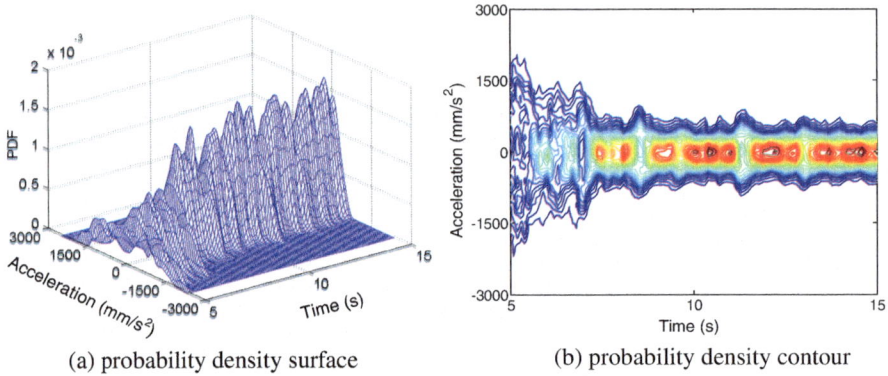

(a) probability density surface                    (b) probability density contour

**Fig. 6.6** Probability density surface and contour of acceleration of controlled oscillator system with strong nonlinearity at typical time interval

oscillator displacement and acceleration and the optimal control force, owing to the logic of feedback control.

The root-mean-square phase plane evolution of the oscillator system with strong nonlinearity with and without controls is shown in Fig. 6.8. It is readily seen that the uncontrolled oscillator moves mainly far from the initial position, which will never come back to the neighborhood of the initial position once it moves out under the external excitation. While the controlled oscillator is able to return to the neighborhood of initial position under the control force even it moves out. This benefit is also seen from the sample phase plane evolution; see Fig. 6.9. Besides, one might see that the uncontrolled oscillator first forms into steady loops at the inner layer domain, and then jumps out to the outer layer domain and moves in circularity. This phenomenon is the so-called bifurcation, a celebrated term in the nonlinear dynamics, which can be derived from the similarities of phase plane evolution among samples. In fact,

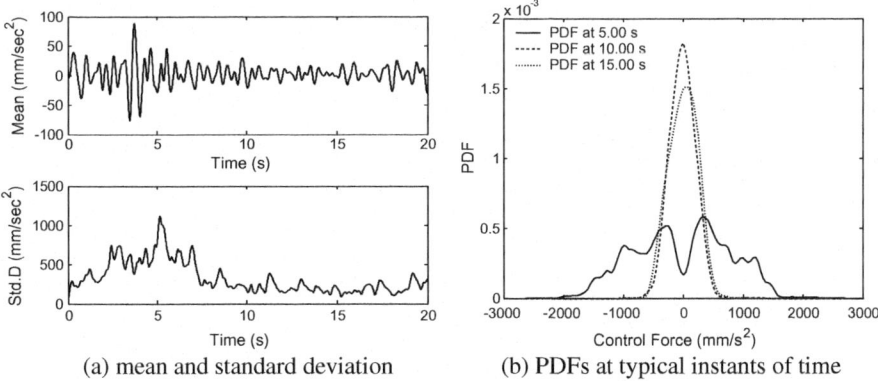

(a) mean and standard deviation                    (b) PDFs at typical instants of time

**Fig. 6.7** Probabilistic characteristics of optimal control force of controlled oscillator system with strong nonlinearity

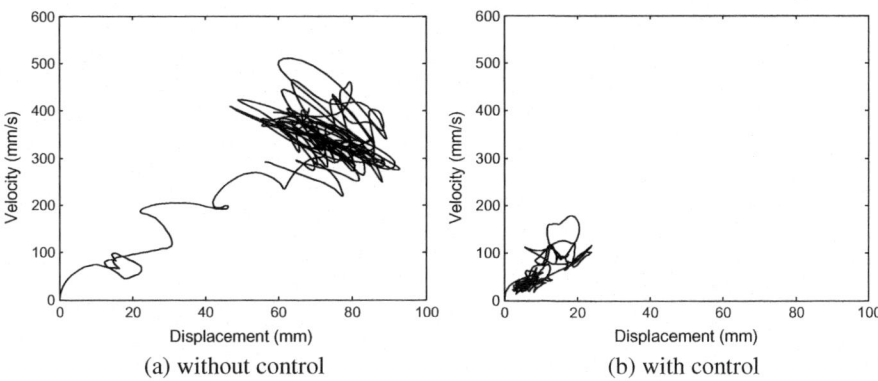

(a) without control                              (b) with control

**Fig. 6.8** Root-mean-square phase plane evolution of oscillator system with strong nonlinearity with and without controls

burfication is an essential feature of Duffing oscillators. Owing to the randomness inherent in external excitations, burfication of Duffing oscillators arises to be stochastic. It is revealed in the sample phase plane evolution of controlled oscillator system that the number of steady loops increases after control and most of the loops are distributed in the neighborhood of initial position.

In order to proceed with a comparative study against the classical stochastic optimal control, the statistical linearization based LQG control is investigated. As to the nonlinear structural system shown in Eq. (6.3.1), the mean-square displacement and the mean-square control force derived from the LQG control are given by (details of deduction refers to Appendix C)

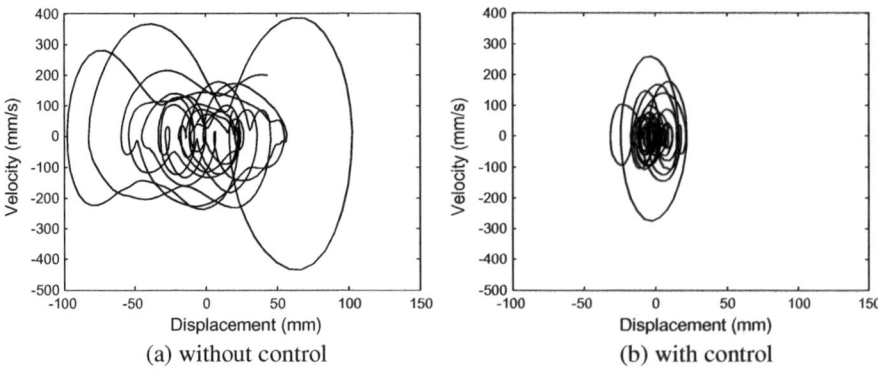

(a) without control                              (b) with control

**Fig. 6.9** Sample phase plane evolution of oscillator system with strong nonlinearity with and without controls

$$E[x^2(t)] = \frac{\sqrt{(\omega_0^2 + \widehat{K})^2 + \frac{12\pi\mu\omega_0^2 S_0}{(2\zeta\omega_0 + \widehat{C})}} - (\omega_0^2 + \widehat{K})}{6\mu\omega_0^2} \qquad (6.3.5)$$

$$E[u^2(t)] = \frac{\pi S_0 \widehat{C}^2}{2\zeta\omega_0 + \widehat{C}} + \frac{\left[\sqrt{(\omega_0^2 + \widehat{K})^2 + \frac{12\pi\mu\omega_0^2 S_0}{(2\zeta\omega_0 + \widehat{C})}} - (\omega_0^2 + \widehat{K})\right]\widehat{K}^2}{6\mu\omega_0^2} \qquad (6.3.6)$$

where $S_0$ denotes the spectral intensity factor of random seismic ground motion; $\widehat{C}$, $\widehat{K}$ denote the numerical damping and numerical stiffness provided by the control force $u(t)$, as shown in Eq. (C.14) in Appendix C in which the cost-function weights have the same design parameters as Table 6.1:

$$\mathbf{Q_Z} = \begin{bmatrix} 0.0 & 0 \\ 0 & 40.0 \end{bmatrix}, \mathbf{R_U} = 1.0 \qquad (6.3.7)$$

Figure 6.10 shows the comparison between root-mean-square quantities of controlled oscillator systems in different nonlinearity levels by the stochastic optimal polynomial control (OPC-PSO) and the statistical linearization based LQG (SL-LQG), involving the equivalent extreme values of displacement and control force. It is seen that when the control force weighting matrix is set as 1.0, the OPC-PSO gains a less displacement than the SL-LQG where their difference reduces along the increasing of nonlinearity level, and the OPC-PSO requires a larger control force. Meanwhile, as to the Duffing system under investigation, the SL-LQG does not exhibit good robustness, and the system response is sensitive to the nonlinearity level: the root-mean-square displacement gradually reduces when the nonlinearity level of oscillator system increases. If the control force weighting matrix of the SL-LQG is set as 0.1, the root-mean-square displacement reduces significantly and is less than the result of the OPC-PSO with control force weighting matrix 1.0, so

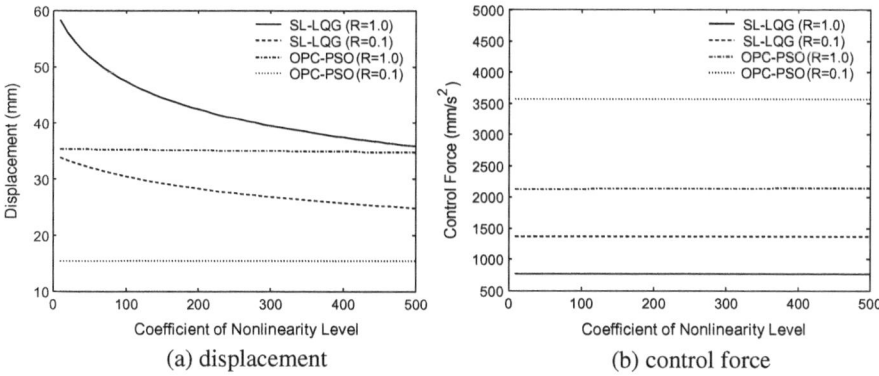

(a) displacement        (b) control force

**Fig. 6.10** Comparison between root-mean-square quantities of controlled oscillator systems in different nonlinearity levels by stochastic optimal polynomial control (OPC-PSO) and statistical linearization based LQG (SL-LQG)

as the control force. When the control force weighting matrix of the OPC-PSO is set as 0.1 as well, a less displacement and larger control force can be gained than the SL-LQG. Similarly, the response of SL-LQG is still sensitive to the nonlinearity level. In summary, the SL-LQG does not exhibit good robustness and underestimates the requisite control force. Therefore, using white Gaussian noise as input cannot logically design the control system of civil engineering structures.

## 6.3.2 Comparative Studies Between Control Criteria

As was mentioned previously, the linear control with the first-order controller can implement the control effectiveness of the nonlinear control with high-order controllers when the criterion of minimizing the performance function in exceedance probability (MPFE) is employed. However, whether other criteria exhibit this feature as well? For validating purposes, the Duffing oscillator with strong nonlinearity is investigated, and two well-visited criteria are employed, e.g., the criterion on system second-order statistics evaluation (SSSE) (Zhang and Xu 2001) and the criterion on Lyapunov asymptotic stability condition (LASC) (Yang et al. 1992).

As to the criterion on SSSE, the constraint quantity is defined by the oscillator displacement, the evaluation quantities include the oscillator displacement, acceleration and control force, and the quantile function is defined as the mean plus one time of standard deviation. The cost-function weights have the formula as follows:

$$\mathbf{Q_Z} = q\begin{bmatrix} 1 & 0 \\ 0 & 1 \end{bmatrix}, \mathbf{R_U} = r, \mathbf{Q_{Z,2}} = \frac{q}{5}\begin{bmatrix} 1 & 0 \\ 0 & 1 \end{bmatrix}, \mathbf{Q_{Z,i}} = \frac{q}{10}\begin{bmatrix} 1 & 0 \\ 0 & 1 \end{bmatrix}, \quad i = 3, 4, \ldots, k$$

(6.3.8)

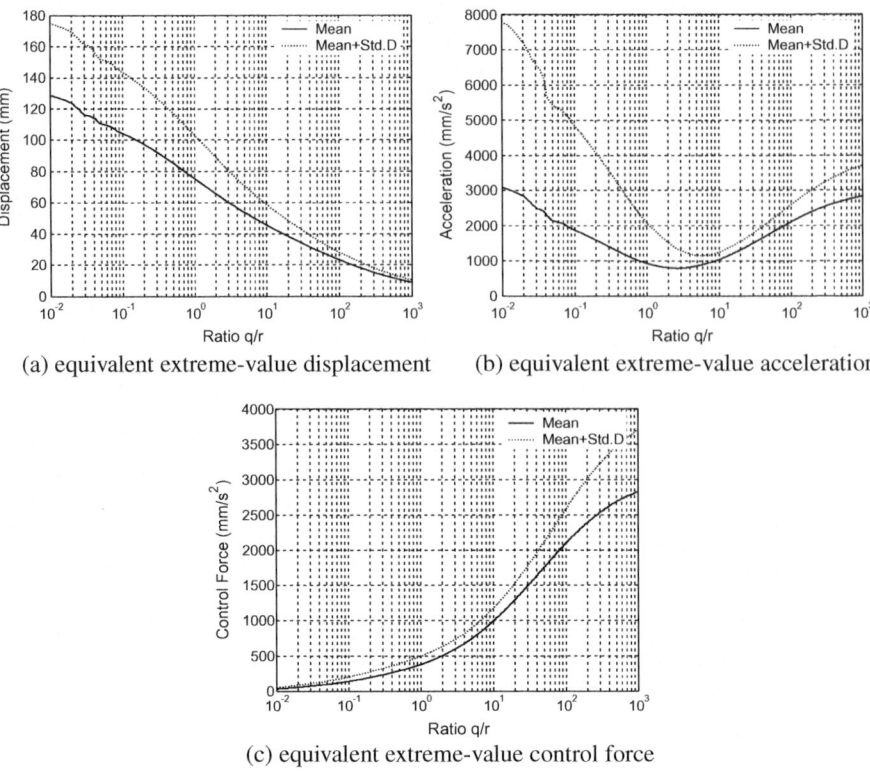

(a) equivalent extreme-value displacement     (b) equivalent extreme-value acceleration

(c) equivalent extreme-value control force

**Fig. 6.11** Relation between mean, quantile of quantities, and coefficient ratio of weighting matrices

Using the first-order controller, the relation between the statistical moments of equivalent extreme values of oscillator responses and the ratio of coefficients of weighting matrices $q/r$ is shown in Fig. 6.11 ($r = 1.0$).

(i) As $q/r \geqslant 20$, the quantile of the displacement constraint is within the threshold 50 mm, and its mean and standard deviation both gradually decrease, meanwhile the mean and standard deviation of control force increase significantly.

(ii) As $q/r = 20$, the mean and standard deviation of the acceleration are both minimum under the constraint condition.

Therefore, the coefficients of the weighting matrices are taken as $q = 20.0$, $r = 1.0$.

As to the criterion on LASC, the cost-function weights have the formulation as follows (Yang et al. 1992):

$$\mathbf{Q_Z} = q \begin{bmatrix} \omega_0^2 & 0 \\ 0 & 1 \end{bmatrix}, \mathbf{R_U} = r, \mathbf{Q}_{\mathbf{Z},2} = \frac{q}{5} \begin{bmatrix} \omega_0^2 & 0 \\ 0 & 1 \end{bmatrix},$$

$$\mathbf{Q}_{\mathbf{Z},i} = \frac{q}{10}\begin{bmatrix} \omega_0^2 & 0 \\ 0 & 1 \end{bmatrix}, \quad i = 3, 4, \ldots, k \tag{6.3.9}$$

According to the cost-function weights defined in the criterion on SSSE, the coefficients of weighting matrices shown in Eq. (6.3.9) are chosen as $q = 10, r = 1.0$.

The control effectiveness of polynomial controllers using the presented three control criteria is listed in Table 6.2. It is seen that the objective function of the linear controller using the criterion of MPFE is less than those of nonlinear controllers using the criteria on SSSE and LASC. The main reason underlying the difference is that the later two criteria involve non-optimization procedures for designing the cost-function weights, and thus at most gain the approximately optimal solution. The former, however, accommodates the minimum failure probabilities of system quantities in the sense of trade-off. Besides, the polynomial controllers with high-order terms have better control effectiveness than the linear controllers with first-order terms using the two criteria involving non-optimization procedures. It is revealed that the linear controller cannot implement the control effectiveness of the nonlinear controller when the criteria on SSSE and LASC are employed, and the nonlinear controller with high-order terms has better control effectiveness. It follows the traditional knowledge that a nonlinear optimal control is more robust and more effective than the counterpart of linear optimal control (Bernstein 1993). Besides, the third-order controller using the criteria on SSSE and LASC achieves the almost same control effectiveness as the fifth-order controller, indicating that the third-order controller is definitely satisfactory to the investigated Duffing oscillator using the criteria involving non-optimization procedures. In summary, the criterion MPFE is a preferable criterion for stochastic optimal control of nonlinear dynamical systems.

It is seen from Table 6.2 that the control criteria have their emphases on the response control of oscillator systems. The first-order controller using the criterion of MPFE, for example, has a superior capacity on displacement control. The third-order controller using the criterion on SSSE has a superior capacity on acceleration control;

**Table 6.2** Control effectiveness of polynomial controllers using three control criteria

| Polynomial controller | Exceedance probabilities | | | | Objective function $J_2$ |
|---|---|---|---|---|---|
| | $P_{f,d}$ | $P_{f,v}$ | $P_{f,a}$ | $P_{f,u}$ | |
| First (MPFE) | 0.0668 | 0.1474 | 0.0771 | $3.601 \times 10^{-7}$ | 0.0161 |
| First (SSSE) | 0.0941 | 0.2276 | 0.0016 | 0.0000 | 0.0303 |
| Third (SSSE) | 0.0870 | 0.1969 | 0.0548 | $3.602 \times 10^{-7}$ | 0.0247 |
| Fifth (SSSE) | 0.0869 | 0.1960 | 0.0579 | $3.604 \times 10^{-7}$ | 0.0247 |
| First (LASC) | 0.1471 | 0.3708 | 0.0028 | 0.0000 | 0.0796 |
| Third (LASC) | 0.0716 | 0.0349 | 0.2339 | $3.603 \times 10^{-7}$ | 0.0305 |
| Fifth (LASC) | 0.0707 | 0.0262 | 0.2458 | $3.601 \times 10^{-7}$ | 0.0330 |
| Uncontrolled | 1.0000 | 0.9968 | 0.4559 | – | 1.1007 |

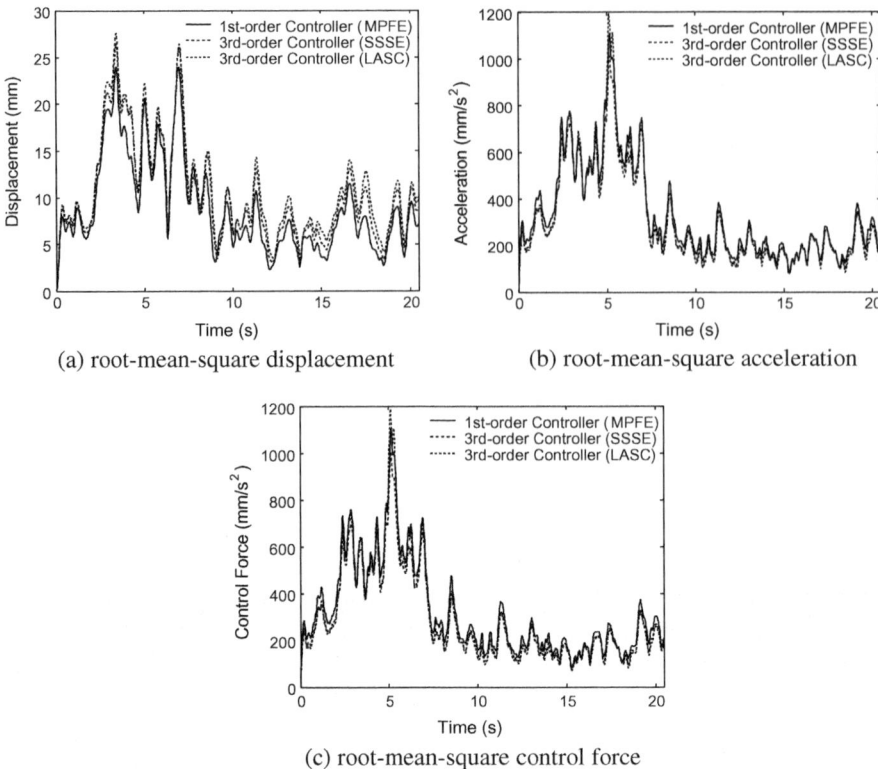

(a) root-mean-square displacement          (b) root-mean-square acceleration

(c) root-mean-square control force

**Fig. 6.12**  Time histories of root-mean-square quantities of controlled oscillator system with strong nonlinearity in terms of three control criteria

while the third-order controller using the criterion on LASC has a superior capacity on velocity control. It is explained that the control effectiveness is closely related to the physical meaning of the applied criterion, which relies upon the components of weighting matrices such as displacement, velocity component and control force, and prompts the balance among system quantities. It is obvious that the criterion of MPFE attains the best trade-off. Figure 6.12 shows the time histories of root-mean-square quantities of controlled oscillator system with strong nonlinearity in terms of the three control criteria, which exposes a consistent result with Table 6.2.

## 6.4  Stochastic Optimal Control of Hysteretic Structures

In conjunction with the generalized optimal control policy addressed in Chap. 5 and the optimal polynomial control method, the stochastic optimal control of nonlinear structural systems with hysteretic components can be readily carried out.

Without loss of generality, the equation of motion of controlled hysteretic structures subjected to random excitation is given by

$$\mathbf{M}\ddot{\mathbf{X}}(t) + \mathbf{C}_t\dot{\mathbf{X}}(t) + \mathbf{R}_t(\mathbf{X}, \mathbf{z}) = \mathbf{B}_s\mathbf{U}(t) + \mathbf{D}_s\mathbf{F}(\Theta, t), \ \mathbf{X}(t_0) = \mathbf{0}, \dot{\mathbf{X}}(t_0) = \mathbf{0} \quad (6.4.1)$$

where $\mathbf{C}_t$ denotes the instantaneous damping matrix; $\mathbf{R}_t(\mathbf{X}, \mathbf{z})$ is the $n$-dimensional column vector denoting the restoring force, including the elastic component and the hysteretic component induced by the hysteretic displacement $\mathbf{z}$. The restoring force can thus be modeled as a combination of an elastic term and a hysteretic term:

$$\mathbf{R}_t(\mathbf{X}, \mathbf{z}) = \alpha \mathbf{K}_0\mathbf{X} + (1 - \alpha)\mathbf{K}_0\mathbf{z} \quad (6.4.2)$$

where $\alpha$ denotes the stiffness ratio between the post-yielding stiffness $\mathbf{K}_1$ and the pre-yielding stiffness $\mathbf{K}_0$.

The function of the hysteretic displacement $\mathbf{z}$ underlies the various formula of the hysteretic model. In this chapter, two families of hysteretic structures are explored, i.e., the Clough hysteretic system and the Bouc–Wen hysteretic system.

Using the Maclaurin series, the equation of motion of controlled hysteretic structures can be rewritten as the formulation of state equation:

$$\dot{\mathbf{Z}}(t) = \mathbf{A}(\mathbf{Z})\mathbf{Z}(t) + \mathbf{B}\mathbf{U}(t) + \mathbf{D}\mathbf{F}(\Theta, t) \quad (6.4.3)$$

$$\mathbf{Z}(t) = \begin{bmatrix} \mathbf{X}(t) \\ \dot{\mathbf{X}}(t) \end{bmatrix}, \mathbf{B} = \begin{bmatrix} \mathbf{0} \\ \mathbf{M}^{-1}\mathbf{B}_s \end{bmatrix}, \mathbf{D} = \begin{bmatrix} \mathbf{0} \\ \mathbf{M}^{-1}\mathbf{D}_s \end{bmatrix},$$

$$\mathbf{A}(\mathbf{Z}) = \begin{bmatrix} \mathbf{0} & \mathbf{I} \\ -\mathbf{M}^{-1}\left(\alpha\mathbf{K}_0 + (1-\alpha)\mathbf{K}_0 \sum\limits_{i=1}^{m} \frac{1}{i!}\frac{\partial^i \mathbf{z}(0,0)}{\partial \mathbf{X}^i} {}^i \mathbf{X}^{i-1}\right) & -\mathbf{M}^{-1}\left(\mathbf{C} + (1-\alpha)\mathbf{K}_0 \sum\limits_{i=1}^{m} \frac{1}{i!}\frac{\partial^i \mathbf{z}(0,0)}{\partial \mathbf{X}^i} {}^i \dot{\mathbf{X}}^{i-1}\right) \end{bmatrix}. \quad (6.4.4)$$

The system matrix can be simplified into the truncated formulation with respect to the zero-order and first-order terms of the Maclaurin series, i.e., the second-order and higher-order terms are ignored:

$$\mathbf{A}(\mathbf{Z}) \doteq \begin{bmatrix} \mathbf{0} & \mathbf{I} \\ -\mathbf{M}^{-1}\left(\alpha\mathbf{K}_0 + (1-\alpha)\mathbf{K}_0 \frac{\partial z(0,0)}{\partial \mathbf{X}}\right) & -\mathbf{M}^{-1}\left(\mathbf{C} + (1-\alpha)\mathbf{K}_0 \frac{\partial z(0,0)}{\partial \mathbf{X}}\right) \end{bmatrix} \quad (6.4.5)$$

As to the optimal polynomial control, the control law exhibits the uniform expression as follows:

$$\mathbf{U}(\Theta, t) = -\mathbf{f}(\mathbf{I}^*, \mathbf{L}^*)\mathbf{f}(\ddot{\mathbf{X}}, \dot{\mathbf{X}}, \mathbf{X}) \quad (6.4.6)$$

Therefore, the essence of solving the generalized optimal control law of the nonlinear structures is still the optimization of the parameters of control law $(\mathbf{I}^*, \mathbf{L}^*)$ so that the structural performance objective can be attained.

The probabilistic criterion of minimizing the performance function in exceedance probability; see Eq. (5.4.1), is applied here for optimizing the parameters of control law. In order to prove the benefits of the generalized optimal control policy in stochastic optimal control of nonlinear structures, a provided number of active tendons is optimized on their control laws, controller parameters and their placements so as to make the control effectiveness maximum. For comparative purposes, the case of the same number of active tendons placed from the bottom acts as the reference, of which controller parameters are simultaneously optimized as the same.

## 6.4.1 Clough Hysteretic System

A ten-story shear frame with nonlinear components represented by Clough hysteretic model subjected to random seismic ground motion is investigated. The story mass and story stiffness are shown in Table 6.3. The structural damping is modeled by the Rayleigh's damping matrix $\mathbf{C} = a\mathbf{M} + b\mathbf{K}_t$, where $a = 0.01$, $b = 0.005$, $\mathbf{K}_t$ denotes the instantaneous stiffness matrix. The natural frequencies of the unyielded structure are 3.46, 10.00, 15.83, 21.26, 26.57, 31.39, 35.25, 38.64, 41.33, and 44.44 rad/s, respectively. The damping ratio of the first-order vibrational mode is thus 1.01%. The physically motivated random seismic ground motion model is employed as the input, and its peak ground acceleration is set as $0.3g$. The restoring force of structural components represented by Clough hysteretic model is shown in Fig. 6.13. The ratio between the post-yielding stiffness and the pre-yielding stiffness of all the interstories $\alpha = K_t / K_0$ is set as 0.1. The initial yielding displacement of interstories $\Delta_y$ is listed in Table 6.3. According to the restoring force relation of structural components, the instantaneous stiffness matrix $\mathbf{K}_t$ can be evaluated. The thresholds of the interstory drift, interstory velocity, story acceleration, and the control force are 15 mm, 150 mm/s, 8000 mm/s$^2$, and 200 kN, respectively. The nonlinear dynamic analysis involved in solving the generalized probability density evolution equation resorts to the Newmark-$\beta$ implicit integral scheme (Clough and Penzien 1993).

Utilizing the generalized optimal control policy addressed in Chap. 5, the controller parameters and placement of the active tendons are optimized. The optimal placement and parameters of weighting matrices of the newly added active tendon using the first-order controller in each sequence are shown in Table 6.4. It is seen that the six tendons are deployed in the tenth interstory, the seventh interstory, the sixth interstory, the eighth interstory, the fifth interstory, and the third interstory in turn. The parameters of generalized optimal control law applied in this numerical example are given by

**Table 6.3** Parameters of ten-story shear frame with nonlinear components

| Story number | 1 | 2 | 3 | 4 | 5 | 6 | 7 | 8 | 9 | 10 |
|---|---|---|---|---|---|---|---|---|---|---|
| Mass ($10^5$ kg) | 1.2 | 1.2 | 1.0 | 1.0 | 1.0 | 1.0 | 1.0 | 1.0 | 0.6 | 0.6 |
| Pre-yielding stiffness (kN/mm) | 48 | 48 | 45 | 45 | 45 | 45 | 45 | 45 | 40 | 40 |
| Yielding displacement (mm) | 10.0 | 10.0 | 8.0 | 8.0 | 8.0 | 8.0 | 8.0 | 8.0 | 6.0 | 6.0 |

**Fig. 6.13** Curve of restoring force of structural components represented by Clough hysteretic model

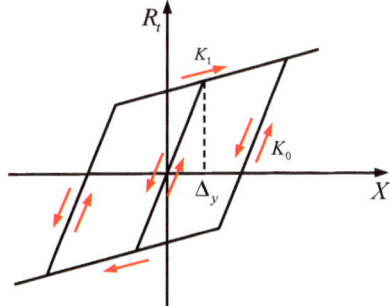

**Table 6.4** Optimal placement and parameters of weighting matrices of newly added tendon in Clough hysteretic system

| Sequence No. | Placement vector | Design parameters[a] | | | | | | |
|---|---|---|---|---|---|---|---|---|
| | | $Q_d$ | $Q_v$ | $Q_{d,2}$ | $Q_{v,2}$ | $Q_{d,3}$ | $Q_{v,3}$ | $R_u$ |
| 0 | $[0\,0\,0\,0\,0\,0\,0\,0\,0\,0]^T$ | – | – | – | – | – | – | – |
| 1 | $[0\,0\,0\,0\,0\,0\,0\,0\,0\,1]^T$ | 187.2 | 0.0 | – | – | – | – | $10^{-10}$ |
| 2 | $[0\,0\,0\,0\,0\,0\,2\,0\,0\,1]^T$ | 84.3 | 217.5 | – | – | – | – | $10^{-10}$ |
| 3 | $[0\,0\,0\,0\,0\,3\,2\,0\,0\,1]^T$ | 207.9 | 0.0 | – | – | – | – | $10^{-10}$ |
| 4 | $[0\,0\,0\,0\,0\,3\,2\,4\,0\,1]^T$ | 0.0 | 532.2 | – | – | – | – | $10^{-10}$ |
| 5 | $[0\,0\,0\,0\,5\,3\,2\,4\,0\,1]^T$ | 73.3 | 0.0 | – | – | – | – | $10^{-10}$ |
| 6 | $[0\,0\,6\,0\,5\,3\,2\,4\,0\,1]^T$ | 0.0 | 62.3 | – | – | – | – | $10^{-10}$ |
| 6 (Third-order) | $[0\,0\,6\,0\,5\,3\,2\,4\,0\,1]^T$ | 0.0 | 62.3 | 0.0 | 0.0 | – | – | $10^{-10}$ |
| 6 (Fifth-order) | $[0\,0\,6\,0\,5\,3\,2\,4\,0\,1]^T$ | 0.0 | 62.3 | 0.0 | 0.0 | 0.0 | 1.8 | $10^{-10}$ |
| Reference | $[1\,1\,1\,1\,1\,1\,0\,0\,0\,0]^T$ | 300.0 | 0.0 | – | – | – | – | $10^{-10}$ |

[a]Initial values of weighting matrices $Q_d = Q_v = 100$, $Q_{d,2} = Q_{v,2} = 5$, $Q_{d,3} = Q_{v,3} = 2$, $R_u = 10^{-10}$

$$(Q_d^*, Q_v^*, R_u^*, L^*) = \begin{bmatrix} 0\,0 & 0.0 & 0 & 73.3 & 207.9 & 84.3 & 0.0 & 0 & 187.2 \\ 0\,0 & 62.3 & 0 & 0.0 & 0.0 & 217.5 & 532.2 & 0 & 0.0 \\ 0\,0 & 10^{-10} & 0 & 10^{-10} & 10^{-10} & 10^{-10} & 10^{-10} & 0 & 10^{-10} \\ 0\,0 & 6 & 0 & 5 & 3 & 2 & 4 & 0 & 1 \end{bmatrix}^{\mathrm{T}} . \quad (6.4.7)$$

The design parameters of high-order terms, i.e., the third-order and the fifth-order, of the polynomial controllers used for the optimally deployed six active tendons are shown in Table 6.4 as well, which are scheduled as the same for all the controllers. It is seen the high-order terms of polynomial controllers have no contributions to the structural control. In fact, the objective function of the high-order controllers remains the same numerical accuracy as that of the first-order control, which has consistent results in four effective figures. It is revealed again that using the criterion of minimizing the performance function in exceedance probability, the linear control with first-order controller can completely implement the control effectiveness of the nonlinear control with high-order controllers.

The exceedance probabilities of system quantities of concern and the objective function are shown in Table 6.5. It is revealed that due to the deployment of the active tendons, the structural performance is enhanced gradually. Figure 6.14 shows the sequences of the tendon placement where the number indicates the order that the tendon is placed. Meanwhile, the reference case with the tendon placed from the bottom has similar control effectiveness to the case of Sequence 3, which indicates that

**Table 6.5** Seismic mitigation by active tendons in Clough hysteretic system

| Sequence No. | Placement vector | Exceedance probabilities | | | | Objective function $J_2$ |
|---|---|---|---|---|---|---|
| | | $P_{f,d}$ | $P_{f,v}$ | $P_{f,a}$ | $P_{f,u}$ | |
| 0 | $[0\,0\,0\,0\,0\,0\,0\,0\,0\,0]^{\mathrm{T}}$ | 0.5016 | 0.6032 | 0.7829 | – | 0.6142 |
| 1 | $[0\,0\,0\,0\,0\,0\,0\,0\,0\,1]^{\mathrm{T}}$ | 0.4942 | 0.5822 | 0.7685 | $3.600 \times 10^{-7}$ | 0.5869 |
| 2 | $[0\,0\,0\,0\,0\,0\,2\,0\,0\,1]^{\mathrm{T}}$ | 0.4355 | 0.5420 | 0.7597 | 0.0268 | 0.5306 |
| 3 | $[0\,0\,0\,0\,0\,3\,2\,0\,0\,1]^{\mathrm{T}}$ | 0.3927 | 0.5152 | 0.7573 | 0.0364 | 0.4973 |
| 4 | $[0\,0\,0\,0\,0\,3\,2\,4\,0\,1]^{\mathrm{T}}$ | 0.3694 | 0.4330 | 0.7534 | 0.1859 | 0.4630 |
| 5 | $[0\,0\,0\,0\,5\,3\,2\,4\,0\,1]^{\mathrm{T}}$ | 0.3580 | 0.4091 | 0.7511 | 0.0326 | 0.4304 |
| 6 | $[0\,0\,6\,0\,5\,3\,2\,4\,0\,1]^{\mathrm{T}}$ | 0.3527 | 0.3918 | 0.7497 | 0.0705 | 0.4225 |
| Reference | $[1\,1\,1\,1\,1\,1\,0\,0\,0\,0]^{\mathrm{T}}$ | 0.3952 | 0.5178 | 0.7596 | 0.0221 | 0.5009 |

**Fig. 6.14**  Schematic of tendon deployments in Clough hysteretic system

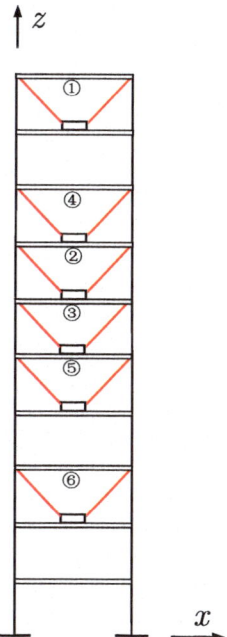

using the criterion of minimum story controllability index gradient, just three tendons in optimal deployment can attain the control effectiveness of six tendons placed from the bottom. A schematic diagram showing the similar control effectiveness between the two schemes of tendon deployments refers to Fig. 6.15.

Figure 6.16 shows the mean and standard deviation of the extreme value of inter-story drift that changes as the deployment of the tendons. It is seen that the statistical moments of extreme values of interstory drift is reduced in sequences. Although the interstory control force acting on the structure cannot be so large and the control effectiveness is limited due to the consideration of nonlinear system stability, the interstories with larger displacement responses gain significant improvement. The mean and standard deviation of extreme values of story acceleration that changes as the deployment of the tendons are shown in Fig. 6.17. Similarly, the story accelerations are reduced gradually in sequences, and the stories with large acceleration response gain significant improvement. However, reduction of the standard deviation of story accelerations is not uniform along the story level, e.g., the low stories exhibit an acceleration reduction but the high stories exhibit an acceleration amplification in sequences. This is due to the fact that the control force input changes the filtering performance of the nonlinear system upon the ground accelerations. Consequently, control effectiveness on the story acceleration is not obvious as on the interstory drift, which is in consistency with the results shown in Table 6.5.

Figures 6.18 shows the hysteretic curves of the first and the tenth interstory components with and without controls subjected to representative ground motion. It is

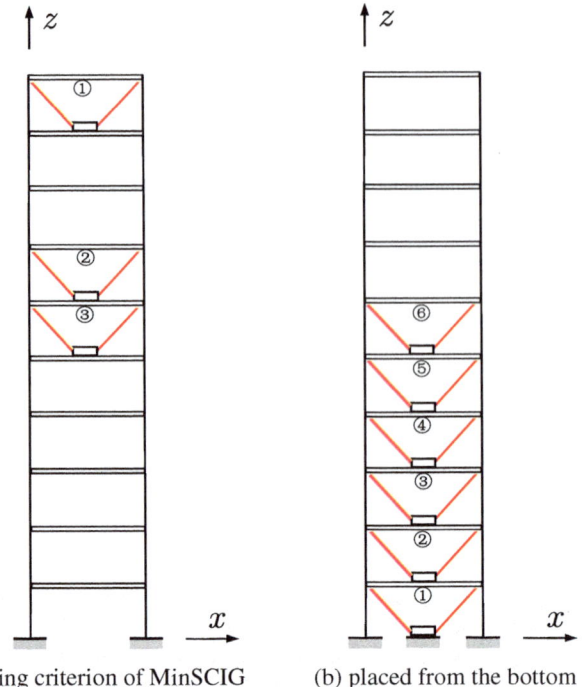

(a) using criterion of MinSCIG          (b) placed from the bottom

**Fig. 6.15**  Schematic of tendon deployments in Clough hysteretic system as different schemes

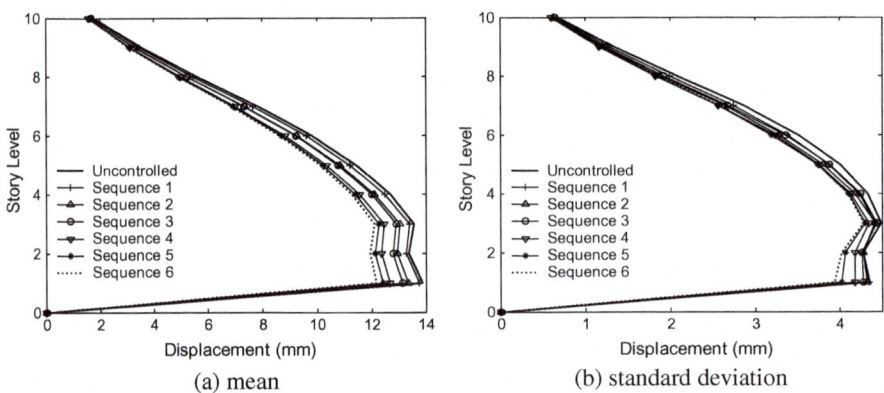

(a) mean                              (b) standard deviation

**Fig. 6.16**  Mean and standard deviation of extreme values of interstory drift that changes as the deployment of tendons in Clough hysteretic system

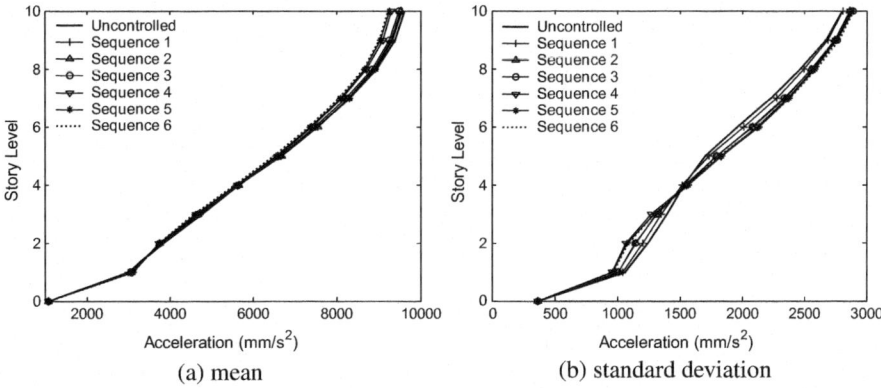

**Fig. 6.17** Mean and standard deviation of extreme values of story acceleration that changes as the deployment of tendons in Clough hysteretic system

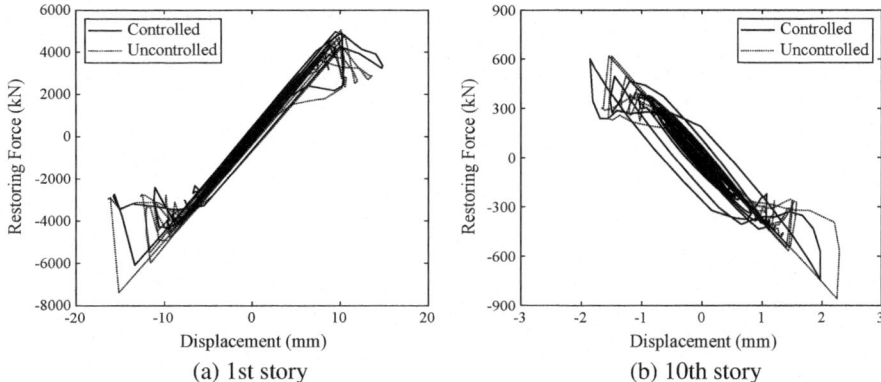

**Fig. 6.18** Hysteretic curves of first and tenth interstory components with and without controls subjected to representative ground motion

seen that the interstory drifts of the structure with control become less than those without control, and the interstory restoring forces become less as well owing to the compensation of additional control forces. Moreover, the profile of hysteretic curves of the interstory components with control has a similarity as that without control; while the interstory component with control moves back and forth in a less range from initial positions. Besides, one might recognize that the first and the tenth interstory components experience opposing motions when subjected to the representative ground motion.

## 6.4.2 Bouc–Wen Hysteretic System

The structural parameters of another exemplified eight-story shear frame are follows: $m_1 = m_2 = 1.0 \times 10^5$ kg, $m_3 = m_4 = 0.9 \times 10^5$ kg, $m_5 = m_6 = 0.9 \times 10^5$ kg, $m_7 = m_8 = 0.8 \times 10^5$ kg; $k_1 = k_2 = 36$ kN/mm, $k_3 = k_4 = 32$ kN/mm, $k_5 = k_6 = 32$ kN/mm, $k_7 = k_8 = 28$ kN/mm. Rayleigh's damping $\mathbf{C} = a\mathbf{M} + b\mathbf{K}$ is employed, where $a = 0.01$, $b = 0.005$. The natural frequencies of the unyielded structure are 3.64, 10.40, 16.46, 22.45, 27.91, 31.89, 34.68, and 36.81 rad/s, respectively. The damping ratio of the first-order vibrational mode is thus 1.05%. The physically motivated random seismic ground motion model is employed as the input, and its peak ground acceleration is set as $0.3g$. The restoring force of structural components is represented by Bouc–Wen hysteretic model. The formulation of Bouc–Wen hysteretic model and the pertinent parameters refer to Appendix A. The thresholds of the interstory drift, interstory velocity, story acceleration and the control force are 30 mm, 300 mm/s, 3000 mm/s$^2$ and 200 kN, respectively. An explicit time integration method is employed in the nonlinear dynamic analysis of the hysteretic system:

$$\ddot{\mathbf{X}}(k+1) = \mathbf{M}^{-1}[\mathbf{B}_s\mathbf{U}(k) + \mathbf{D}_s\mathbf{F}(\mathbf{\Theta}, k) - \mathbf{C}\dot{\mathbf{X}}(k) - \mathbf{R}_t(\mathbf{X}(k), \mathbf{z}(k))] \quad (6.4.8)$$

$$\dot{\mathbf{X}}(k+1) = \dot{\mathbf{X}}(k) + (1 - \gamma_a)\ddot{\mathbf{X}}(k)\Delta t + \gamma_a\ddot{\mathbf{X}}(k+1)\Delta t \quad (6.4.9)$$

$$\mathbf{X}(k+1) = \mathbf{X}(k) + \dot{\mathbf{X}}(k)\Delta t + (\tfrac{1}{2} - \beta_a)\ddot{\mathbf{X}}(k)\Delta t^2 + \beta_a\ddot{\mathbf{X}}(k+1)\Delta t^2 \quad (6.4.10)$$

It has been proved (Chung and Lee 1994) that the integral scheme is of second-order accuracy when $\gamma_a = \tfrac{3}{2}$, and it is unconditionally stable when $1 \leq \beta_a \leq \tfrac{28}{27}$. In this study, the control parameters $\gamma_a = 1.5$, $\beta_a = 1.0$. Besides, the hysteretic displacement $\mathbf{z}$ is solved employing the fourth-order Runge–Kutta method.

Table 6.6 shows the optimal placement and parameters of weighting matrices of the newly added active tendon using the first-order controller in each sequence. It is seen that the five tendons are deployed in the first interstory, the second interstory, the third interstory, the fourth interstory, and the sixth interstory in turn. The parameters of generalized optimal control law applied in this numerical example are given by

$$(Q_d^*, Q_v^*, R_u^*, L^*) = \begin{bmatrix} 642.5 & 100.2 & 777.9 & 57.1 & 0 & 225.5 & 0 & 0 \\ 73.0 & 47.9 & 485.7 & 1072.0 & 0 & 193.1 & 0 & 0 \\ 10^{-9} & 10^{-9} & 10^{-9} & 10^{-9} & 0 & 10^{-9} & 0 & 0 \\ 1 & 2 & 3 & 4 & 0 & 5 & 0 & 0 \end{bmatrix}^{\mathrm{T}} \quad (6.4.11)$$

In order to evaluate the control effectiveness of the first-order control, the nonlinear controls with high-order controllers on the hysteretic system are also investigated. The parameters of weighting matrices of third-order and fifth-order terms are shown in Table 6.6 as well. It is seen that the parameters of weighting matrices of high-order terms are almost all equal to zero, indicating once again that the high-order terms have no contributions to improve the structural performance, and the linear control

**Table 6.6** Optimal placement and parameters of weighting matrices of newly added tendon in Bouc–Wen hysteretic system

| Sequence No. | Placement vector | Design parameters[a] | | | | | | |
|---|---|---|---|---|---|---|---|---|
| | | $Q_d$ | $Q_v$ | $Q_{d,2}$ | $Q_{v,2}$ | $Q_{d,3}$ | $Q_{v,3}$ | $R_u$ |
| 0 | $[0\,0\,0\,0\,0\,0\,0\,0]^T$ | – | – | – | – | – | – | – |
| 1 | $[1\,0\,0\,0\,0\,0\,0\,0]^T$ | 642.5 | 73.0 | – | – | – | – | $10^{-9}$ |
| 2 | $[1\,2\,0\,0\,0\,0\,0\,0]^T$ | 100.2 | 47.9 | – | – | – | – | $10^{-9}$ |
| 3 | $[1\,2\,3\,0\,0\,0\,0\,0]^T$ | 777.9 | 485.7 | – | – | – | – | $10^{-9}$ |
| 4 | $[1\,2\,3\,4\,0\,0\,0\,0]^T$ | 57.1 | 1072.0 | – | – | – | – | $10^{-9}$ |
| 5 | $[1\,2\,3\,4\,0\,5\,0\,0]^T$ | 225.5 | 193.1 | – | – | – | – | $10^{-9}$ |
| 5 (Third-order) | $[1\,2\,3\,4\,0\,5\,0\,0]^T$ | 225.5 | 193.1 | 0.0 | 0.0 | – | – | $10^{-9}$ |
| 5 (Fifth-order) | $[1\,2\,3\,4\,0\,5\,0\,0]^T$ | 225.5 | 193.1 | 0.0 | 0.0 | 0.0 | 0.6 | $10^{-9}$ |
| Reference | $[1\,1\,1\,1\,1\,0\,0\,0]^T$ | 0.0 | 1027.0 | – | – | – | – | $10^{-9}$ |

[a]Initial values of weighting matrices $Q_d = Q_v = 100$, $Q_{d,2} = Q_{v,2} = 20$, $Q_{d,3} = Q_{v,3} = 10$, $R_u = 10^{-9}$

with first-order controller can implement the control effectiveness of the nonlinear control with high-order controllers when the criterion of minimizing the performance function in exceedance probability is employed.

Table 6.7 shows the exceedance probabilities of system quantities of concern and the objective function in each sequence. It is readily seen that the exceedance probabilities of system quantities are gradually reduced as the optimal deployment and design of active tendons; the objective function attains the minimum until all the five controllers are placed. The orders of active tendons deployed in the structural interstories are shown in Fig. 6.19. The reference case with the five active tendon placed from the bottom has similar control effectiveness to the case of Sequence 3. It is indicated that using the criterion of minimum story controllability index gradient, just three tendons in optimal deployment can attain the control effectiveness of five tendons placed from the bottom. A schematic diagram showing similar control effectiveness between the two tendon deployments refers to Fig. 6.20.

**Table 6.7** Seismic mitigation by active tendons in Bouc–Wen hysteretic system

| Sequence no. | Placement vector | Exceedance probabilities | | | | Objective function $J_2$ |
|---|---|---|---|---|---|---|
| | | $P_{f,d}$ | $P_{f,v}$ | $P_{f,a}$ | $P_{f,u}$ | |
| 0 | $[0\,0\,0\,0\,0\,0\,0\,0]^T$ | 0.8354 | 0.1334 | 0.8441 | – | 0.7141 |
| 1 | $[1\,0\,0\,0\,0\,0\,0\,0]^T$ | 0.5714 | 0.0186 | 0.6867 | 0.0000 | 0.3992 |
| 2 | $[1\,2\,0\,0\,0\,0\,0\,0]^T$ | 0.5833 | 0.0210 | 0.6252 | $3.603 \times 10^{-7}$ | 0.3658 |
| 3 | $[1\,2\,3\,0\,0\,0\,0\,0]^T$ | 0.5279 | 0.0585 | 0.3917 | $3.602 \times 10^{-7}$ | 0.2178 |
| 4 | $[1\,2\,3\,4\,0\,0\,0\,0]^T$ | 0.4777 | 0.0919 | 0.1846 | $4.276 \times 10^{-7}$ | 0.1354 |
| 5 | $[1\,2\,3\,4\,0\,5\,0\,0]^T$ | 0.4286 | 0.0756 | 0.1740 | $3.603 \times 10^{-7}$ | 0.1098 |
| Reference | $[1\,1\,1\,1\,1\,0\,0\,0]^T$ | 0.7125 | 0.1372 | 0.1345 | 0.0041 | 0.2723 |

**Fig. 6.19** Schematic of tendon deployments in Bouc–Wen hysteretic system

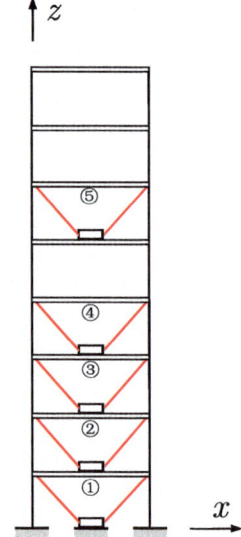

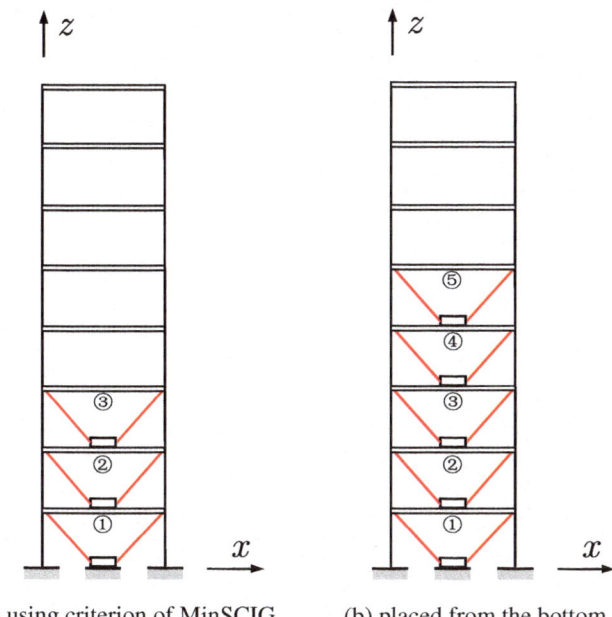

(a) using criterion of MinSCIG             (b) placed from the bottom

**Fig. 6.20**   Schematic of tendon deployments in Bouc–Wen hysteretic system as different schemes

   The mean and standard deviation of extreme values of interstory drift varying along the height of the structure as the placement of active tendons are shown in Fig. 6.21. It is seen that the mean and standard deviation of the extreme values of interstory drift changes nonuniformly along the height of the structure as the tendon deployment, and the displacement of medium interstories even occurs to be amplified, and they are smoother along the height of the structure. One might understand that the stability of nonlinear systems is a critical performance argument for evaluating the optimal tendon deployment. Nevertheless, the global performance of the hysteretic structure with control is better than that without control. Besides, the control effectiveness on story acceleration is better than the interstory drift; see Fig. 6.22 that shows the mean and standard deviation of the extreme values of story acceleration varying along the height of the structure as the placement of active tendons. It is seen that the upper stories with larger acceleration before control have a significant performance enhancement after the control.
   The hysteretic curves of the first and the eighth interstory components with and without controls subjected to a representative ground motion are shown in Fig. 6.23. It is seen that the hysteretic curves of the uncontrolled components and controlled components feature the Bouc–Wen prosperities, i.e., the strength deterioration, the stiffness degradation, and the pinching effect. The interstory drifts of the controlled structure become smaller than those of the uncontrolled structure, and meanwhile, the stiffness degradation of components gains an alleviation after the control. Besides, the

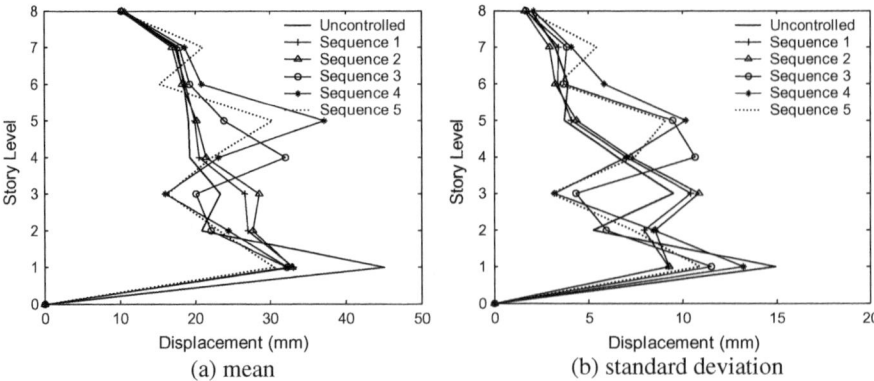

**Fig. 6.21** Mean and standard deviation of extreme values of interstory drift that changes as tendon deployments in Bouc–Wen hysteretic system

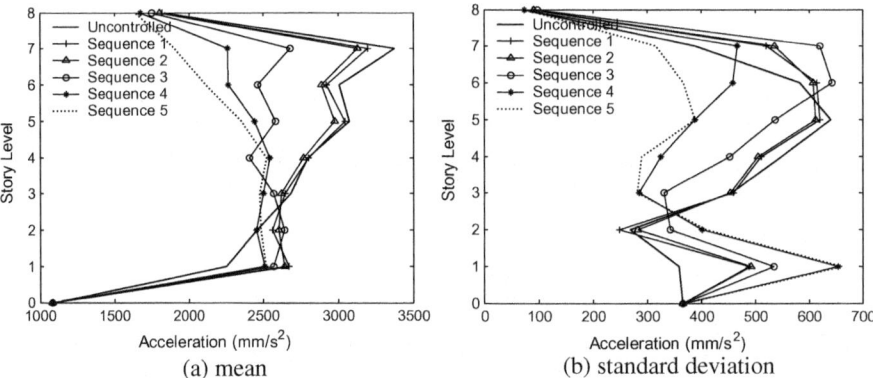

**Fig. 6.22** Mean and standard deviation of extreme values of story acceleration that changes as tendon deployments in Bouc–Wen hysteretic system

restoring forces of components become smaller due to the compensation of control forces.

The time histories of root-mean-square hysteretic energy dissipation of the first and the eighth interstory components with and without controls are shown in Fig. 6.24. It is seen that the energy dissipations of components at different interstories are reduced significantly after the control, and eventually become into an approximate stationary processes, which indicates a better energy-dissipation behavior pertaining to the hysteretic system.

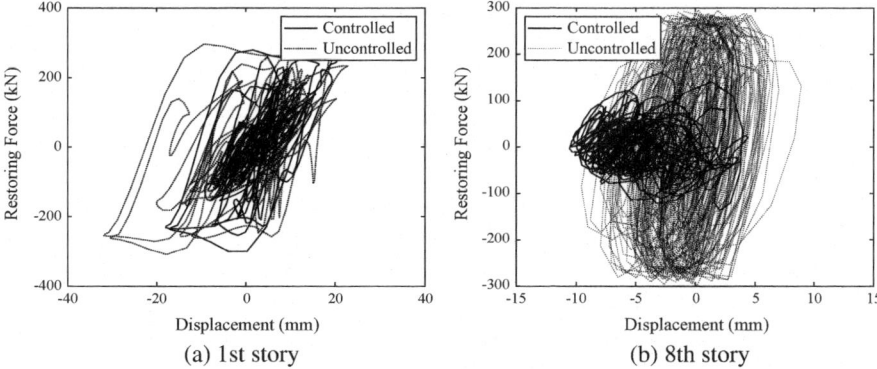

**Fig. 6.23** Hysteretic curves of first and eighth interstory components with and without controls subjected to representative ground motion

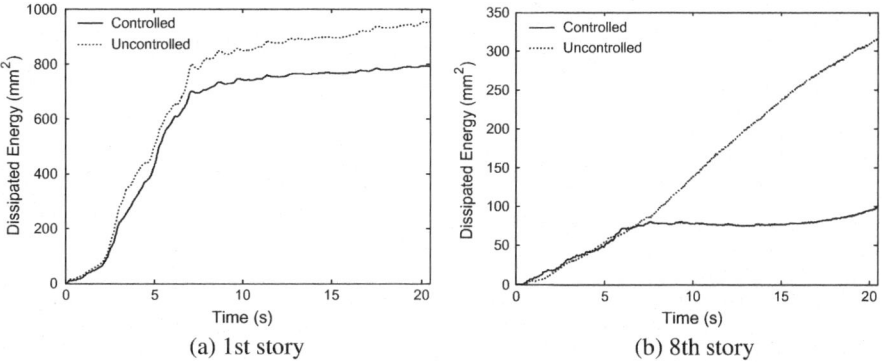

**Fig. 6.24** Time histories of root-mean-square hysteretic energy dissipation of first and eighth interstory components with and without controls

## 6.5  Discussions and Summaries

In conjunction with the scheme of optimal polynomial control, the physically based stochastic optimal control has been applied to the performance enhancement of nonlinear and hysteretic structures. The Duffing oscillator systems, the ten-story shear frame with nonlinear components represented by Clough hysteretic model and the eight-story shear frame with nonlinear components represented by Bouc–Wen hysteretic model are investigated.

The stochastic optimal control of Duffing oscillator systems with variant nonlinearity levels shows that using the exceedance probability criterion on energy trade-off, the linear control with a first-order controller can implement the control effectiveness of nonlinear control with high-order controllers, which bypasses the structural instability associated with nonlinear control systems due to the time delay and unpredictable feedback errors. However, other control criteria such as the system second-order statistics evaluation (SSSE) and the Lyapunov asymptotic stability condition (LASC) have no such benefit. This finding exhibits practical significance. Besides, in comparison with the statistical linearization based LQG, the stochastic optimal polynomial control exhibits good robustness which is not sensitive to the nonlinearity level of Duffing oscillator systems.

It is also indicated in the stochastic optimal control of hysteretic structures that the first-order controller can implement the control effectiveness of nonlinear control with high-order controllers. With the optimal control, the Clough hysteretic systems and the Bouc–Wen hysteretic systems both attain a certain degree of performance enhancement, and their hysteretic behaviors and energy-dissipation capacities gain an obvious improvement. Meanwhile, the application of the generalized optimal control policy with the criterion of minimum story controllability index gradient can accommodate an structural performance objective, and implement a maximum control effectiveness with a minimum control cost.

# References

Anderson Brian DO, Moore J (1990) Optimal control: linear quadratic methods. Prentice-Hall, Englewood Cliffs

Baber TT, Noori MN (1985) Random vibration of degrading, pinching systems. ASCE J Eng Mech 111(8):1010–1027

Baber TT, Wen YK (1981) Random vibration of hysteretic degrading systems. ASCE J Eng Mech Div 107(6):1069–1087

Bernstein DS (1993) Nonquadratic cost and nonlinear feedback control. Int J Robust Nonlinear Control 3:211–229

Bouc R (1967) Forced vibration of mechanical system with hysteresis. In: Proceedings of 4th conference on nonlinear oscillations, Prague, Czechoslovakia

Chung J, Lee JM (1994) A new family of explicit time integration methods for linear and nonlinear structural dynamics. Int J Numer Meth Eng 37:3961–3976

Clough RW, Johnson SB (1966) Effects of stiffness degradation on earthquake ductility requirements. In: Proceedings of 2nd Japan national earthquake engineering conference, Tokyo, Japan, pp 227–232

Clough RW, Penzien J (1993) Dynamics of structures, 2nd edn. McGraw-Hill, New York

Housner GW, Bergman LA, Caughey TK, Chassiakos AG, Claus RO, Masri SF, Skelton RE, Soong TT, Spencer BF Jr, Yao James TP (1997) Structural control: past, present, and future. ASCE J Eng Mech 123(9):897–971

Iwan WD (1961) The dynamic response of bilinear hysteretic system. Doctoral Thesis, California Institute of Technology, Pasadena, California

Peng YB, Li J (2011) Exceedance probability criterion based stochastic optimal polynomial control of Duffing oscillators. International Journal of Non-Linear Mechanics 46(2):457–469

Sekar P, Narayanan S (1994) Periodic and chaotic motions of a square prism in cross-flow. J Sound Vib 170(1):1–24

Shefer M, Breakwell JV (1987) Estimation and control with cubic nonlinearities. J Optim Theory Appl 53:1–7

Suhardjo J, Spencer BF Jr, Sain MK (1992) Nonlinear optimal control of a duffing system. Int J Non-Linear Mech 27(2):157–172

Wen YK (1976) Method for random vibration of hysteretic systems. ASCE J Eng Mech Div 102(2):249–263

Yang JN, Li Z, Liu SC (1992) Stable controllers for instantaneous optimal control. ASCE J Eng Mech 118(8):1612–1630

Yang JN, Li Z, Vongchavalitkul S (1994) Stochastic hybrid control of hysteretic structures. Probab Eng Mech 9(1–2):125–133

Yang JN, Agrawal AK, Chen S (1996) Optimal polynomial control for seismically excited non-linear and hysteretic structures. Earthq Eng Struct Dyn 25:1211–1230

Zhang WS, Xu YL (2001) Closed form solution for along-wind response of actively controlled tall buildings with LQG controllers. J Wind Eng Ind Aerodyn 89:785–807

Zhu WQ, Ying ZG, Ni YQ, Ko JM (2000) Optimal nonlinear stochastic control of hysteretic systems. ASCE J Eng Mech 126(10):1027–1032

# Chapter 7
# Stochastic Optimal Control of Wind-Resistant Structures with Viscous Dampers

## 7.1 Preliminary Remarks

The high-rise buildings experience alongwind-induced motions or acrosswind-induced motions when they are subjected to wind actions. The occupants would feel uncomfortable if these motions reach a certain amplitude. The structural performance, in this case, is generally denoted by the habitability, which is often measured by the wind-induced acceleration of structures (Chan and Chui 2006). The serviceability design problem of high-rise buildings associated with habitability enhancement is one of the most concerned issues, especially during typhoon seasons. The reports relevant to typhoon events in the past years often mentioned that the strong vibration of high-rise buildings results in discomfort or even dazzling state to occupants. Numerical investigations of TMD deployed in the building "Taipei 101" indicated that the vibration of the structure subjected to frequently occurring wind actions with a half-year return period would exceed 30% of the design maximum acceleration if removing the control device (Chung et al. 2013). Therefore, the serviceability-based control and design retain a practical significance to the high-rise buildings.

The structural control for mitigating wind-induced vibration can be largely categorized into the passive and active modalities. The former is a widely applied means due to its practical feasibility (Housner et al. 1997). A most efficient measure for mitigating the wind-induced vibration of structures is the damping reinforcement. The viscous dampers are proved to be an effective proposal of implementing the damping reinforcement owing to their many technical advantages (Housner et al. 1997; Patil and Jangid 2011), e.g., being insensitive to the working temperature (with steady behaviors from −40 centidegrees to 70 centidegrees) (Symans and Contantinou 1998), remaining a visco-response in a wide frequency domain (Soong and Constantinou 1994), exhibiting a damper force out of phase with displacement (Soong and Dargush 1997), and providing considerable damper force even in case of low structural velocity. Another highlighting feature of the viscous damper is its benefit

© Springer Nature Singapore Pte Ltd. and Shanghai Scientific and Technical Publishers 2019
Y. Peng and J. Li, *Stochastic Optimal Control of Structures*,
https://doi.org/10.1007/978-981-13-6764-9_7

in acceleration control owing to a less stiffness and a larger damping that the damper supplies. Therefore, the viscous damper has been proved to the best control device for the wind-induced comfortability control of high-rise and ultra high-rise buildings. The parameter definition and placement optimization of viscous dampers are the critical issues associated with the stochastic optimal control of wind-induced structural comfortability. However, the viscous damper is a velocity-relevant damper, and exhibits a strong nonlinearity. The structure deployed with viscous dampers becomes a nonlinear system in essence. The primary task for the stochastic optimal control of viscously damped structures is thus seeking for a solving procedure with sufficient efficiency and accuracy.

In the present chapter, the equivalent linearization techniques with respect to the structural system attached with nonlinear viscous dampers are addressed first. The probabilistic criteria and numerical methods for the optimal design of viscous dampers used in randomly wind-induced vibration control of structures are provided. For validating purposes, the optimal control of wind-induced comfortability of a high-rising building is investigated, of which the randomness inherent in wind excitations is included.

## 7.2   Equivalent Linearization of Viscously Damped Systems

The stochastic optimal control of viscously damped structural systems involves the design of damper parameters and the optimization of damper deployments. This features practical significance for the high-rise building with available limited space. As mentioned in Chap. 1, the optimization methods for damper deployments can be categorized into three classes, i.e., the sequential method (Zhang and Soong 1992), the gradient method (Takewaki 1997; Peng et al. 2013), and the genetic algorithm based method (Singh and Moreschi 2002). These three classes of optimization methods all involve the iterative solution of viscously damped structures with nonlinearity. For this reason, a highly efficient method which allows for solving the viscously damped structures underlies the optimization and design of the viscous dampers.

The analysis methods for the damper control of structural wind-induced vibration are mainly classified into frequency-domain and time-domain methods. The frequency-domain method is widely used in practice due to the simple algorithm and the rigorous principle (Davenport 1961). The time-domain method can accurately secure the response details of structures even with nonlinear behaviors, which has been paid extensive attention in recent years. As to the reliability-based control for wind-induced vibration mitigation of structures, the time-domain method is usually required. It is revealed in previous investigations (Chen et al. 2017) that the equation of motion of viscously damped structural systems with low-velocity exponent dampers often refers to stiff problem, which belongs to a family of strong nonlinearities. This issue results in that the traditional equivalent linearization techniques such as the energy-dissipation equivalent linearization method and statistical linearization technique remain a challenge.

## 7.2.1 Stiff Differential Equation for Viscously Damped Systems

Consider a single-degree-of-freedom (SDOF) structural system attached with a viscous damper, the equation of motion of the controlled structure subjected to time-varying load $F(t)$ is given by

$$m\ddot{x}(t) + c\dot{x}(t) + kx(t) - F_D(\dot{x}(t)) = F(t) \tag{7.2.1}$$

where $m$, $c$, and $k$ denote the mass, damping, and stiffness of the structure, respectively; $x(t)$, $\dot{x}(t)$ and $\ddot{x}(t)$ denote the displacement, velocity, and acceleration of the structure, respectively, which exhibit an opposing displacement and velocity as the piston; $F(t)$ denotes the external excitation; $F_D(\cdot)$ denotes the damper force exerted by the viscous damper:

$$F_D(\dot{x}(t)) = -c_D \mathrm{sgn}(\dot{x}(t))|\dot{x}(t)|^{\alpha} \tag{7.2.2}$$

where $c_D$ denotes the damping coefficient; $\alpha$ denotes the velocity exponent, which is a positive quantity between 0 and 1, being closer to 0 implying stronger nonlinearity. The velocity exponent is usually valued in the range $0.3 \leq \alpha \leq 0.5$ for the case of building control, and is usually valued in the range $0.15 \leq \alpha \leq 0.3$ for the case of bridge control. The sign $\mathrm{sgn}(\cdot)$ is the signum function, taking value 1 for positive argument, $-1$ for negative argument, and otherwise 0.

The damper force of the viscous damper represented by Eq. (7.2.2) for different $\alpha$ and the same unit of damping coefficient $c_D = 1.0$ is shown in Fig. 7.1, in which the loaded sine wave of displacement exhibits unit amplitude and unit circular frequency. It is seen clearly that in the case $\alpha = 1.0$, it reduces to a linear damping; whereas in the case $\alpha = 0.0$, it becomes a dry-friction force. When $0 < \alpha < 1.0$,

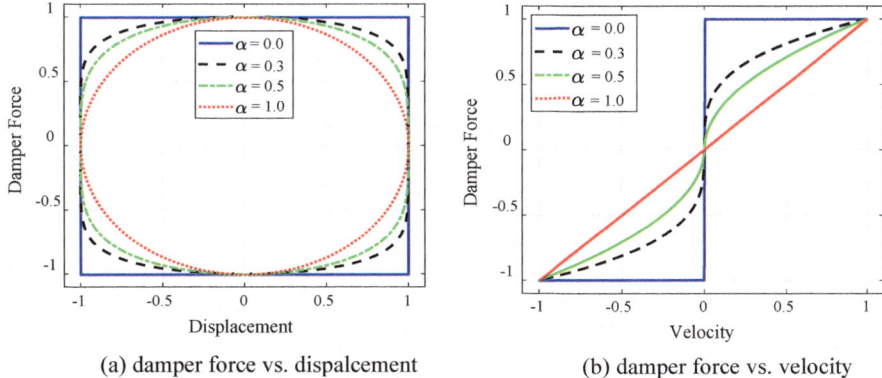

(a) damper force vs. dispalcement       (b) damper force vs. velocity

**Fig. 7.1** Damper force of viscous damper in cases of different velocity exponents

the smaller of the velocity exponent $\alpha$, the stronger of nonlinearity that the viscous damper exhibits. Moreover, the hysteretic curve of the damper involves from the elliptical shape to the rectangular shape as the velocity exponent decreases. In the case of same maximum damper force and same maximum displacement, the smaller of the velocity exponent, the more similar to a rectangular shape with a larger area that the hysteretic curve gives rise to, which indicates a stronger energy-dissipation capacity of the viscous damper and a better control effectiveness on the viscously damped structure. However, the change of the damper force as the velocity arises to fast-varying and slow-varying behaviors. For instance, the damper force varies very fast in the case of low velocity close to zero if the velocity exponent $\alpha = 0.3$; while it varies mildly in other velocity domains. Such coexistence of fast-varying and low-varying behaviors of viscous dampers leads to a typical stiff system of the controlled structure attached with viscous damper. A stiff system can be revealed by its stiff differential equation.

To further clarify this problem, consider the case of free vibration, i.e., $F(t) = 0$ in Eq. (7.2.1). Substituting Eq. (7.2.2) into Eq. (7.2.1), and letting $y_1(t) = x(t)$, $y_2(t) = \dot{x}(t)$ lead to the state equation as follows:

$$\begin{cases} \dot{y}_1(t) = y_2(t) \\ \dot{y}_2(t) = -\frac{c}{m}y_2(t) - \frac{k}{m}y_1(t) - \frac{c_D}{m}\text{sgn}(y_2(t))|y_2(t)|^\alpha \end{cases} \tag{7.2.3}$$

In a vector form

$$\dot{\mathbf{Y}} = \mathbf{A}(\mathbf{Y}, t) \tag{7.2.4}$$

where

$$\mathbf{Y} = (y_1(t), y_2(t))^T \mathbf{A} = (A_1, A_2)^T A_1(y_1(t), y_2(t)) = y_2(t)$$

$$A_2(y_1(t), y_2(t)) = -\frac{c}{m}y_2(t) - \frac{k}{m}y_1(t) - \frac{c_D}{m}\text{sgn}(y_2(t))|y_2(t)|^\alpha \tag{7.2.5}$$

The Jacobian matrix for Eq. (7.2.5) can thus be obtained by

$$\mathbf{J} = \begin{bmatrix} \frac{\partial A_1}{\partial y_1} & \frac{\partial A_1}{\partial y_2} \\ \frac{\partial A_2}{\partial y_1} & \frac{\partial A_2}{\partial y_2} \end{bmatrix} = \begin{bmatrix} 0 & 1 \\ -\frac{k}{m} & \left(-\frac{c}{m} - \alpha\frac{c_D}{m}|y_2(t)|^{\alpha-1}\right) \end{bmatrix} \tag{7.2.6}$$

The characteristic equation can be obtained as

$$\lambda^2 + \left(\frac{c}{m} + \alpha\frac{c_D}{m}|\dot{x}(t)|^{\alpha-1}\right)\lambda + \frac{k}{m} = 0 \tag{7.2.7}$$

where $\lambda$ denotes the eigenvalues of the Jacobian matrix. If Eq. (7.2.7) satisfies the conditions as follows:

$$\begin{cases} \mathrm{Re}(\lambda_j) < 0, \quad j = 1, 2, \cdots, m \\ s := \max_{1 \le j \le m} |\mathrm{Re}(\lambda_j)| / \min_{1 \le j \le m} |\mathrm{Re}(\lambda_j)| \gg 1 \end{cases} \tag{7.2.8}$$

where $\mathrm{Re}(\cdot)$ denotes the real counterpart of the complex eigenvalues. Equation (7.2.1) is then defined as a stiff differential equation, where $s$ denotes the stiff ratio.

For illustrative purposes, consider Eq. (7.2.1) as the first modal equation of a high-rise building structure, where the modal mass $m = 3.75 \times 10^7$ kg, the modal damping ratio $\zeta = 0.01$, and the fundamental period $T_n = 4.94$ s. Since the stiffness $k = \frac{4\pi^2 m}{T_n^2}$ and the damping coefficient $c = \frac{4\pi\zeta m}{T_n}$, Eq. (7.2.7) becomes

$$\lambda^2 + \left( \frac{4\pi\zeta}{T_n} + \alpha \frac{c_D}{m} |\dot{x}|^{\alpha-1} \right) \lambda + \frac{4\pi^2}{T_n^2} = 0 \tag{7.2.9}$$

The two roots of Eq. (7.2.9) are given by

$$\lambda_{1,2} = -\frac{1}{2} \left( \frac{4\pi\zeta}{T_n} + \alpha \frac{c_D}{m} |\dot{u}|^{\alpha-1} \right) \pm \frac{1}{2} \sqrt{ \left( \frac{4\pi\zeta}{T_n} + \alpha \frac{c_D}{m} |\dot{u}|^{\alpha-1} \right)^2 - \frac{16\pi^2}{T_n^2} } \tag{7.2.10}$$

For notation convenience, let $\frac{4\pi\tilde{\zeta}_D}{T_n} = \alpha \frac{c_D}{m} |\dot{x}|^{\alpha-1}$, such that

$$\tilde{\zeta}_D = \alpha \frac{c_D T_n}{4\pi m} |\dot{x}|^{\alpha-1} \tag{7.2.11}$$

It is indicated that the argument defined in Eq. (7.2.11) is an instantaneous damping ratio due to its relevance with velocity. Then

$$\lambda_{1,2} = -\frac{2\pi}{T_n} \left( \zeta_t \pm \sqrt{\zeta_t^2 - 1} \right) \tag{7.2.12}$$

where $\zeta_t = \zeta + \tilde{\zeta}_D$, which could be noted as the instantaneous total damping ratio.

The following special cases are of interest for addressing the stiff ratio:

Case 1: If $c_D \equiv 0$, i.e., there is no viscous dampers in the system, then $\tilde{\zeta}_D \equiv 0$, and the two roots in Eq. (7.2.12) are thus reduced to $\lambda_{1,2} = -\frac{2\pi}{T_n} \left( \zeta \pm \sqrt{\zeta^2 - 1} \right)$. The damping ratio of the structure itself is usually far less than 1.0. For instance, it is usually less than 5% for concrete structures and in the range of 1%–3% for steel structures. The two roots are thus conjugate complex numbers $\lambda_{1,2} = -\frac{2\pi}{T_n} \left( \zeta \pm i\sqrt{1 - \zeta^2} \right)$, where $i = \sqrt{-1}$ denotes the imaginary unit. In this case, the stiff ratio $s = |\mathrm{Re}(\lambda_1)| / |\mathrm{Re}(\lambda_2)| = 1$.

Case 2: If $\alpha = 1$, i.e., the viscous damper is reduced to a linear damper, there is $\tilde{\zeta}_D = \frac{c_D T_n}{4\pi m}$. Generally, in engineering practice the additional "equivalent" damping ratio $\tilde{\zeta}_D$ owing to the installation of damping devices might be in the range of 2%–4%, and thus the total damping ratio of the controlled structure might be in the range of

4%–7%, which is still much smaller than 1.0. In this case, the two roots in Eq. (7.2.12) are conjugate complex numbers, and the stiff ratio is 1 as well.

Case 3: As the damping coefficient $c_D$ increases and the other quantities keep fixed, the instantaneous damping ratio $\tilde{\zeta}_D$ will increase according to Eq. (7.2.11). When $c_D$ is large enough, $\zeta_t^2 - 1 \geq 0$ will occur, i.e., the system becomes an instantaneous over-damped system. In this case, the stiff ratio $s = \frac{\text{Re}(\lambda_1)}{\text{Re}(\lambda_2)} = \frac{\zeta_t + \sqrt{\zeta_t^2 - 1}}{\zeta_t - \sqrt{\zeta_t^2 - 1}} = \left(\zeta_t + \sqrt{\zeta_t^2 - 1}\right)^2$, which tends to $s \approx 4\zeta_t^2$, and the stiff ratio will increase very fast against the instantaneous total damping ratio, and in turn against the increase of $c_D$.

Case 4: As the fundamental period $T_n$ increases when the other quantities are kept fixed, the instantaneous damping ratio $\tilde{\zeta}_D$ will increase. When $T_n$ is large enough, $\zeta_t^2 - 1 \geq 0$ might occur. Similar to Case 3, the stiff ratio will be given also by $s = \left(\zeta_t + \sqrt{\zeta_t^2 - 1}\right)^2$. Thus, in this case, the stiff ratio will increase very fast against the instantaneous total damping ratio, and in turn against the increase of $T_n$. Generally, high-rise buildings have longer fundamental periods. Therefore, the stiff ratio of high-rise building structures might be large. This is just the case that the wind-induced vibration of high-rise buildings shall be suppressed by the control systems consisting of viscous dampers.

Case 5: As the velocity exponent $\alpha$ decreases and the other quantities are fixed, the instantaneous damping ratio $\tilde{\zeta}_D$ will first increase and then decrease from some "turning point". The derivative of $\tilde{\zeta}_D$ with respect to $\alpha$ is given by $\frac{\partial \tilde{\zeta}_D}{\partial \alpha} = \frac{c_D T_n}{4\pi m}|\dot{x}|^{\alpha-1}[1 + \alpha \ln(|\dot{x}|)]$, and thus the "turning point" will occur when $\frac{\partial \tilde{\zeta}_D}{\partial \alpha} = 0$, i.e., $\alpha = -\frac{1}{\ln(|\dot{x}|)}$, which is related to the velocity. Therefore, with the decrease of $\alpha$ from one to zero, the stiff ratio $s$ first equals 1 when $\zeta_t^2 - 1 \leq 0$, and then quadratically increase to its peak at the "turning point" following the same pattern as Case 3 and Case 4, afterward it decreases to 1 again in a sharp manner when $\alpha = 0$. One might recognize that the "turning point" is straightforwardly related to the velocity that the smaller the velocity is, the closer the "turning point" approaches to zero.

To reveal the influence parameters on the stiff ratio of the structural system intuitively, case studies are carried out. Since the damper force changes very fast in the case of a small velocity, the velocity is set as $\dot{x} = 0.1$ mm/s. Shown in Fig. 7.2 are the stiff ratios against different parameters of viscously damped structural systems, including the damping coefficient of viscous dampers, the fundamental period of the structural system, and the velocity exponent. It is observed that the stiff ratio quadratically increases by the order of magnitudes against the increase of damping coefficient and the fundamental period; see Fig. 7.2a. As the velocity exponent decreases from 1 down to 0, however, the stiff ratio first experiences a stage with value 1.0, then quadratic increase to the "turning point", and decreases sharply; see Fig. 7.2b. This result is in agreement with the discussions in Case 1–Case 5. One might recognize that a higher damping ratio indicates a larger damper force. Clearly, if the damping ratio is too small, the viscous damper has little effects on the response of the system, and it could be expected that the stiff ratio will also change slightly. Therefore, only

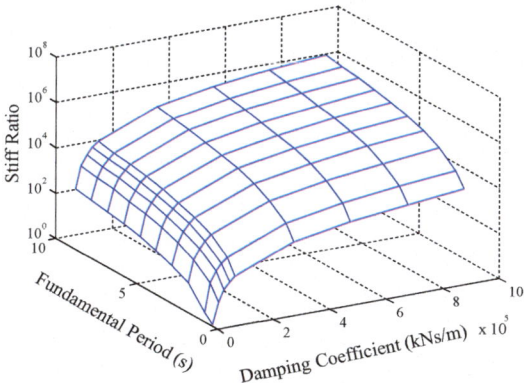

(a) stiff ratio vs. damping coefficient and fundamental period (velocity exponent $\alpha = 0.5$)

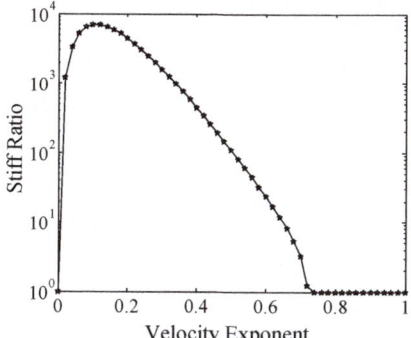

(b) stiff ratio vs. velocity exponent (damping coefficient $c_D = 10000\text{kN} \cdot (\text{s/m})^{\alpha}$, fundamental period

$$T_n = 4.94 \text{ s})$$

**Fig. 7.2**   Stiff ratio against different parameters of viscously damped systems

for relatively large damping coefficient its effect on stiff ratio becomes significant. The stiff ratio being relatively large for longer fundamental period is also physically reasonable, since a longer fundamental period indicates a lower predominant frequency in the response. A higher frequency content might be compensated by the sharp change of the damper force of viscous dampers with a moderate velocity exponent $\alpha$. This is extremely significant for high-rise buildings installed with viscous damper, owing to their fundamental periods are usually longer than 2 s even as high as nearly 10 s, and the damping component of viscous dampers is usually smaller than 0.5 down to 0.3.

Generally, a system is denoted by a stiff differential equation if its stiff ratio is greater than $10^p$ ($p \geqslant 1$). It is seen from Fig. 7.2 that high-rise buildings with the fundamental period longer than 2 s, attached with viscous dampers exhibiting the damping coefficient more than $10{,}000 \text{ kN}(\text{s/m})^{\alpha}$ and the velocity exponent 0.3–0.5, are seriously stiff since their stiff ratios are typically in the order of magnitude of $10^2$ or higher.

## 7.2.2  Solution of Viscously Damped Systems

The equation of motion of an MDOF structure attached with viscous dampers can be denoted by

$$\mathbf{M}\ddot{\mathbf{X}}(t) + \mathbf{C}\dot{\mathbf{X}}(t) + \mathbf{K}\mathbf{X}(t) + \mathbf{F}_D(\dot{\mathbf{X}}(t)) = \mathbf{F}(\mathbf{\Theta}, t) \tag{7.2.13}$$

where $\mathbf{M}$, $\mathbf{C}$, and $\mathbf{K}$ are the mass, damping, and stiffness matrices, respectively; $\mathbf{X}$, $\dot{\mathbf{X}}$, and $\ddot{\mathbf{X}}$ are the structural displacement, velocity, and acceleration, respectively; $\mathbf{F}(\mathbf{\Theta}, t)$ denotes the random wind load; $\mathbf{\Theta}$ is the random vector denoting the randomness inherent in the random wind load; $\mathbf{F}_D(\dot{\mathbf{X}}(t))$ denotes the attached damper force:

$$\mathbf{F}_D(\dot{\mathbf{X}}(t)) = \mathbf{S} \cdot \left( |\mathbf{T}_{\dot{\mathbf{X}}}|^\alpha \times \mathrm{sgn}(\mathbf{T}_{\dot{\mathbf{X}}}) \right) =
\begin{bmatrix}
c_{D,1} & -c_{D,2} & 0 & \cdots & 0 & 0 \\
0 & c_{D,2} & -c_{D,2} & \cdots & 0 & 0 \\
0 & 0 & c_{D,3} & \cdots & 0 & 0 \\
\vdots & \vdots & \vdots & \vdots & \vdots & \vdots \\
0 & 0 & 0 & \cdots & c_{D,n-1} & -c_{D,n} \\
0 & 0 & 0 & \cdots & 0 & c_{D,n}
\end{bmatrix}$$

$$\cdot \left( \left| \begin{bmatrix} 1 & 0 & 0 & \cdots & 0 & 0 \\ -1 & 1 & 0 & \cdots & 0 & 0 \\ 0 & -1 & 1 & \cdots & 0 & 0 \\ \vdots & \vdots & \vdots & \vdots & \vdots & \vdots \\ 0 & 0 & 0 & \cdots & 1 & 0 \\ 0 & 0 & 0 & \cdots & -1 & 1 \end{bmatrix} \begin{Bmatrix} \dot{X}_1 \\ \dot{X}_2 \\ \dot{X}_3 \\ \vdots \\ \dot{X}_{n-1} \\ \dot{X}_n \end{Bmatrix} \right|^\alpha \times \mathrm{sgn}\left( \begin{bmatrix} 1 & 0 & 0 & \cdots & 0 & 0 \\ -1 & 1 & 0 & \cdots & 0 & 0 \\ 0 & -1 & 1 & \cdots & 0 & 0 \\ \vdots & \vdots & \vdots & \vdots & \vdots & \vdots \\ 0 & 0 & 0 & \cdots & 1 & 0 \\ 0 & 0 & 0 & \cdots & -1 & 1 \end{bmatrix} \begin{Bmatrix} \dot{X}_1 \\ \dot{X}_2 \\ \dot{X}_3 \\ \vdots \\ \dot{X}_{n-1} \\ \dot{X}_n \end{Bmatrix} \right) \right)$$

$$\tag{7.2.14}$$

where $\mathbf{S}$ denotes the damping coefficient matrix; $c_{D,j}$ denotes the total damping coefficient of the $j$th interstory with viscous damper; $\mathbf{T}_{\dot{\mathbf{X}}}$ denotes the interstory velocity vector, and $|\mathbf{T}_{\dot{\mathbf{X}}}|^\alpha$ indicates taking the component-wise power, i.e., $|\mathbf{T}_{\dot{\mathbf{X}}}|^\alpha = \left( |T_{\dot{X},1}|^\alpha, |T_{\dot{X},2}|^\alpha, \cdots, |T_{\dot{X},n}|^\alpha \right)^{\mathrm{T}}$, where $T_{\dot{X},j}$ denotes the $j$th component of the vector $\mathbf{T}_{\dot{\mathbf{X}}}$.

It is found quite often that the equation of motion of a high-rise building installed with nonlinear viscous dampers is surprisingly difficult to solve by conventional numerical schemes, just due to its stiff problem as addressed previously. Most of the widely used time integral schemes, such as the Newmark method and the Wilson method, suffer from instability or spurious numerical spikes. In fact, as to solving ordinary differential equations, a family of backward differentiation formulae

(BDF) is demonstrated to be efficient. The extended formulation of BDF is written as (Shampine and Reichelt 1997)

$$\sum_{j=1}^{k} \frac{1}{j} \nabla^j z_{n+1} = hL(t_{n+1}, z_{n+1}) + \kappa \gamma_k \left( z_{n+1} - z_{n+1}^{(0)} \right) \tag{7.2.15}$$

where

$$\gamma_k = \sum_{j=1}^{k} \frac{1}{j}, \quad z_{n+1} - z_{n+1}^{(0)} = \nabla^{k+1} z_{n+1} \tag{7.2.16}$$

in which $\nabla^j z_n = \nabla^{j-1} z_n - \nabla^{j-1} z_{n-1}$ denotes the operator of backward differentiation, $\nabla^0 z_n = z_n$; $z_n$ denotes the system state; $k$ denotes the computational order; $h$ denotes the step-length of differentiation; $L$ denotes the system operator of initial value problems, i.e. $\dot{z} = L(t, z)$, $z(t_0) = z_0$; $\kappa$ is a scalar parameter. Equation (7.2.15) reduces to the standard BDF in the case $\kappa = 0$.

It is seen from Eq. (7.2.15) that although the backward differentiation formulae for solving the nonlinear systems are accurate, their solutions involve multiple-step schemes and are not suitable for the iterative optimization and design of viscously damped structural systems. In practice, the equivalent linearization techniques are usually employed which transfers the original nonlinear system into a linearized system as a certain equivalent criterion resulting in the responses between original nonlinear and linearized systems to be the same or in an acceptable error range. Among those equivalent linearization techniques, the widely used is the energy-dissipation equivalent linearization method.

The equivalent criterion of the energy-dissipation equivalent linearization method is that the energy dissipations of attached viscous dampers to the equivalent linearized system and to the original nonlinear system are equal. The additional equivalent damping ratio as the energy-dissipation equivalent criterion $\zeta_k^{(E-E)}$ is given by (Seleemah and Costantinou 1997):

$$\zeta_k^{(E-E)} = \frac{T_k^{2-\alpha} \sum_j c_{D,j} \lambda \left[ (u_{k,j} - u_{k,j-1}) \cos(\theta_j) \right]^{1+\alpha}}{(2\pi)^{3-\alpha} A_k^{1-\alpha} \sum_i m_i u_{k,i}^2} \tag{7.2.17}$$

where

$$\lambda = 2^{2+\alpha} \frac{\Gamma^2(1 + \alpha/2)}{\Gamma(2 + \alpha)} \tag{7.2.18}$$

where $T_k$ denotes the period of the $k$th vibrational mode; $\theta_j$ denotes the angle between the story and viscous damper in the $j$th interstory; $u_{k,j}$ denotes the modal displacement the $j$th story of the $k$th vibrational mode; $m_i$ denotes the mass of the $j$th story;

$A_k$ denotes the roof displacement amplitude of the $k$th vibrational mode in unit of modal displacement $u_{k,j}$; $\Gamma(\cdot)$ denotes the Gamma function $\Gamma(z) = \int_0^\infty t^{z-1}/e^t \mathrm{d}t$.

It is seen that the formulation of the equivalent damping ratio $\zeta_k^{(\text{E-E})}$ includes the roof displacement amplitude of the $k$th vibrational mode $A_k$, indicating that the solution of equivalent damping ratio needs a known roof displacement amplitude. However, solving the roof displacement amplitude needs to have the equivalent damping ratio in advance. Therefore, the energy-dissipation equivalent linearization method involves an iterative scheme.

It is noted that in the energy-dissipation equivalent linearization method, the structural response is assumed to be a harmonic process, which is inconsistent with the response characteristics of engineering structures subjected to dynamic excitations such as seismic ground motion and wind load. Another route to solve the nonlinear system is the stochastic equivalent linearization method, i.e., the statistical linearization technique (Roberts and Spanos 1990). In the statistical linearization technique, the structural response is assumed to be a Gaussian stationary process, and the difference between the linearized system and original nonlinear system is minimized in the sense of mean square.

By virtue of the statistical linearization technique, the equivalent damping ratio of a viscously damped structural system can be denoted by (see Appendix D)

$$\zeta_k^{(\text{S-E})} = \eta_k \rho(\alpha) \left( \frac{G_{\tilde{F}_k(t)}(\omega)}{\left(\zeta_k^{(\text{S-E})}+\zeta_k\right)\omega_k} \right)^{(\alpha-1)/2} \tag{7.2.19}$$

where $\rho(\alpha) = \Gamma(1+\alpha/2)\sqrt{2^{3-\alpha}\pi^{\alpha-2}}$; $G_{\tilde{F}_k(t)}(\omega)$ denotes the one-sided power spectral density of the generalized excitation $\tilde{F}_k(t) = \phi_k^{\mathrm{T}}\mathbf{F}(\mathbf{\Theta}, t)/\left(\phi_k^{\mathrm{T}}\mathbf{M}\phi_k\right)$ of the $k$th vibrational mode; $\eta_k = q_k/(2\bar{m}_k\omega_k)$, $q_k = \sum\limits_{j=1}^{n} \left(c_{D,j}\Delta_j^{(k)} + c_{D,j+1}\Delta_{j+1}^{(k)}\right)u_{k,j}^2 - 2\sum\limits_{j=2}^{n} c_{D,j}\Delta_j^{(k)}u_{k,j}u_{k,j-1}$, $\Delta_j^{(k)} = \left|u_{k,j} - u_{k,j-1}\right|^{\alpha-1}$; $\bar{m}_k$ denotes the modal mass of the $k$th vibrational mode; $\zeta_k$, $\omega_k$ denote the damping ratio and the circular frequency of the $k$th vibrational mode, respectively.

To verify the effectiveness and accuracy of equivalent linearization techniques and the BDF, a 20-story shear frame controlled by viscous dampers is investigated. The basic information of the structure is as follows: the structural height is 72 m with story height of 3.6 m and depth–width ratio of 2.2; the mass of each story is $m_1 = m_2 = \cdots = m_{20} = 2.0 \times 10^5$ kg, and the interstory stiffness is $k_1 = \cdots = k_4 = 1.7 \times 10^5$ kN/m, $k_5 = \cdots = k_{10} = 1.5 \times 10^5$ kN/m, $k_{11} = \cdots = k_{16} = 1.2 \times 10^5$ kN/m, $k_{17} = \cdots = k_{20} = 1.0 \times 10^5$ kN/m. The damping ratios of the first two vibrational modes are both 0.01. The structural damping is represented by Rayleigh damping matrix $\mathbf{C} = a\mathbf{M} + b\mathbf{K}$. The circular frequencies of the first ten vibrational modes are 2.09, 5.88, 9.74, 13.45, 17.27, 20.94, 24.28, 27.81, 30.90, and 33.93 rad/s, receptively. The wind-induced vibration control of the structure is carried out using the viscous

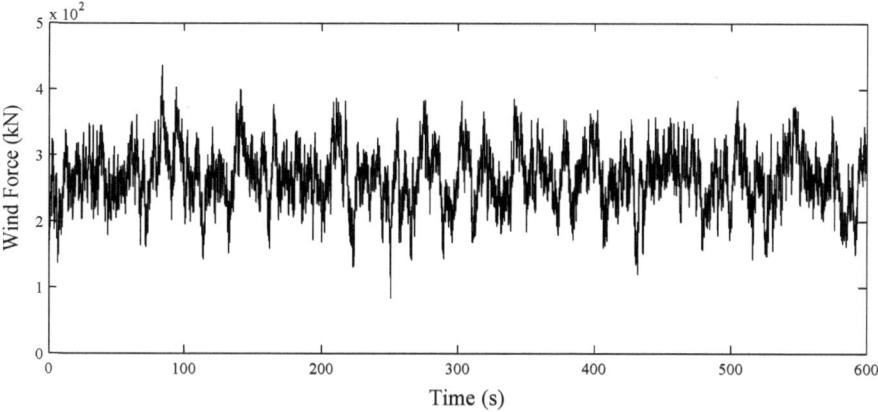

**Fig. 7.3**  Time history of representative wind force on the structural roof

dampers. For illustrative purposes, the viscous dampers are deployed uniformly in the interstories along the story level. The additional damping coefficient at each interstory is $500 \text{ kN}(\text{s}/\text{m})^\alpha$, where the velocity exponent is valued by $\alpha = 1.0, 0.5, 0.3$.

The Newmark-$\beta$ scheme on the linearized system by the energy-dissipation equivalent linearization method (EEN), the Newmark-$\beta$ scheme on the linearized system by the stochastic equivalent linearization method (SEN), and the backward differentiation formulae on the original nonlinear system (BDF) are employed to carry out the time-domain analysis. The wind excitation is represented by the spatial fluctuating wind-velocity field model addressed in Sect. 2.5.2. The three basic random parameters included in the wind velocity Fourier spectrum model are valued as: the 10-min mean wind velocity $\bar{U}_{10}$ at the standard height 10 m is assumed to follow the extreme-value type I distribution with mean 39.33 m/s and coefficient of variation 0.1; the surface roughness length $z_0$ is assumed to follow the log-normal distribution with mean 0.2 m and coefficient of variation 0.2; the zero-phase evolution time $T_e$ is assumed to follow the Gamma distribution with mean $0.902 \times 10^9$ s and coefficient of variation 0.1. A representative time history of the random wind excitation on the structural roof is shown in Fig. 7.3. The approaching flow is assumed to be perpendicular to the building surface. Under the representative wind excitation, the roof displacement and roof acceleration of the viscously damped structure with variant velocity exponents and using different numerical schemes are shown in Figs. 7.4, 7.5, 7.6.

It is seen that the structural responses by the three schemes match well with each other in the case of velocity exponent 1.0. With the decreasing of velocity exponent, the result of the energy-dissipation equivalent linearization method deviates with that of the BDF to a larger extent, in comparison with the stochastic equivalent linearization method. In the case of velocity exponent 0.3, this tendency becomes much more significant, that is, the result of the stochastic equivalent linearization method is much closer to the BDF, in comparison with the energy-dissipation equivalent

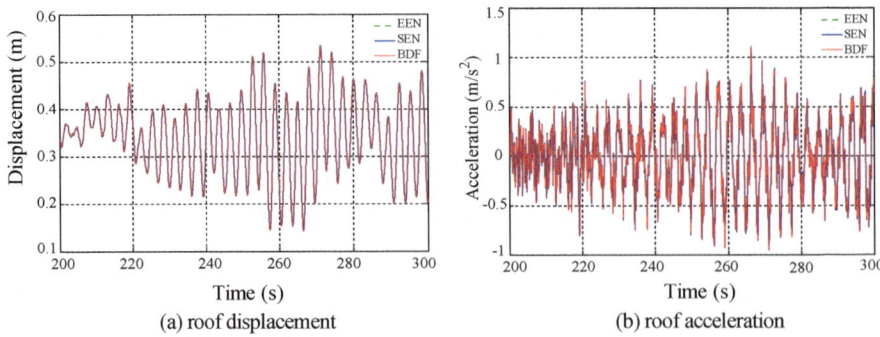

**Fig. 7.4** Comparison of roof responses using different schemes in the case of velocity exponent 1.0

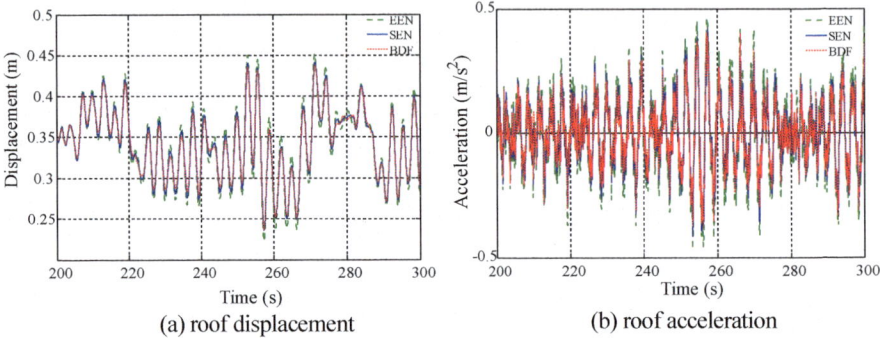

**Fig. 7.5** Comparison of roof responses using different schemes in the case of velocity exponent 0.5

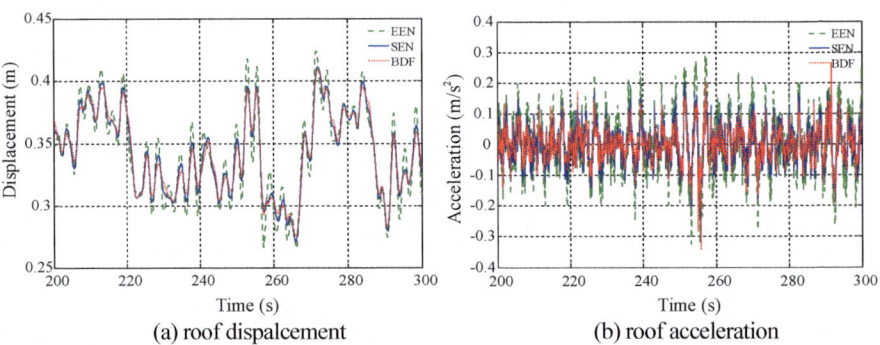

**Fig. 7.6** Comparison of roof responses using different schemes in the case of velocity exponent 0.3

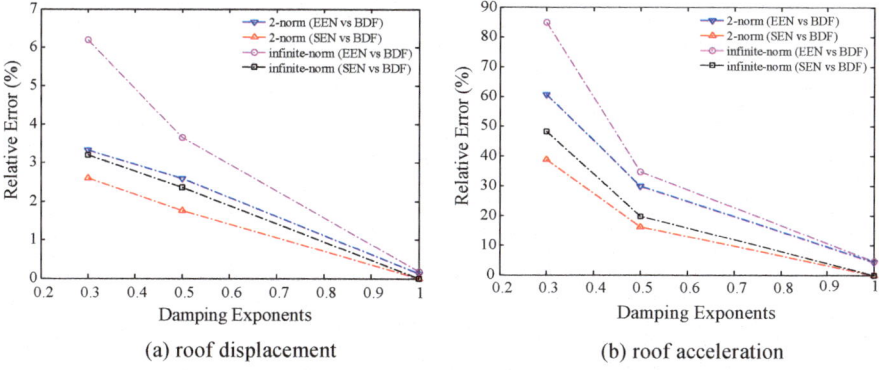

(a) roof displacement          (b) roof acceleration

**Fig. 7.7** Relative errors in 2-norm and infinite-norm of roof responses with different velocity exponents and using different schemes

linearization method. It is thus demonstrated that the stochastic equivalent linearization method has a considerable accuracy even when the nonlinearity of the viscously damped structural system is relatively strong, and the energy-dissipation equivalent linearization method is shown to be conservative.

The relative errors in 2-norm and infinite-norm of roof responses with different velocity exponents using the measure of relative error in 2-norm using the three schemes "EEN" and "SEN" compared to the "BDF" are shown in Fig. 7.7. It is well recognized that it is smaller than that in infinite-norm both in terms of the roof displacement and roof acceleration; the relative errors of acceleration are more than those of displacement by over 10 times, indicating a computational challenge inherent in the accurate solution of the structural acceleration. It is also seen that the relative error would increase rapidly with the reduction of velocity exponent. Meanwhile, the relative errors of structural responses between the energy-dissipation equivalent linearization method and the BDF are always larger than those between the stochastic equivalent linearization method and the BDF, no matter in 2-norm or in infinite-norm.

In summary, the stochastic equivalent linearization method is an elegant scheme since it reveals the stochastic essence of structural responses to some extent by invoking the Gaussian-process assumption of structural responses, and gains the equivalent modal damping ratio through minimizing the mean-square error between the linearized system and the original nonlinear system. In comparison with the energy-dissipation equivalent linearization method, the stochastic equivalent linearization method attains a more accurate solution that is more close to the result of the BDF. A solution with considerable accuracy and efficiency can thus be obtained by performing frequency-domain and time-domain analysis upon the linearized structural system, especially upon the high-rise buildings structures, where intermediate nonlinear dampers are the most common ones to be installed for performance enhancement. The stochastic equivalent linearization method is thus used to facilitate the serviceability-based optimal design of viscously damped structures as both the criteria of minimizing the standard deviation and of minimizing the exceedance probability of roof acceleration.

## 7.3  Optimal Deployment of Viscous Dampers

As mentioned in Sect. 5.3, the comfortability is one of the critical arguments representing the performance of structural systems, which is usually measured by the acceleration. As to the high-rise buildings, the roof acceleration is generally far larger than other stories. In practice, the roof acceleration control is thus an efficient means for enhancing the structural comfortability.

Since the structural responses under the random excitations are random processes, the probabilistic criteria similar to the proposal in Chap. 4 can be employed in the present investigation. The probabilistic criterion of the conventional serviceability design is usually defined as the standard deviation or the peak value of wind-induced roof acceleration of structures (Huang et al. 2011). Considering the fundamental value of reliability in the optimization and design of structural performance, two families of serviceability criteria are thus employed with the minimization of single-objective performance function as follows:

(i)   serviceability criterion 1 (SC-1): minimizing the standard deviation of roof acceleration.

$$c_{D,i}^* = \arg\min_{c_{D,i}}\{J_2\} = \arg\min_{c_{D,i}}\left\{\sigma_{\ddot{X}_n}\,\Big|\,\Big\{\sum_i c_{D,i} = C_{D,\text{total}}\Big\}\right\}, \quad i = 1, 2, \ldots, n$$

(7.3.1)

where $\sigma_{\ddot{X}_n}$ denotes the standard deviation of roof acceleration; $c_{D,i}^*$ denotes the optimal damping coefficient of the viscous dampers allocated in the $i$th interstory; $C_{D,\text{total}}$ denotes the target of total cost in terms of the sum of damping coefficient of the viscous dampers allocated in all the structural interstories. It is thus initiated that using the serviceability criterion shown in Eq. (7.3.1), the optimal distribution of damping coefficients related to the damper sizes and placements can be attained.

(ii)  serviceability criterion 2 (SC-2): minimizing the exceedance probability of roof acceleration

$$c_{D,i}^* = \arg\min_{c_{D,i}}\{J_2\}$$

$$= \arg\min_{c_{D,i}}\left\{\Pr\left\{\bigcup_{t\in[0,T]}\left(\left|\ddot{X}_n(t)\right| > \ddot{X}_{\text{thd}}\right)\right\}\,\Big|\,\Big\{\sum_i c_{D,i} = C_{D,\text{total}}\Big\}\right\},$$

$$i = 1, 2, \ldots, n$$

(7.3.2)

where $\ddot{X}_n(t)$ denotes the roof acceleration in the time interval $[0, T]$; $\ddot{X}_{\text{thd}}$ denotes threshold of the roof acceleration; $\Pr\{\cdot\}$ denotes the probability of random event. It is indicated that by virtue of the generalized optimal control policy, the serviceability criterion shown in Eq. (7.3.2) can accommodate the optimal design of damper parameters and placement so as to guarantee sufficient structural comfortability. One might recognize that this serviceability criterion is constructed according to the first-passage problem, of which the exceedance probability of roof acceleration can be readily solved via the generalized probability density evolution equations and the equivalent extreme-value event criterion.

The stochastic equivalent linearization method has been proved to exhibit high efficiency and accuracy for solving nonlinear multi-degree-of-freedom systems, which is thus first applied to carry out the linearization of nonlinear structural systems attached with viscous dampers. As to the linearized system, the modal superposition method in frequency-domain and the probability density evolution method with time-domain analysis by Newmark-β schemes are employed to solve the performance function $J_2$ in the serviceability criteria shown in Eqs. (7.3.1) and (7.3.2), respectively.

By virtue of the statistical linearization technique shown in Eq. (7.2.19), the additional equivalent damping ratio of viscously damped structures can be gained, and the nonlinear structural system is readily transferred into a linearized structural system. In modal space, the linearized system can be decomposed to a series of single-degree-of-freedom systems with independent equations of motion as follows:

$$\ddot{u}_j(t) + 2\zeta_j^{(e)}\omega_j\dot{u}_j(t) + \omega_j^2 u_j(t) = \tilde{F}_j(t) \tag{7.3.3}$$

where $\tilde{F}_j(t) = \phi_j^{\mathrm{T}}\mathbf{F}(\Theta, t)\big/\bar{m}_j$ denotes the generalized wind load of the $j$th vibrational mode; $\omega_j = \sqrt{\bar{k}_j/\bar{m}_j}$ denotes the circular frequency of the $j$th vibrational mode; $\zeta_j^{(e)} = \zeta_j^{(\text{S-E})} + \zeta_j$ denotes the sum of the inherent damping ratio and the equivalent damping ratio of the $j$th vibrational mode; $\bar{m}_j$, $\bar{k}_j$ denote the generalized mass and generalized stiffness of the $j$th vibrational mode, respectively; $\phi_j^{\mathrm{T}}$ denotes the $j$th modal vector.

According to the modal superposition method, the cross-power spectral density of generalized wind load between modes $i$ and $j$ is represented by

$$S_{\tilde{F}_j\tilde{F}_k}(\omega) = \frac{1}{\bar{m}_j\bar{m}_k}\phi_j^{\mathrm{T}}\mathbf{S}_{\mathbf{F}}(\omega)\phi_k \tag{7.3.4}$$

where $\mathbf{S}_{\mathbf{F}(t)}$ denotes the power spectral density matrix of wind load. The power spectral density of generalized structural responses $u_j(t)$ can be obtained by integrating the frequency response transfer function of systems and the power spectral density of generalized wind load:

$$S_{U_j}(\omega) = |H_j(\omega)|^2 S_{\tilde{F}_j}(\omega) \tag{7.3.5}$$

where $H_j(\omega)$ denotes the frequency response transfer function:

$$|H_j(\omega)|^2 = \frac{1}{\left(\omega_j^4 - 2\omega_j^2\omega^2 + \omega^4\right) + 4\left[\zeta_j^{(e)}\right]^2\omega_j^2\omega^2} \tag{7.3.6}$$

The power spectral density of the $i$th story displacement then can be written as

$$S_{X_j}(\omega) = \sum_{k=1}^{n} \phi_{jk}^2 S_{U_j}(\omega) = \sum_{k=1}^{n} \phi_{jk}^2 |H_j(\omega)|^2 S_{\tilde{F}_j}(\omega) \tag{7.3.7}$$

where $\phi_{jk}$ denotes the $k$th component of the $j$th modal vector of structural systems.

According to the relation between the power spectral densities of the structural responses and their differentiated arguments, the power spectral densities of story acceleration and those of story displacement have the relation function as follows:

$$S_{\ddot{X}_j}(\omega) = \omega^4 S_{X_j}(\omega) \tag{7.3.8}$$

The mean-square roof acceleration of the structure is then given by

$$\sigma_{\ddot{X}_j}^2 = \int_{-\infty}^{\infty} \omega^4 S_{X_j}(\omega)\mathrm{d}\omega = \sum_{k=1}^{n} \phi_{jk}^2 \int_{-\infty}^{\infty} \omega^4 |H_j(\omega)|^2 S_{\tilde{F}_j}(\omega)\mathrm{d}\omega \tag{7.3.9}$$

According to the equivalent extreme-value event criterion, the extreme value of roof acceleration $\ddot{X}_n$ in the time interval $[0, T]$ is defined by

$$W(\mathbf{\Theta}, T) = \max_{t\in[0,T]} \left(|\ddot{X}_n(\mathbf{\Theta}, t)|\right) \tag{7.3.10}$$

Introducing a pseudo random process, there is

$$Z(\tau) = \varphi(W(\mathbf{\Theta}, T), \tau) \tag{7.3.11}$$

$$Z(\tau)|_{\tau=\tau_0} = 0, \ Z(\tau)|_{\tau=\tau_c} = W(\mathbf{\Theta}, T) \tag{7.3.12}$$

According to the probability preservation principle, the joint probability density function $p_{Z\Theta}(z, \boldsymbol{\theta}, \tau)$ of $(Z(\tau), \mathbf{\Theta})$ satisfies the generalized probability density evolution equation as follows (Li and Chen 2009):

$$\frac{\partial p_{Z\Theta}(z, \boldsymbol{\theta}, \tau)}{\partial \tau} + \dot{Z}(\tau)\frac{\partial p_{Z\Theta}(z, \boldsymbol{\theta}, \tau)}{\partial z} = 0 \tag{7.3.13}$$

where $\tau$ denotes the generalized time. The associated initial condition is given by

$$p_{Z\Theta}(z, \boldsymbol{\theta}, \tau_0) = \delta(z - z_0)p_{\Theta}(\boldsymbol{\theta}) \qquad (7.3.14)$$

The joint probability density function $p_{Z\Theta}(z, \boldsymbol{\theta}, \tau)$ can be then obtained by solving Eq. (7.3.13), in view of the numerical procedure solving the generalized probability density evolution equation addressed in Sect. 2.3.3. The probability density function of extreme value of the roof acceleration $Z(\tau_c)$ is then obtained:

$$p_Z(z, \tau_c) = \int_{\Omega_\Theta} p_{Z\Theta}(z, \boldsymbol{\theta}, \tau_c)d\boldsymbol{\theta} \qquad (7.3.15)$$

where $\Omega_\Theta$ is the distribution space of $\Theta$.

The dynamic reliability of structures is then given by

$$R(T) = \Pr\{W(\Theta, T) \in \Omega_s\} = \int_0^{\ddot{X}_{\text{thd}}} P_Z(z, \tau_c)dz \qquad (7.3.16)$$

where $P_Z(z, \tau_c)$ denotes probability density of the virtual random process $Z(\tau)$ at the instant of time $\tau = \tau_c$.

The failure probability of roof acceleration $\ddot{X}_n$ in the time interval $[0, T]$ is then given by

$$\Pr\left\{ \bigcup_{t \in [0,T]} \left( \left| \ddot{X}_n(t) \right| > \ddot{X}_{\text{thd}} \right) \right\} = 1 - \int_0^{\ddot{X}_{\text{thd}}} P_Z(z, \tau_c)dz \qquad (7.3.17)$$

As mentioned in Sect. 7.2, the optimization methods for the damper deployment include the case-sequential scheme, the minimum gradient scheme, and the genetic algorithm. The former two schemes are both explicit strategies aiming at approaching the performance objective, which provide feasibility for the decision maker who is able to readily define the optimal parameters and placements of the viscous dampers in steps according to the structural performance. In comparison with the case-sequential scheme, the minimum gradient method exhibits the capacity with more expeditious convergence (Peng et al. 2013). However, the genetic algorithm is an implicit strategy aiming at minimizing the objective function. Although a repeat optimization might be incurred once the structural performance objective changes and the constraint on the optimization needs to be redefined, the genetic algorithm still has been widely used in the optimization and design of viscous dampers due to its good adaptability and excellent global optimization capability (Silvestri and Trombetti 2007). As to the issue of wind-induced comfortability control of high-rise buildings, an updated scheme for gaining a higher convergence velocity is developed by integrating the genetic algorithm and the minimum gradient criterion where the standard deviation and exceedance probability of the roof acceleration at the searching points and their change rates, i.e., gradient, are included.

The genetic algorithm is an iterative procedure following the rule of fittest to survive. In each generation of the population, the individuals are first selected according to the evaluation results of the individual fitness values in the addressed problem. The individual with larger fitness value exhibits a larger possibility of being selected. Then the crossover and mutation similar to the genetic operator in the genetics are carried out to yield the next generation of the population. This process results in that the next generation has a stronger fitness in the problem domain. The individuals in the last generation can be viewed as the optimal solution to the problem. The genetic algorithm involves a three-step procedure, i.e., selection, crossover, and mutation. As to the problem of viscous damper deployment with respect to the minimization of exceedance probability of roof acceleration, the fitness evaluations of individuals in each generation of population all involves solving the generalized probability density evolution equations, which incurs an unacceptable computational cost. In order to reduce the calculation efforts, the neural network algorithm is utilized.

The neural network algorithm aims at building the nonlinear mapping model exhibiting memory and prediction capacities through the training and predicting on the provided data (Rojas 1996). Utilizing the nonlinear mapping model, the computational cost of the objective function is saved, and the computational efficiency can be enhanced significantly.

The support vector machine (SVM) is employed serving as the tool for modeling of the neural network. The SVM has a distinguished ability for efficient prediction of the objective function of individuals, which can significantly reduce the computational cost (Haykin 2007). The flowchart of the SVM-based genetic algorithm is shown in Fig. 7.8. It is seen that the individual fitness of genetic algorithm relies upon the SVM, which thus plays a critical role in enhancing the accuracy and efficiency of the optimization procedure.

## 7.4 Case Studies

As a practical application of the optimal design of viscous dampers in the wind-induced comfortability control of high-rise buildings, a 58-story steel structure subjected to random wind excitations is studied. The height of the structure is 249 m, and the building area is 1.25 km$^2$. According to the Chinese Code for Design Loads of Building Structures (GB50009-2012), the structural basic wind pressure is 0.75 kN/m$^2$, the occupant-comfortability validation wind pressure is 0.45 kN/m$^2$, and the ground surface roughness belongs to type A.

Using the software PKPM to carry out the finite-element modeling and the analysis, the structural wind-induced responses can be readily derived from the formulae shown in the Chinese Code for Steel Structure of High-Rise Buildings (JGJ99-1998) and in the Chinese Code for Design Loads of Building Structures (GB50009-2012): the maximum crosswind roof accelerations along $Y$ direction are 0.426 m/s$^2$, 0.381 m/s$^2$, respectively; the maximum crosswind roof accelerations along $X$ direction are 0.399 m/s$^2$, 0.308 m/s$^2$, respectively. However, the threshold of structural

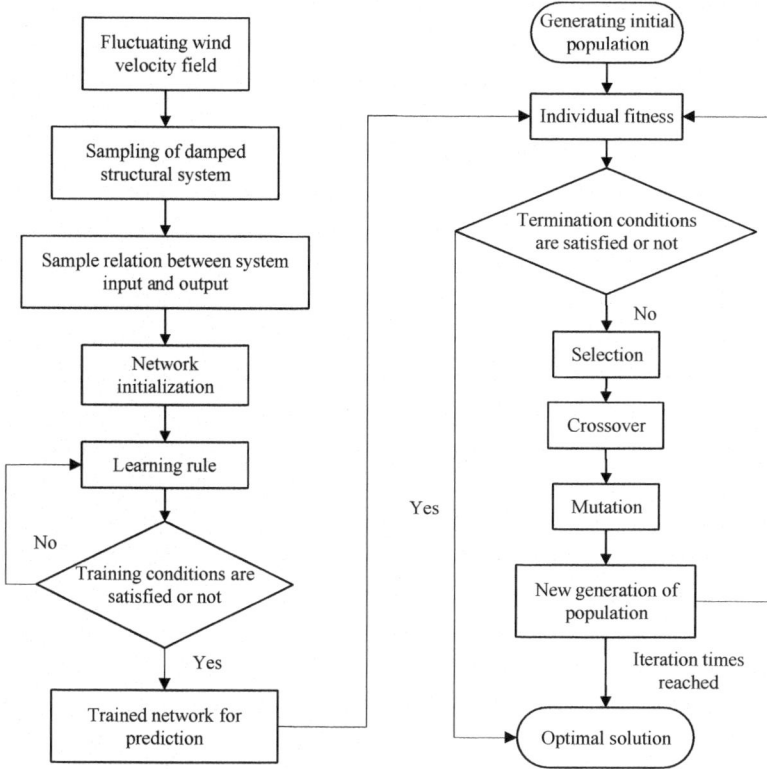

**Fig. 7.8** Flowchart of SVM-based genetic algorithm

roof acceleration are defined as 0.28 m/s$^2$ for public buildings, 0.20 m/s$^2$ for apartment buildings, respectively, in terms of the Chinese code JGJ99-1998. It is seen that the structural wind-induced roof acceleration surpasses the threshold around 50%. Therefore, considering a successful structural control strategy such that the wind-induced comfortability satisfies with the provisions is a critical task of structural design.

## 7.4.1 Dimension-Reduced Model of High-Rise Building

The software SAP2000 is employed to perform accurate finite-element modeling as shown in Fig. 7.9. In view of the mass matrix and stiffness matrix derived from the finite-element model, the structural parameters such as the interstory stiffness

**Fig. 7.9** SAP finite-element
model of a high-rise building

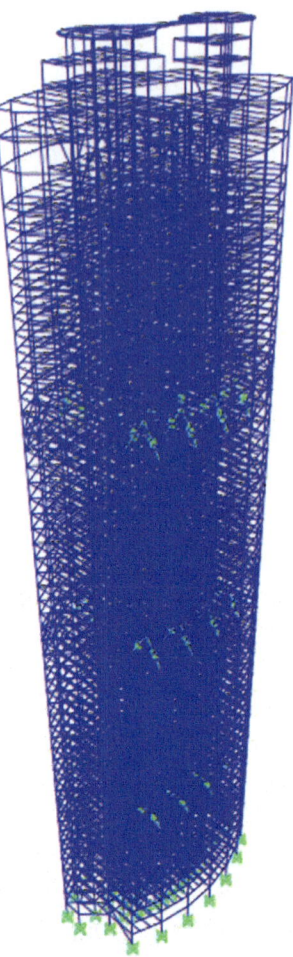

and story mass of the two-dimensional mass-lumped system can be attained; see
Table 7.1. The damping ratio of the first two vibrational modes of the mass-lumped
system along $X$ and $Y$ directions are both 0.01. The Rayleigh damping hypothesis, i.e.,
$\mathbf{C} = a\mathbf{M} + b\mathbf{K}$, is employed in this study. Table 7.2 shows the fundamental periods
and the roof acceleration of the finite-element model and of the mass-lumped system.
In the comparative study, the basic wind pressure 0.75 kN/m$^2$ is used.

It is seen that differences of fundamental periods and roof displacements between
the simplified mass-lumped system and SAP finite-element model are both in a
range of acceptable error. This simplified model underlies the feasibility of stochastic
analysis and optimal control of the high-rise building structure.

**Table 7.1**  Interstory stiffness and story mass of mass-lumped system

| Story level | Interstory stiffness along $X$ ($10^3$ kN/m) | Interstory stiffness along $Y$ ($10^3$ kN/m) | Story mass ($10^3$ kg) |
|---|---|---|---|
| 1 | 9400 | 22,300 | 1954.6 |
| 2 | 5450 | 10,700 | 2478.5 |
| 3 | 4850 | 8530 | 2443.3 |
| 4 | 4590 | 7210 | 2475.0 |
| 5 | 4560 | 6800 | 2493.6 |
| 6 | 6160 | 8780 | 2210.8 |
| 7 | 6090 | 8150 | 2191.9 |
| 8 | 6030 | 7400 | 2192.0 |
| 9 | 6050 | 6910 | 2199.4 |
| 10 | 6960 | 7730 | 3550.1 |
| 11 | 9420 | 9120 | 3052.3 |
| 12 | 6450 | 6680 | 2209.4 |
| 13 | 5000 | 5510 | 2142.7 |
| 14 | 4620 | 5040 | 2143.3 |
| 15 | 4450 | 4950 | 2131.2 |
| 16 | 4330 | 4980 | 2176.1 |
| 17 | 4240 | 4660 | 2143.5 |
| 18 | 4160 | 4410 | 2144.3 |
| 19 | 4110 | 4180 | 2144.4 |
| 20 | 4040 | 4070 | 2151.5 |
| 21 | 4020 | 4000 | 2123.4 |
| 22 | 4020 | 3860 | 2124.0 |
| 23 | 4090 | 3700 | 2124.1 |
| 24 | 4680 | 3780 | 3469.2 |
| 25 | 6550 | 3870 | 2842.9 |
| 26 | 7560 | 4420 | 2949.6 |
| 27 | 4340 | 3540 | 2109.0 |
| 28 | 3400 | 3120 | 2037.5 |
| 29 | 3150 | 2930 | 2037.5 |
| 30 | 3030 | 2820 | 2038.5 |
| 31 | 2960 | 2730 | 2041.2 |
| 32 | 2910 | 2670 | 2040.9 |
| 33 | 2870 | 2640 | 2073.4 |
| 34 | 2840 | 2590 | 2040.8 |
| 35 | 2800 | 2500 | 2017.9 |
| 36 | 2800 | 2410 | 2018.3 |

(continued)

**Table 7.1** (continued)

| Story level | Interstory stiffness along $X$ ($10^3$ kN/m) | Interstory stiffness along $Y$ ($10^3$ kN/m) | Story mass ($10^3$ kg) |
|---|---|---|---|
| 37 | 2840 | 2300 | 1997.1 |
| 38 | 3180 | 2340 | 2591.8 |
| 39 | 5080 | 2710 | 3429.3 |
| 40 | 4000 | 2140 | 3014.0 |
| 41 | 3550 | 2310 | 1905.1 |
| 42 | 2720 | 2030 | 1938.8 |
| 43 | 2480 | 1920 | 1938.9 |
| 44 | 2350 | 1810 | 1939.0 |
| 45 | 2260 | 1750 | 1954.8 |
| 46 | 2170 | 1690 | 1952.1 |
| 47 | 2090 | 1620 | 1939.0 |
| 48 | 2010 | 1550 | 1938.9 |
| 49 | 1900 | 1460 | 1930.5 |
| 50 | 1680 | 1290 | 1951.8 |
| 51 | 1560 | 1180 | 1934.9 |
| 52 | 1410 | 1070 | 1935.6 |
| 53 | 599 | 514 | 1731.3 |
| 54 | 742 | 565 | 2790.3 |
| 55 | 559 | 398 | 1558.8 |
| 56 | 459 | 301 | 311.4 |
| 57 | 425 | 243 | 325.0 |
| 58 | 330 | 177 | 634.5 |

**Table 7.2** Fundamental periods and roof displacements of SAP finite-element model and mass-lumped system

| Models | Fundamental period along $X$ direction | Fundamental period along $Y$ direction | Alongwind roof displacement along $X$ direction (m) | Alongwind roof displacement along $Y$ direction (m) |
|---|---|---|---|---|
| SAP model | 4.99 | 5.26 | 0.336 | 0.480 |
| Mass-lumped | 4.91 | 4.94 | 0.255 | 0.506 |
| Error (%)[a] | 1.6 | 6.1 | 5.7 | 5.4 |

[a]Error = abs (Mass-lumped system − SAP model)/SAP model

## 7.4.2 Dynamics Analysis of Model

Using the formulae shown in the Chinese code JGJ99-1998 and in the Chinese code GB50009-2012, the validation of roof acceleration of mass-lumped system subjected to wind load with the occupant-comfortability validation wind pressure 0.45 kN/m$^2$ is carried out. It is seen from Table 7.3 that using the Newmark-$\beta$ integral scheme, the dynamic analysis of mass-lumped system gains similar results with the formulae in provisions, and the roof acceleration along $Y$ direction is always larger than that along $X$ direction no matter subjected to alongwind or subjected to crosswind loads. In this study, the alongwind load is simulated by the spatial fluctuating wind velocity model addressed in Sect. 2.5.2. The three basic random parameters included in the wind velocity Fourier spectrum model are valued as: the 10-min mean wind velocity $\bar{U}_{10}$ at the standard height 10 m is assumed to follow the extreme-value type I distribution with mean 26.83 m/s and coefficient of variation 0.1; the surface roughness length $z_0$ is assumed to follow the log-normal distribution with mean 0.16 m and coefficient of variation 0.2; the zero-phase evolution time $T_e$ is assumed to follow the Gamma distribution with mean $0.902 \times 10^9$ s and coefficient of variation 0.1. The crosswind load is simulated by the spectral representation method in conjunction with the experimental crosswind force spectrum and coherence function (Liang et al. 2002). Time histories of representative roof alongwind and crosswind forces of high-rise building are shown in Fig. 7.10. In fact, the building widths along $X$ and $Y$ directions are 37.26 m and 63.34 m, and the aspect ratios are 6.68 and 3.99, respectively. It is indicated that the crosswind effects are of the main concern since the aspect ratios in the two main directions are larger than 3.0 (Liang et al. 2002).

The one-dimensional mass-lumped system along $Y$ direction is investigated, of which the wind-induced vibration and comfortability control are carried out. The circular frequencies of the first ten vibrational modes are denoted by 1.27, 3.15, 4.86, 6.78, 8.20, 10.02, 11.37, 13.23, 14.84, and 17.27 rad/s, respectively.

**Table 7.3** Validation of roof acceleration of mass-lumped system subjected to wind load

| Cases | Mass-lumped (m/s$^2$) | PKPM (JGJ99-1998) (m/s$^2$) | PKPM (GB50009-2002) (m/s$^2$) |
|---|---|---|---|
| Alongwind along $X$ | 0.154 | 0.080 | 0.115 |
| Alongwind along $Y$ | 0.255 | 0.132 | 0.183 |
| Crosswind along $X$ | 0.361 | 0.399 | 0.308 |
| Crosswind along $Y$ | 0.413 | 0.426 | 0.381 |

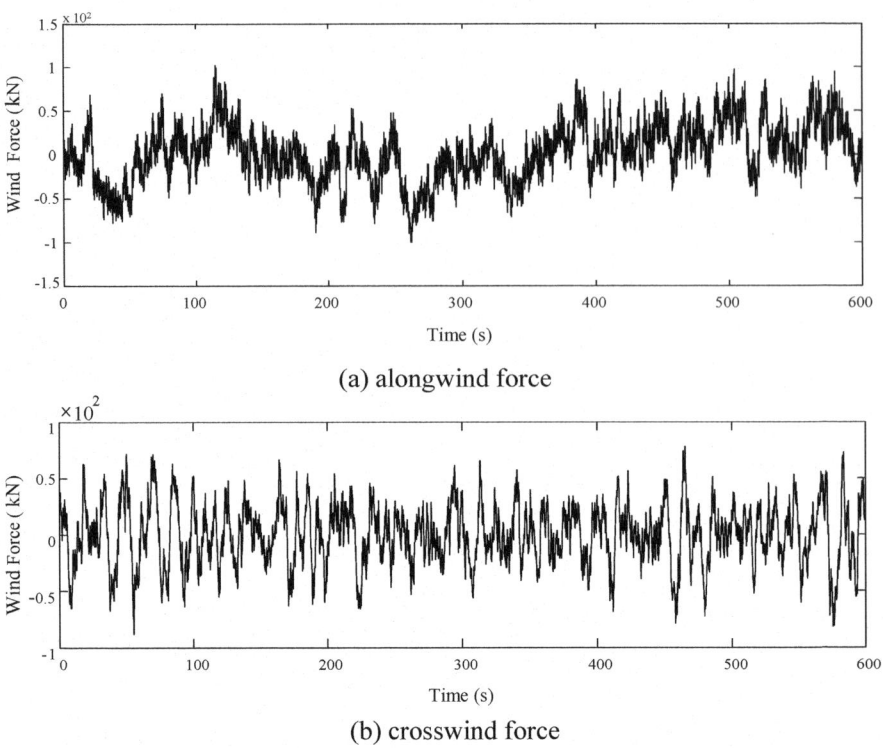

(a) alongwind force

(b) crosswind force

**Fig. 7.10**   Time histories of representative roof wind forces of a high-rise building

### 7.4.3   Wind-Induced Comfortability Control

The desired total damping coefficient is first defined through evaluating the structural system with uniformly deployed viscous dampers along story level. It is assumed that the velocity exponents of all the viscous dampers are the same and set as $\alpha = 0.5$. The total damping coefficient is tentatively set as $C_{D,\text{total}} = 8 \times 10^4 \text{ kN(s/m)}^{0.5}$. In this case, the mean of additional damping ratios of the first three vibrational modes is 0.77%, and a comparative result of the roof accelerations with and without viscous dampers is shown in Fig. 7.11. It is seen that with the viscous damper control, the roof acceleration of structure subjected to a representative wind force decreases significantly, of which the maximum acceleration is around 0.21 m/s$^2$ and reduced to an acceptable range defined by the provisions. Therefore, the total damping coefficient used for the proceeding optimization of parameters and deployments of viscous dampers is set as $C_{D,\text{total}} = 8 \times 10^4 \text{kN(s/m)}^{0.5}$.

The parameter and placement optimizations of viscous dampers as both the serviceability criteria SC-1 and SC-2 are carried out. The optimization as the criterion SC-1 employs the genetic algorithm; while the optimization as the criterion SC-

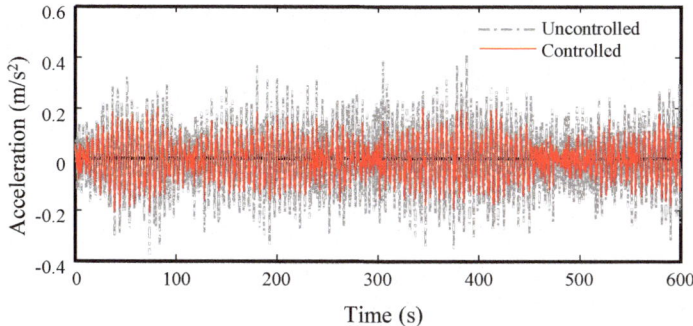

**Fig. 7.11** Roof acceleration of structure with and without viscous dampers subjected to representative crosswind force

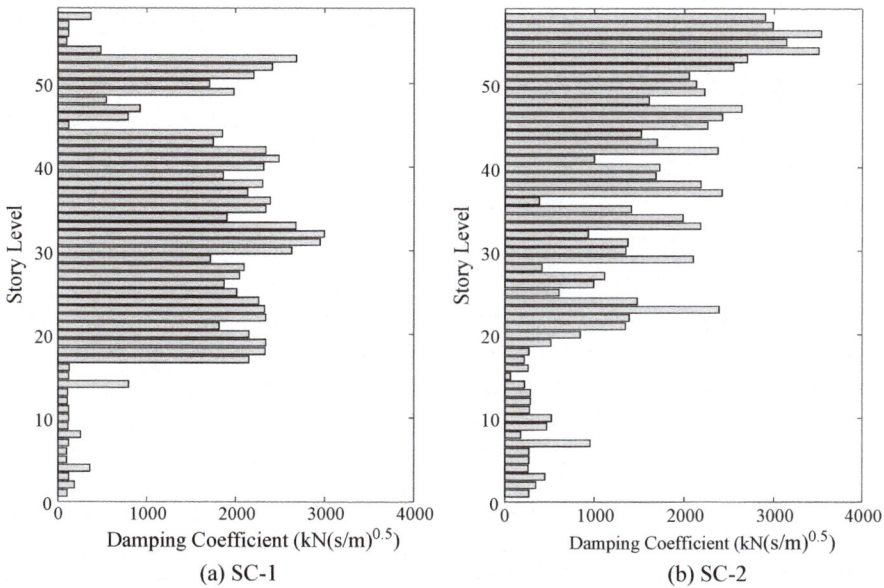

(a) SC-1                                       (b) SC-2

**Fig. 7.12** Schematic diagrams of viscous damper deployments as serviceability criteria SC-1 and SC-2

2 employs the SVM-based genetic algorithm. Both optimizations involve a same parameter group of the genetic algorithm: the size of initial population is 1024, the size of other populations is 200, the number of genetic generations is 300, number of variable dimensions is 58, and the parameters k1, k2, k3, k4 for the adaptive crossover and mutation are 0.5, 0.3, 0.7, 0.5, respectively.

With the optimization of the genetic algorithm, the schematic diagrams of viscous damper deployments as the serviceability criteria SC-1 and SC-2 are shown in Fig. 7.12. It is seen that if the traditional criteria on the optimization of mean-

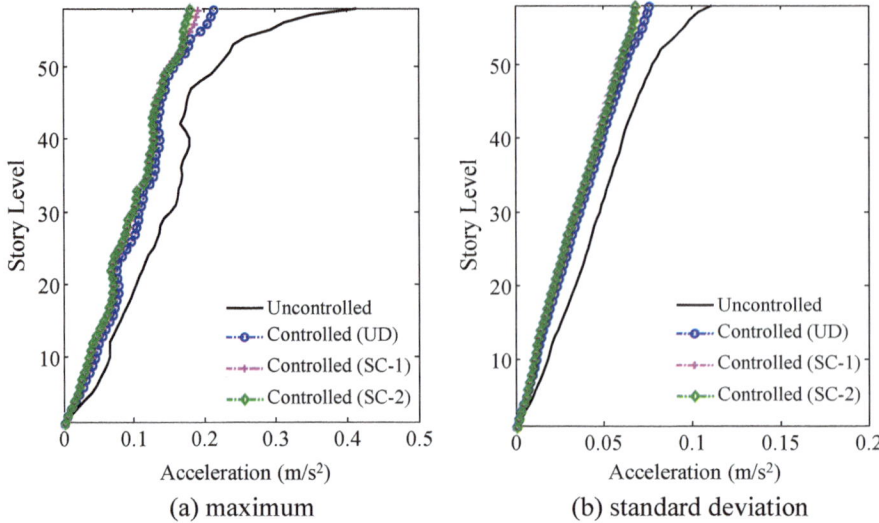

**Fig. 7.13** Maximum and standard deviation of story acceleration along story level subjected to representative wind force

square responses such as SC-1 is employed, the middle-level stories; see the stories 17–44 and stories 49–53, need more number of viscous dampers. However, if the exceedance probability based serviceability criteria SC-2 is employed, the high-level stories need more number of viscous dampers, especially at the interstories with a large requirements of viscous dampers such as the stories 50–58.

In order to analyze the control effectiveness of the viscous dampers deployed in the structural system, a comparative study between the wind-induced vibration control by virtue of the optimally deployed viscous dampers as criterion SC-2 and that of the optimally deployed viscous dampers as criterion SC-1 is carried out. Figure 7.13 shows the maximum and standard deviation of story accelerations with and without viscous damper deployments when subjected to a representative wind force. In the figure, the case of viscous dampers uniformly deployed along story level is labeled as Controlled (UD); the cases of optimally deployed viscous dampers as criteria SC-1 and SC-2 are labeled as Controlled (SC-1) and Controlled (SC-2), respectively. All the case calculations resort to the stochastic equivalent linearization method and Newmark-$\beta$ integral scheme. It is seen that the story accelerations without control are far larger than those with control, which proves the effectiveness of the viscous dampers in the wind-induced comfortability control. One might recognize that the story accelerations of the controlled structure in design as criteria SC-1 and SC-2 are both less than that in the design of uniformly deployed viscous dampers, indicating that the viscous damper deployment as serviceability criteria can gain better control effectiveness, and optimization of viscous dampers exhibits a good trade-off.

**Table 7.4**  Dynamic reliabilities of roof acceleration of structure subjected to different thresholds

| Cases | Thresholds | | | | | | | | |
|---|---|---|---|---|---|---|---|---|---|
| | 0.2 | 0.21 | 0.22 | 0.23 | 0.24 | 0.25 | 0.26 | 0.27 | 0.28 |
| UD | 0.0026 | 0.0177 | 0.0315 | 0.0476 | 0.0814 | 0.1654 | 0.6099 | 0.6855 | 0.7600 |
| SC-1 | 0.0057 | 0.0158 | 0.0328 | 0.0490 | 0.0897 | 0.4321 | 0.6497 | 0.7329 | 0.7918 |
| SC-2 | 0.0779 | 0.1420 | 0.1999 | 0.5127 | 0.7595 | 0.8269 | 0.8704 | 0.9015 | 0.9326 |

Units of threshold are $(m/s^2)$

In fact, the means of the additional former three-order damping ratios of viscously damped structures as the criteria SC-1 and SC-2 are 0.79% and 1.32%, respectively, which are both larger than that of the case with uniformly deployed viscous dampers; i.e., 0.77%. Moreover, the maximum roof accelerations in the cases of optimally deployed viscous dampers as serviceability criteria SC-1 and SC-2 are reduced to 10.6% and 15.7%, respectively, against the case with uniformly deployed viscous dampers. The standard deviations of roof acceleration in the former two cases are reduced to 8.8% and 10.1%, respectively, against the later case. It is well understood that the serviceability criterion using the exceedance probability as the objective argument accommodates better control effectiveness.

The PDF and CDF of equivalent extreme-value roof acceleration can be obtained using the probability density evolution method, as shown in Figs. 7.14 and 7.15. It is seen that the PDF and CDF of uncontrolled roof acceleration incline to the right, and those of controlled roof acceleration as the criteria SC-1 and SC-2 incline to the left to a large extent, by comparison with the case with uniformly deployed viscous dampers. It is also shown that the exceedance probability based serviceability criterion secures better structural habitability. Besides, as shown in Fig. 7.14, the 95% quantile in the case of optimally deployed viscous dampers as criterion SC-2 is much less than those in other two cases. In order to quantitatively assess the differences from the cases, the dynamic reliabilities of roof acceleration subjected to different thresholds are shown in Table 7.4.

It is seen from Figs. 7.14 and 7.15 that the probability density of uncontrolled roof acceleration has a significant difference from those of controlled roof acceleration, which proves again the effectiveness of viscous damper control. In the cases of viscous damper control, the extreme value of roof acceleration as the criterion SC-2 arises to be minimum, then as the criterion SC-1, and the non-optimized case ND has the largest roof acceleration. A straightforward comparison between the case without control and the cases with control can be seen from the 95% quantile. Table 7.4 further shows that in the condition of the same total damping coefficient, the optimal deployment of viscous dampers as the exceedance probability based serviceability criterion can attain the best wind-induced comfortability.

**Fig. 7.14** PDFs of equivalent extreme-value roof acceleration and 95% quantiles

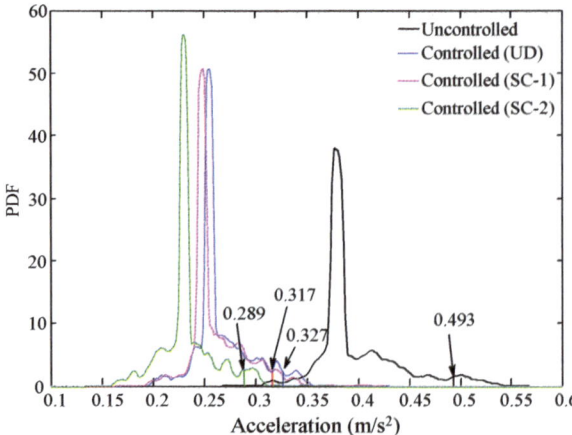

**Fig. 7.15** CDFs of equivalent extreme-value roof acceleration

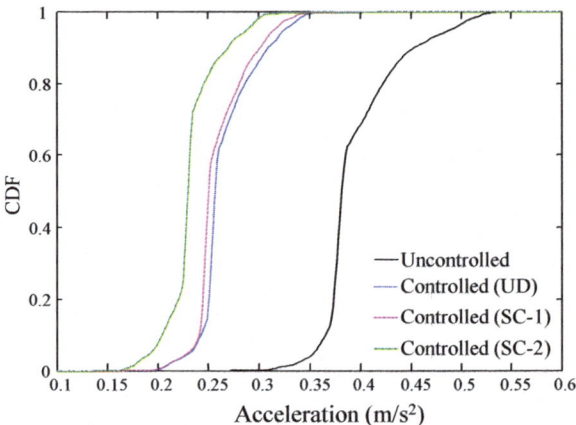

## 7.5 Discussions and Summaries

The present chapter addresses the stiff ratio of nonlinear structural systems with viscous dampers, in the context of practical challenges on the wind-induced comfortability control of high-rise buildings. The classical equivalent linearization methods including the energy-dissipation equivalent linearization method and the stochastic equivalent linearization technique are investigated. Two families of probabilistic criteria for the optimal design of viscous dampers deployed in the structural system are provided. For validating purposes, the reliability based stochastic optimal control of wind-induced comfortability of a high-rise building in practice is detailed. Some concluding remarks are drawn as follows:

(1) The damper force of nonlinear viscous dampers arises a fast-changing and slow-changing behavior along with the variation of piston velocity, which results in

a stiff problem inherent in the viscously damped structural system. This stiff problem becomes more significant with the increasing of damping coefficient and fundamental period of structures.

(2) Due to the essential nonlinearities inherent in viscously damped structures, the conventional energy-dissipation equivalent linearization method fails to derive an acceptable linearized system. The stochastic equivalent linearization technique is verified to have sufficient accuracy and efficiency in the case of the wind-induced vibration mitigation, which enables the modal superposition method to be used in the highly efficient optimization of nonlinear viscous dampers allocated in high-rise buildings.

(3) The damper allocations as the serviceability criteria of minimum standard deviation and of minimum exceedance probability of roof acceleration have the benefit to reduce the wind-induced vibration significantly, which gain a similar control effectiveness in the case of a same total damping coefficient, and both exhibit a better trade-off than the non-optimized case with uniformly deployed viscous dampers. However, the traditional optimization of viscous damper deployments based on the serviceability criterion of minimum standard deviation of roof acceleration is a deterministic scheme in essence, which has limitation of enhancing the wind-induced comfortability of high-rise building, by comparison with the optimization of viscous damper deployments based on the serviceability criterion of minimum exceedance probability of roof acceleration.

# References

Chan CM, Chui JKL (2006) Wind-induced response and serviceability design optimization of tall steel buildings. Eng Struct 28(4):503–513

Chen JB, Zeng XS, Peng YB (2017) Probabilistic analysis of wind-induced vibration mitigation of structures by fluid viscous dampers. J Sound Vib 409:287–305

Chung LL, Wu LY, Yang CSW, Lien KH, Lin MC, Huang HH (2013) Optimal design formulas for viscous tuned mass dampers in wind-excited structures. Struct Control Health Monit 20(3):320–336

Davenport AG (1961) The spectrum of horizontal gustiness near the ground in high winds. Q J R Meteorol Soc 87:194–211

Haykin S (2007) Neural networks: a comprehensive foundation, 3rd edn. Prentice-Hall, Inc

Housner GW, Bergman LA, Caughey TK, Chassiakos AG, Claus RO, Masri SF, Skelton RE, Soong TT, Spencer BF Jr, Yao James TP (1997) Structural control: past, present, and future. ASCE J Eng Mech 123(9):897–971

Huang MF, Chan CM, Kwok KCS (2011) Occupant comfort evaluation and wind-induced serviceability design optimization of tall buildings. Wind Struct 14(6):559–582

Li J, Chen JB (2009) Stochastic dynamics of structures. Wiley, Singapore

Liang SG, Liu S, Li QS, Ming G (2002) Mathematical model of acrosswind dynamic loads on rectangular tall buildings. J Wind Eng Ind Aerod 90(12):1757–1770

Patil VB, Jangid RS (2011) Response of wind-excited benchmark building installed with dampers. Struct Des Tall Spec 20(4):497–514

Peng YB, Ghanem R, Li J (2013) Generalized optimal control policy for stochastic optimal control of structures. Struct Control Health Monit 20(2):187–209

Roberts JB, Spanos PD (1990) Random vibration and statistical linearization. Wiley, West Sussex

Rojas R (1996) Neural networks: a systematic introduction. Springer Science & Business Media

Seleemah AA, Constantinou MC (1997) Investigation of seismic response of buildings with linear and nonlinear fluid viscous dampers. State University of New York at Buffalo, New York

Shampine LF, Reichelt MW (1997) The matlab ode suite. SIAM J Sci Comput 18(1):1–22

Silvestri S, Trombetti T (2007) Physical and numerical approaches for the optimal insertion of seismic viscous dampers in shear-type structures. J Earthq Eng 11(5):787–828

Singh MP, Moreschi LM (2002) Optimal placement of dampers for passive response control. Earthq Eng Struct Dyn 31(4):955–976

Soong TT, Constantinou MC (1994) Passive and active structural vibration control in civil engineering. Springer, New York

Soong TT, Dargush GF (1997) Passive energy dissipation systems in structural engineering. Wiley, New York

Symans MD, Constatinou MC (1998) Passive fluid viscous passive systems for seismic energy dissipation. J Earthq Technol 35(4):185–206

Takewaki I (1997) Optimal damper placement for minimum transfer functions. Earthq Eng Struct Dyn 26(11):1113–1124

Zhang RH, Soong TT (1992) Seismic design of viscoelastic dampers for structural applications. ASCE J Struct Eng 118(5):1375–1391

# Chapter 8
# Stochastic Optimal Control of Seismic Structures with MR Dampers

## 8.1 Preliminary Remarks

Although the active structural control can attain a desired structural performance, the power supply system for implementing the structural control might suffer from a serious damage when subjected to hazardous dynamic excitations (Patten et al. 1998). Moreover, the entire structural system tends to instability due to the inevitable modeling error, measurement noise, and time delay (Soong 1990). A refined means is to actualize the logical combination between the active and passive control modalities so as to carry out a semiactive structural control. This modality exhibits a less demand of external energy and a low risk of system dynamic instability, which has thus received extensive attention in practice (Chu et al. 2005; Dan et al. 2015).

Owing to the perfect dynamic damping behaviors, the magnetorheological (MR) damper is regarded as one of the most promising control devices for implementing the semiactive structural control (Casciati et al. 2006). It has been an active area of research worldwide in the past two decades. The relevant topics include semiactive control algorithms and strategies (Jansen and Dyke 2000; Yoshioka et al. 2002; Nagarajaiah and Narasimhan 2006; Li et al. 2007; Xu and Guo 2008; Hogsberg 2011), modeling and dynamic performance of MR dampers (Spencer et al. 1997; Yang et al. 2002; Tsang et al. 2006; Boada et al. 2011; Xu et al. 2012; Chae et al. 2013), novel materials and technologies (Carlson and Jolly 2000; Tse and Chang 2004; Jung et al. 2010; Imaduddin et al. 2013), real-time hybrid simulations (Carrion et al. 2009; Cha et al. 2013; Asai et al. 2015), etc; while a few attempts, in the theoretical framework of stochastic optimal control, have been carried out for the design and optimization of MR damped structures. For instance, Dyke et al. proposed a LQG clipped-optimal control strategy implemented by MR dampers for strengthening the seismic safety of structures (Dyke et al. 1996). Ni et al. developed a neural network controller with MR damper, which achieved the similar gain to the LQG clipped-optimal controller (Ni et al. 2002). Ying et al. proposed a non-clipped strategy of semiactive stochastic optimal control for nonlinear structural systems with MR dampers based

© Springer Nature Singapore Pte Ltd. and Shanghai Scientific and Technical Publishers 2019
Y. Peng and J. Li, *Stochastic Optimal Control of Structures*,
https://doi.org/10.1007/978-981-13-6764-9_8

on the stochastic averaging method and stochastic dynamic programming (Ying et al. 2009). Using the modal-based LQG control algorithm and MR dampers, a smart system design was addressed to enhance the seismic performance of base-isolated buildings (Wang and Dyke 2013).

As a semiactive control device, the MR damper needs accurate dynamic models in the application of civil engineering, so as to online predict the control law, i.e., input current, in each time step in view of the relation between the dynamical model, the expected semiactive control force, and the structural state. However, the dynamic constitutive relation of the magnetorheological fluid arises to be complicated since its mechanical behaviors hinge upon a series of factors such as the magnetic field intensity and the shear rate driven by the damper piston. The complex behaviors of the MR fluid bring a challenging issue for accurate modeling of MR dampers. The dynamic test shows as well that the hysteretic behaviors of MR dampers indeed give rise to significant nonlinearity. Therefore, the accurate, simple, and feasible mechanical models ought to be established so as to fulfill the performance of MR dampers and guarantee the real-time effectiveness of the semiactive control strategy.

In this chapter, the method of stochastic optimal control using MR dampers is first introduced. Dynamic modeling, input current identification, and microstructured suspension behaviors of the MR damper are then addressed. For illustrative purposes, the semiactive stochastic optimal control of an MR damped structural system subjected to random seismic ground motion is carried out.

## 8.2 Semiactive Stochastic Optimal Control Using MR Dampers

A lot of semiactive control algorithms and control strategies have been developed in recent years to fulfill the dynamic performance of MR dampers. For instance, Jansen and Dyke investigated the effectiveness of classical semiactive control algorithms including the Lyapunov stability theory, the LQG clipped-optimal control, the decentralized Bang–Bang control, the modulated homogenous friction, and the maximum energy dissipation (Jansen and Dyke 2000). Chae et al. proposed an updated Maxwell nonlinear slider model for predicting the two-state control modalities of MR dampers subjected to random displacements, i.e., Passive-off and Passive-on, and the variant current and damper outputs (Chae et al. 2013). A semiactive stochastic optimal control in the theoretical framework of the physically based stochastic optimal control was developed (Peng et al. 2017). In conjunction with the bound Hrovat algorithm, the proposed strategy of semiactive stochastic optimal control exhibits the benefits of simplicity and effectiveness.

## 8.2.1  Bound Hrovat Algorithm

For a randomly excited linear structural system attached with MR dampers, the equation of motion is given by

$$\mathbf{M}\ddot{\mathbf{X}}(t) + \mathbf{C}\dot{\mathbf{X}}(t) + \mathbf{K}\mathbf{X}(t) = \mathbf{B}_s\mathbf{U}_s(t) + \mathbf{D}_s\mathbf{F}(\mathbf{\Theta}, t) \qquad (8.2.1)$$

where $\mathbf{M}, \mathbf{C}, \mathbf{K}$ are $n \times n$ mass, damping, and stiffness matrices, respectively; $\mathbf{X}$ is the $n$-dimensional column vector denoting system displacement; $\mathbf{B}_s$ is the $n \times r$ matrix denoting the location of MR dampers; $\mathbf{U}_s$ is the $r$-dimensional column vector denoting control forces pertaining to MR dampers; $\mathbf{D}_s$ is the $n \times p$ matrix denoting the location of external random excitations; and $\mathbf{F}$ is the $p$-dimensional column vector denoting random excitation.

The control force of MR dampers typically consists of two parts: the passive damping force that cannot be regulated by the control law and the variable damping force that can be regulated by the control law. Considering a shear-valve mode MR damper that is often applied in practice, the term related to the MR damper force in Eq. (8.2.1) can thus be denoted by

$$\mathbf{B}_s\mathbf{U}_s(t) = -\mathbf{B}_s\mathbf{C}_D\dot{\mathbf{X}}(t) - \mathbf{B}_s\mathbf{U}_{dc}(t) \qquad (8.2.2)$$

where $\mathbf{B}_s\mathbf{C}_D\dot{\mathbf{X}}(t)$ denotes the passive damping force and $\mathbf{B}_s\mathbf{U}_{dc}(t)$ denotes the variable Coulombic force which can be regulated through changing the input current and the associated magnetic field intensity which influences the yield strength of the MR fluid. The input current is determined by system state and damper models allowing for implementation of the expected damper force as a certain semiactive control algorithm.

Substituting Eq. (8.2.2) into Eq. (8.2.1), one has

$$\mathbf{M}\ddot{\mathbf{X}}(t) + (\mathbf{C} + \mathbf{B}_s\mathbf{C}_D)\dot{\mathbf{X}}(t) + \mathbf{K}\mathbf{X}(t) = -\mathbf{B}_s\mathbf{U}_{dc}(t) + \mathbf{D}_s\mathbf{F}(\mathbf{\Theta}, t) \qquad (8.2.3)$$

Introducing the extended state vector $\mathbf{Z}(t) = [\mathbf{X}^T(t)\ \dot{\mathbf{X}}^T(t)]^T$, Eq. (8.2.3) becomes

$$\dot{\mathbf{Z}}(t) = \mathbf{A}\mathbf{Z}(t) + \mathbf{B}\mathbf{U}_{dc}(t) + \mathbf{D}\mathbf{F}(\mathbf{\Theta}, t) \qquad (8.2.4)$$

where $\mathbf{A}$ is the $2n \times 2n$ system matrix; $\mathbf{B}$ is the $2n \times r$ matrix denoting the location of MR dampers; and $\mathbf{D}$ is the $2n \times p$ matrix denoting the location of random excitation:

$$\mathbf{A} = \begin{bmatrix} \mathbf{0} & \mathbf{I} \\ -\mathbf{M}^{-1}\mathbf{K} & -\mathbf{M}^{-1}(\mathbf{C} + \mathbf{B}_s\mathbf{C}_D) \end{bmatrix}, \mathbf{B} = \begin{bmatrix} \mathbf{0} \\ -\mathbf{M}^{-1}\mathbf{B}_s \end{bmatrix}, \mathbf{D} = \begin{bmatrix} \mathbf{0} \\ \mathbf{M}^{-1}\mathbf{D}_s \end{bmatrix} \quad (8.2.5)$$

In order to attain a good agreement with the dynamic behaviors of MR damper, a simple and efficient control strategy based on the Hrovat algorithm (Hrovat et al.

**Fig. 8.1** Relation between
MR damping force and
damper velocity at a certain
instant of time in the case of
a sample excitation

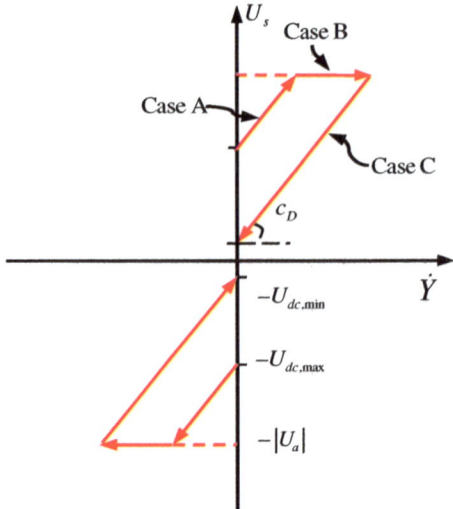

1983) is developed for the MR damper control, of which the component formulation
is given by

$$
U_s(\boldsymbol{\Theta}, t) = \begin{cases}
c_D\dot{Y}(\boldsymbol{\Theta}, t) + U_{dc,\max}\,\text{sgn}(\dot{Y}(\boldsymbol{\Theta}, t)), & \text{Case A: } U_a\dot{Y} < 0 \text{ and } |U_a| > U_{d,\max} \\
|U_a|\text{sgn}(\dot{Y}(\boldsymbol{\Theta}, t)), & \text{Case B: } U_a\dot{Y} < 0 \text{ and } |U_a| < U_{d,\max} \\
c_D\dot{Y}(\boldsymbol{\Theta}, t) + U_{dc,\min}\,\text{sgn}(\dot{Y}(\boldsymbol{\Theta}, t)), & \text{Case C: } U_a\dot{Y} > 0
\end{cases}
\tag{8.2.6}
$$

where $U_s(\boldsymbol{\Theta}, t)$ denotes the semiactive stochastic optimal control force executed
by the MR damper; $U_a(\boldsymbol{\Theta}, t)$ denotes the reference active stochastic optimal con-
trol force; $U_{d,\max}(\boldsymbol{\Theta}, t) = c_D|\dot{Y}(\boldsymbol{\Theta}, t)| + U_{dc,\max}$ denotes the changeable maximum
damping force of MR damper; $U_{dc,\max}$, $U_{dc,\min}$ denote the maximum and minimum
Coulombic forces of MR damper; $c_D$ denotes the viscous damping coefficient of MR
damper; and $\dot{Y}(\boldsymbol{\Theta}, t)$ denotes the damper velocity, i.e., the motion velocity of piston
relative to the damper cylinder which is opposite to the interstory drift between the
stories with the MR damper. In these parameters, $U_{dc,\max}$, $U_{dc,\min}$, $c_D$ are the design
parameters pertaining to the MR damper.

Figure 8.1 shows the relation between the MR damper force $U_s(\boldsymbol{\theta}, t)$ and damper
velocity $\dot{Y}(\boldsymbol{\theta}, t)$ at a certain instant of time in the case of a sample excitation $\boldsymbol{\theta}$.
The control force represented by Eq. (8.2.6) can be realized through driving the
calculated current into the MR damper. The expected input current is typically an
inverse solution of MR damper models. In application, the control effectiveness of
MR dampers highly hinges upon the accuracy and computational cost of the inverse
solution. Herein, the input current is assumed to fully implement the control gain in
demand with Eq. (8.2.6).

Substituting the formulation of control force as shown in Eq. (8.2.6) into the
equation of motion of stochastic dynamical system; say Eq. (8.2.4), one can attain

the solutions of the state vector and the control force. Clearly, the quantities of concern including the component of story velocity $\dot{X}(\Theta, t)$, the component of semiactive stochastic optimal control force $U_s(\Theta, t)$, and the active stochastic optimal control force $U_a(\Theta, t)$ are functions of $\Theta$. Similar to Eqs. (3.2.4) and (3.2.5), these quantities satisfy with the generalized probability density evolution equations (GDEEs), respectively, as follows:

$$\frac{\partial p_{\dot{X}\Theta}(\dot{x}, \theta, t)}{\partial t} + \ddot{X}(\theta, t)\frac{\partial p_{\dot{X}\Theta}(\dot{x}, \theta, t)}{\partial \dot{x}} = 0 \tag{8.2.7}$$

$$\frac{\partial p_{U_s\Theta}(u_s, \theta, t)}{\partial t} + \dot{U}_s(\theta, t)\frac{\partial p_{U_s\Theta}(u_s, \theta, t)}{\partial u_s} = 0 \tag{8.2.8}$$

$$\frac{\partial p_{U_a\Theta}(u_a, \theta, t)}{\partial t} + \dot{U}_a(\theta, t)\frac{\partial p_{U_a\Theta}(u_a, \theta, t)}{\partial u_a} = 0 \tag{8.2.9}$$

Under the provided initial conditions

$$p_{\dot{X}\Theta}(\dot{x}, \theta, t)|_{t=0} = \delta(\dot{x} - \dot{x}_0)p_\Theta(\theta) \tag{8.2.10}$$

$$p_{U_s\Theta}(u_s, \theta, t)|_{t=0} = \delta(u_s - u_{s0})p_\Theta(\theta) \tag{8.2.11}$$

$$p_{U_a\Theta}(u_a, \theta, t)|_{t=0} = \delta(u_a - u_{a0})p_\Theta(\theta) \tag{8.2.12}$$

one can attain the probability density functions of the quantities of concern at any instant of time as follows:

$$p_{\dot{X}}(\dot{x}, t) = \int_{\Omega_\Theta} p_{\dot{X}\Theta}(\dot{x}, \theta, t)d\theta \tag{8.2.13}$$

$$p_{U_s}(u_s, t) = \int_{\Omega_\Theta} p_{U_s\Theta}(u_s, \theta, t)d\theta \tag{8.2.14}$$

$$p_{U_a}(u_a, t) = \int_{\Omega_\Theta} p_{U_a\Theta}(u_a, \theta, t)d\theta \tag{8.2.15}$$

where $\dot{x}_0, u_{s0}, u_{a0}$ denotes the initial deterministic values of $\dot{X}(t), U_s(t), U_a(t)$.

### 8.2.2 Parameter Design of MR Damper

In order to gain a similar control effectiveness as the reference active stochastic optimal control, an MR damper design can be proceeded to facilitate the semiactive control system. The design principle lies in that the maximum output of the MR damper including the viscous damping force equals to the maximum active optimal control force, i.e., the extreme value of active optimal control force.

Assume that the MR damper control and the active optimal control have a similar control effectiveness, for instance, a same interstory velocity at the moment of the maximum active optimal control force:

$$U_{s,\max}(\boldsymbol{\Theta}) = c_D \left| \dot{Y}_{s|U_{s,\max}(\boldsymbol{\Theta})} \right| + |U_{dc,\max}| = c_D \left| \dot{Y}_{a|U_{a,\max}(\boldsymbol{\Theta})} \right| + |U_{dc,\max}| = U_{a,\max}(\boldsymbol{\Theta}) \quad (8.2.16)$$

Since the output of the MR damper can be continuously tuned by the current-driven magnetic field, there is

$$U_{s,\max}(\boldsymbol{\Theta}) = c_D \left| \dot{Y}_{s|U_{s,\max}(\boldsymbol{\Theta})} \right| + U_{dc,\max} = s \left( c_D \left| \dot{Y}_{s|U_{s,\max}(\boldsymbol{\Theta})} \right| + U_{dc,\min} \right) \quad (8.2.17)$$

where $s$ denotes the tunable times of damper force.

Assuming that the minimum Coulombic force $U_{dc,\min} = 0$, then one has

$$U_{s,\max}(\boldsymbol{\Theta}) = s c_D \left| \dot{Y}_{a|U_{a,\max}(\boldsymbol{\Theta})} \right| = U_{a,\max}(\boldsymbol{\Theta}) \quad (8.2.18)$$

The viscous damping coefficient is thus denoted by

$$c_D = \frac{U_{a,\max}(\boldsymbol{\Theta})}{s \left| \dot{Y}_{a|U_{a,\max}}(\boldsymbol{\Theta}) \right|} \quad (8.2.19)$$

and the maximum Coulombic force can be derived from Eqs. (8.2.17) to (8.2.19)

$$U_{dc,\max} = (s-1) c_D \left| \dot{Y}_{a|U_{a,\max}(\boldsymbol{\Theta})} \right| \quad (8.2.20)$$

It is shown in Eq. (8.2.20) that owing to the randomness inherent in the external excitation, the system state and the associated optimal control force are random processes. In this context, the parameters of control law exhibit uncertainties due to the dispersion over the sampling space. For example, the design parameters $U_{dc,\max}$, $c_D$ both rely upon $\boldsymbol{\Theta}$. However, the parameter design and optimization of control law is a deterministic scheme, i.e., the design parameters of structural control ought to be constant regardless of samples of random excitations.

It is revealed that the first step of MR damper control of structures in practice is to gain the expected damper force for the response reduction of structures, and then calculate the input current according to the dynamic model of MR dampers and the real-time system state, i.e., the so-called control law for regulation of MR dampers. In this process, the desired structural performance controlled by the semiactive modality can be precisely derived in theory if the real output of MR dampers just relies upon the amplitude of input current and the real-time state of structural system. This situation, however, is retained under two provided conditions: (i) no measurement noise during

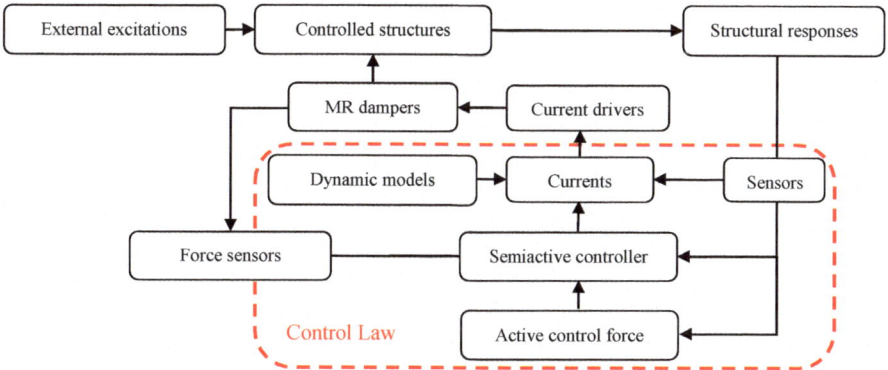

**Fig. 8.2**   Flowchart of implementing MR damper control of structures

the structural state monitoring and no errors inherent in the current calculation and signal delivery to the damper; (ii) no time delays at each step from state measurement, semiactive control force analysis, input current calculation, and signal delivery to the MR damper. However, there still exist differences between the output of MR damper and the expected semiactive optimal control force even the provided conditions are satisfied, due to the physical constraints of MR dampers such as the complicated rheological behaviors of MR fluids. In fact, the MR damping system belongs to a family of feedback control systems in logic. The associated measurement noise and time delay would exist in the control system. Moreover, the calculated current might exhibit a large diverse from the expected due to the modeling error of MR damper. Therefore, a set of measure system is often used in practice to monitor the real-time output of MR damper, thereby a current compensation strategy is thus proposed to better the control effectiveness of MR damper.

The flowchart of implementing the MR damper control of structures is shown in Fig. 8.2. It is seen that the active control force-based semiactive controller design and the dynamic modeling of MR dampers underly the design and optimization of control law of semiactive control.

## 8.3   Dynamic Modeling of MR Dampers

### 8.3.1   Parameterized Model

The dynamic models of MR dampers are mainly classified into parameterized and nonparameterized models (Yang et al. 2013). The parameterized models are mostly the mathematical formulation of damper force derived from the fitted curves of relationships between the damper force and damper displacement or velocity, of which the data is collected from the performance test of MR dampers. The parameterized models usually consist of a collection of mechanical elements such as the spring element, the viscous damping element, and the Coulomb friction element in configuration of serial and parallel systems (Spencer et al. 1997). The widely used parameterized models of MR dampers are mainly the Bingham model, Gamota-Filisko model, nonlinear bi-viscous hysteretic model, Bouc–Wen hysteretic model, their modified versions, etc. Similarly, the nonparameterized models are derived from the data of performance test of MR dampers, and are formulated by the intelligent algorithms such as neural network and fuzzy logic (Chang and Roschke 1998; Xu and Guo 2008), while these two families of modelings both are built toward the phenomenology and match with the accurate description of macroscale dynamic behaviors of MR dampers. However, the parameterized models exhibit a more feasible extension and a better applicability in practice.

Among the parameterized models, the modified Bouc–Wen hysteretic model is a preferable formulation for dynamic modeling of MR dampers since it not only reveals the hysteretic behavior inherent in the relation between damper force and velocity but also improves the slipperiness of piecewise functional curves. This model was first proposed by Bouc (1967), and later modified by Wen (1976). It has been widely used in modeling of hysteretic structural systems owing to its simplicity and feasibility. However, the Bouc–Wen hysteretic model cannot simulate the roll-off characteristics in the relation curves between damper force and velocity in the case that the acceleration and velocity turn direction and the velocity amplitude are very low. For this reason, a modified Bouc–Wen hysteretic model was then developed by Spencer et al. (1997). The schematic of a shear-valve mode MR damper and its modified Bouc–Wen hysteretic model are shown in Fig. 8.3.

The modified version consists of an original Bouc–Wen hysteretic model in series of a damping element and then in parallel of a spring element, which has the formulation with respect to the output of MR dampers as follows:

$$F_D = c_1 \dot{y} + k_1(x - x_0) \tag{8.3.1}$$

$$\dot{y} = \frac{1}{c_0 + c_1}[\alpha z + k_0(x - y) + c_0 \dot{x}] \tag{8.3.2}$$

$$\dot{z} = -\gamma|\dot{x} - \dot{y}|z|z|^{n-1} - \beta(\dot{x} - \dot{y})|z|^n + A(\dot{x} - \dot{y}) \tag{8.3.3}$$

where $F_D$ denotes the damper force; $\dot{y}$ denotes the piston velocity; $z$ denotes the hysteretic component; $k_1$ denotes the equivalent axial spring stiffness of accumulators; $c_0$ denotes the viscous damping coefficient of MR dampers in the case of large damper velocity; $c_1$ denotes the damping coefficient of MR dampers in the case of small damper velocity; $k_0$ denotes the axial stiffness of MR dampers in the case of high damper velocity; $x_0$ denotes the initial displacement of accumulator spring $k_1$; $\alpha$ denotes a stiffness parameter defined by the control current and the MR fluid; and $\gamma$, $\beta$, and $A$ are defined to govern the smoothing of damper force–velocity curves.

For illustrative purposes, the dynamic modeling of MR dampers using the experimental data is carried out, which was derived from a dynamic test of MR damper with

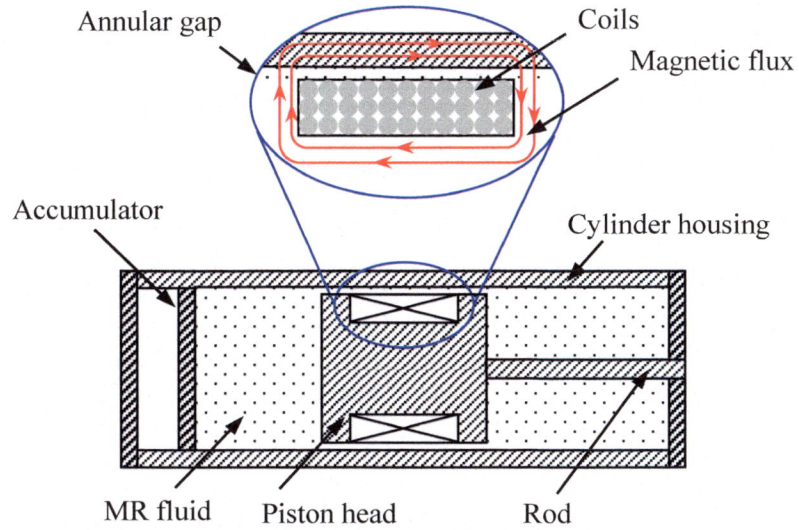

(a) schematic of shear-valve mode MR damper

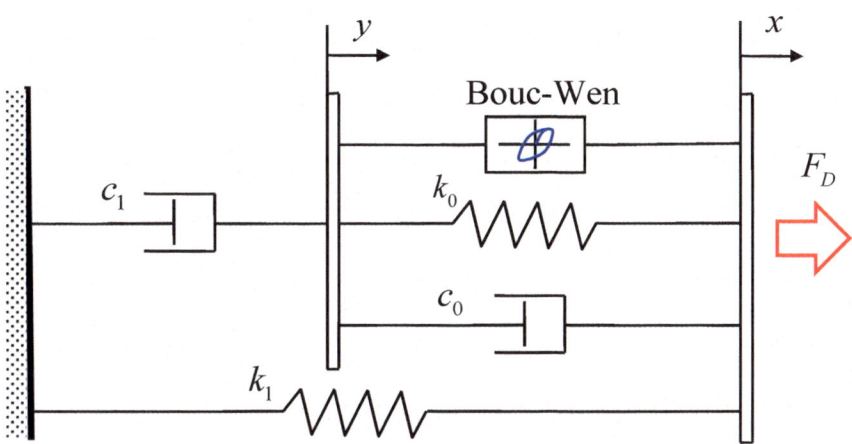

(b) schematic of modified Bouc–Wen hysteretic model

**Fig. 8.3** Schematic of shear-valve mode MR damper and its modified Bouc–Wen hysteretic model

specification MRD-100-10 (Peng et al. 2018). The specimens of the MR damper are shown in Fig. 8.4. The two specimens are labeled by MRD-A and MRD-B, respectively. This specification of MR dampers consists of cylinder, piston, MR fluid, and coils, which is a typical single-rod damper, as shown in Fig. 8.3. When the piston moves back and forth relative to the cylinder, the MR fluid passes through the annual gap between the piston and the cylinder, and yields damping force. The damping force can be readily regulated by changing the density of magnetic flux circumfused around the coils, which is able to be carried out using varying currents driven into MR dampers. The design parameters of the MR damper are listed as follows: maximum output 10 kN, the outer diameter of cylinder 100 mm, the fabrication length 670 mm, the length of stroke ±55 mm, the rated current 2.0 A, and the energy consumption 20 W.

The dynamic test was carried out on an electrohydraulic and servo-controlled material testing machine; see Fig. 8.5. During the test, the active clamp drives the motion of the piston of MR damper so that the piston executes a harmonic motion with specified frequency and amplitude relative to the cylinder. The input current to the MR damper, in four different levels 0.0, 0.5, 1.0, and 1.5 A, is implemented by a DC stabilized power supply. The experimental cases with different displacement amplitudes, excitation frequencies and input currents are proceeded to test the dynamic performance of the MR damper. These experimental cases are listed in Table 8.1.

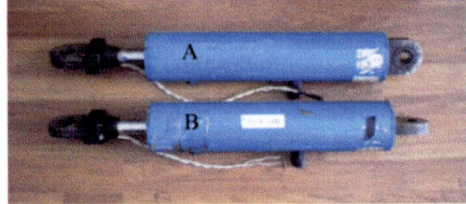

**Fig. 8.4** Two specimens of MR damper with specification MRD-100-10

**Fig. 8.5** Setup of dynamic test of MR dampers

**Table 8.1** Experimental cases of MR damper with specification MRD-100-10

| Amplitudes (mm) | Frequencies (Hz) | | | | | Currents (A) |
|---|---|---|---|---|---|---|
| | 0.25 | 0.5 | 0.75 | 1.0 | 1.5 | |
| 5 | | ✓ | | | | 0.0, 0.5, 1.0, 1.5 |
| 10 | | ✓ | | | | 0.0, 0.5, 1.0, 1.5 |
| 15 | ✓ | ✓ | ✓ | ✓ | ✓ | 0.0, 0.5, 1.0, 1.5 |
| 20 | | ✓ | | | | 0.0, 0.5, 1.0, 1.5 |
| 25 | | ✓ | | | | 0.0, 0.5, 1.0, 1.5 |

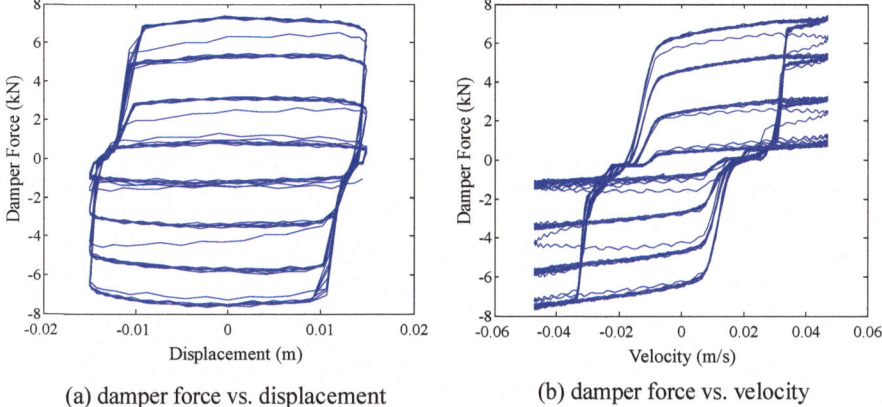

(a) damper force vs. displacement          (b) damper force vs. velocity

**Fig. 8.6**  Testing curves of MRD-A in typical loading conditions with displacement amplitude 15 mm and excitation frequency 0.5 Hz

The testing curves of the MR dampers, i.e., MRD-A and MRD-B, in typical loading conditions with displacement amplitude 15 mm and excitation frequency 0.5 Hz are shown in Figs. 8.6 and 8.7, respectively. Since the input currents are loaded in levels step by step, the curve loops from the inner to the outer are referred to the current level 0.0 A, 0.5 A, 1.0 A, and 1.5 A, respectively. Meanwhile, the relative velocity of the piston of MR damper to the cylinder is calculated through the numerical differential of displacement data monitored in the test.

It is seen that the damper force increases with the enhancement of input current, and the maximum of damper force in each loop arises to be of linear correlation with the input current (the saturation of damper force does not happen since the input current is less than the rated current of MR damper in the experimental cases). In the case of input current 0.0 A, the MR damper possesses viscous behaviors, e.g., the relation curve between damper force and piston displacement approaches to be elliptical, and the relation curve between damper force and piston velocity approaches

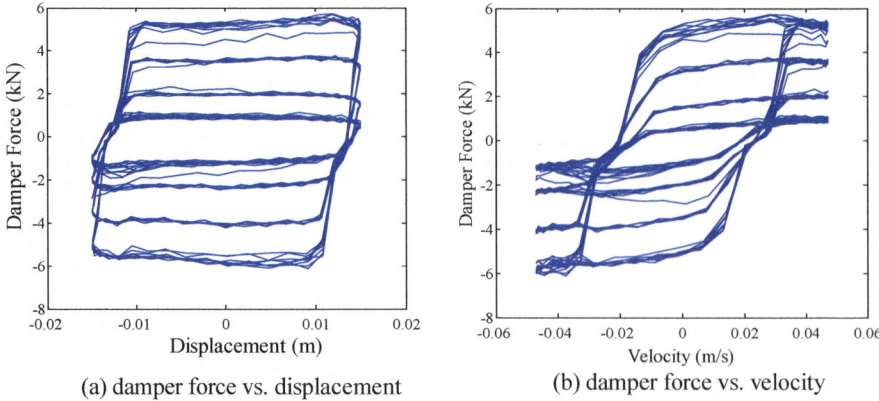

(a) damper force vs. displacement          (b) damper force vs. velocity

**Fig. 8.7** Testing curves of MRD-B in typical loading conditions with displacement amplitude 15 mm and excitation frequency 0.5 Hz

to be S shape. With the increasing of input current, however, the shear stress of MR fluid becomes stronger as well, thereby the MR damper features a viscous–plastic mechanics. The relation curves between damper force and piston velocity arise to be a complicated hysteretic behavior. In the range of high velocity, the linear relation between damper force and velocity is observed, while in the range of low velocity, a nonlinear relation between damper force and piston velocity is observed where the distinguished hysteretic behaviors occur. These findings are in good agreement with the dynamic model of MR dampers; see Fig. 8.3b.

## 8.3.2  Parameter Identification of Model

It is seen from Eqs. (8.3.1) to (8.3.3) that the modified Bouc–Wen hysteretic model involves the coupling of differential equations and includes the terms with high orders, which brings about an extremely difficult for the parameter identification of the model. There are three families of schemes for the parameter identification of Bouc–Wen hysteretic models of MR damper. The first refers to traditional optimization methods such as the nonlinear optimization scheme and least squares method with additional constraints (Spencer et al. 1997; Dyke et al. 1998). The second refers to the mathematical analysis of the relation between parameters and forced-limit cycles of model (Ikhouane and Rodellar 2005). The third refers to the genetic algorithm-based parameter identification (Charalampakis and Koumousis 2008). Comparing with the first two families of schemes, the third family of schemes has a better efficiency and accuracy for the optimization of multiple parameters, and is thus applied in this study.

For illustrative purposes, the MR damper MRD-A is considered for the parameter identification of the modified Bouc–Wen hysteretic model. The testing cases of concern are set as follows: Case 1, loading frequency 0.25 Hz, loading amplitude 15 mm, and loading current 0.0–1.5 A; Case 2, loading frequency 0.50 Hz, loading amplitude 15 mm, and loading current 0.0–1.5 A; and Case 3, loading frequency 1.00 Hz, loading amplitude 15 mm, and loading current 0.0–1.5 A.

Since the modified Bouc–Wen hysteretic model involves the differential equations, a four-order Runge–Kutta method is used to solve the damping force. Meanwhile, a genetic algorithm is employed for parameter identification, which can be readily implemented by the MATLAB toolbox function *ga*. In each step of loop, the assignment and optimization of model parameters are proceeded. The population size, generations, and stall generations in the genetic algorithm toolbox are set as 100, 200, and 50, respectively. The limit of fitness variation between optimized seeds at two neighbor generations is set as 0.001. The solution of differential equations can be derived from the case that all the 10 model parameters are valued through building the logical relation shown in Eqs. (8.3.1)–(8.3.3) with respect to the damper velocity $\dot{y}$ and hysteretic rate $\dot{z}$. For the ready conjunction with the genetic algorithm, nine parameters except the initial displacement $x_0$ are set as *input*. The input values of the parameters are controlled in real time by the prescribed iterative scheme of genetic algorithm toolbox. The damper force $F_D$ is set as *out*. All the identification values of these parameters are evaluated by the experimental or simulated data. An index pertaining to the degree of fitness is defined as follows (Spencer et al. 1997):

$$Fitness = \frac{\sqrt{\frac{1}{n}\sum_{i=1}^{n}(F_{D,i}^{\text{exp}} - F_{D,i}^{\text{fit}})^2}}{\sqrt{\frac{1}{n}\sum_{i=1}^{n}(F_{D,i}^{\text{exp}} - \frac{1}{n}(\sum_{i=1}^{n}F_{D,i}^{\text{exp}}))^2}} \tag{8.3.4}$$

where $n$ denotes the number of data points in the experiments or simulations; $F_{D,i}^{\text{exp}}$ denotes the damper force of the $i$th data point; and $F_{D,i}^{\text{fit}}$ denotes the damper force of the $i$th fitted point.

The modified Bouc–Wen hysteretic model exhibits 10 parameters which might result in a high computational cost and a low accuracy if no constraints are posed upon the parameters. In this study, the initial displacement $x_0$ of accumulator is set as 0.2 m. The upper and lower bounds of the remaining nine parameters $c_1$, $k_1$, $c_0$, $\alpha$, $k_0$, $\gamma$, $n$, $\beta$, and $A$ are denoted by $[10^4, 10^7]$, $[10^2, 10^4]$, $[10^2, 10^5]$, $[10^3, 10^5]$, $[10, 10^4]$, $[10^2, 10^5]$, $[1, 5]$, $[10^2, 10^5]$, and $[10, 10^3]$, respectively.

In consideration of the complexity of solving the differential equation, the four-order Runge–Kutta method is implemented by the solver *ode4* of MATLAB/Simulink. The time interval of the solver is fixed at each step so as to derive the damper force at the setting instant of time. In this study, the time interval is set as 0.0001 s, and the time length is set as 20 s.

In the semiactive control modality, an inverse calculation is usually required so as to regulate the damper force through changing the input current or voltage. The analysis of current relevance of model parameters is a critical step, aiming at the determination of sensitive parameters and their functional relation with the input current. The relation curves between parameters and input current reveal that except

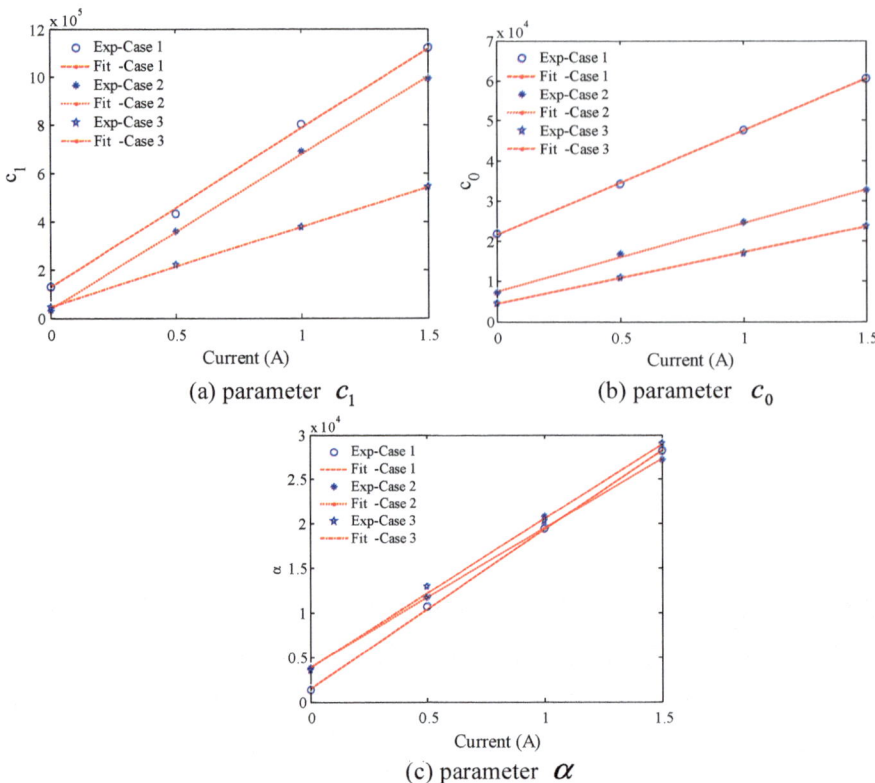

**Fig. 8.8** Fitted curves between control parameters and input current

the parameters $c_1$, $c_0$, and $\alpha$, the relation between the remaining six parameters and input current is not clear. The parameters $c_1$, $c_0$, and $\alpha$ are the so-called critical parameters for the dynamic model of the MR damper, which change linearly along with the increasing of the current; see Fig. 8.8.

According to Eqs. (8.3.1)–(8.3.3), the identification of the remaining six parameters is carried out again through fixing the functional relation of $c_1$, $c_0$, and $\alpha$ with input current. In order to reduce the computational cost, the upper and lower bounds of the parameters to be identified are valued, respectively, by the individual minimum and maximum in the first identification. With an iterative process, the six parameters are eventually defined by Case 1: $[k_1, k_0, \gamma, n, \beta, A] = [782.16, 3007.79, 61530.17, 4.35, 30746.83, 164.20]$, where the fitness is $0.1416$; by Case 2: $[k_1, k_0, \gamma, n, \beta, A] = [413.68, 9562.50, 1010.11, 2.00, 1011.73, 105.87]$, where the fitness is $0.1326$; and by Case 3: $[k_1, k_0, \gamma, n, \beta, A] = [3609.01, 5012.80, 1944.34, 2.00, 1660.78, 173.43]$, where the fitness is $0.1410$.

Using the optimized values of parameters, comparative studies between the modified Bouc–Wen hysteretic model and the experimental data associated with input

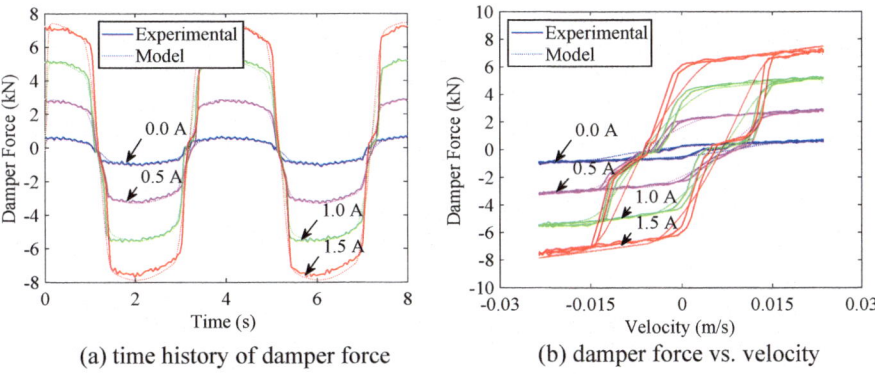

**Fig. 8.9** Comparison between modified Bouc–Wen hysteretic model and experimental data in the case of displacement amplitude 15 mm and excitation frequency 0.25 Hz (Case 1)

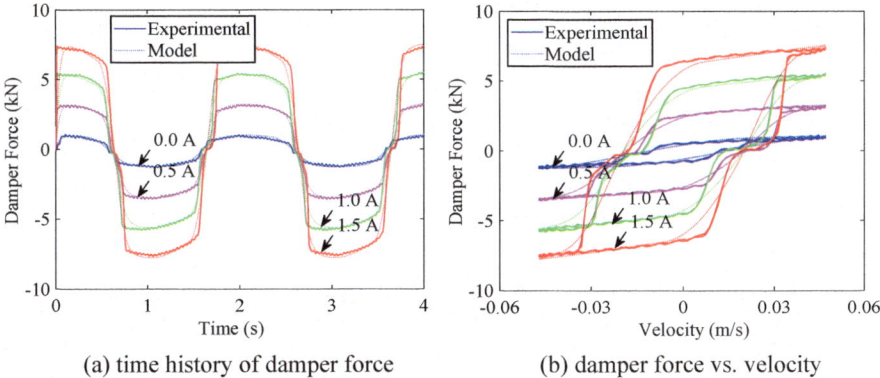

**Fig. 8.10** Comparison between modified Bouc–Wen hysteretic model and experimental data in the case of displacement amplitude 15 mm and excitation frequency 0.50 Hz (Case 2)

currents are proceeded. The model result (labeled as 'Model') and experimental result (labeled as 'Experimental') in the concerned three cases are shown in Figs. 8.9, 8.10 and 8.11, respectively. One might see that the parameterized model secures a sound fitting accuracy with the experimental data in different levels of input current, indicating that identified values of model parameters are satisfied for the individual case. Throughout the three cases, meanwhile, the control parameters, i.e., $c_1$, $c_0$, and $\alpha$, are viewed as the same and all submit to the linear function of input current.

According to the identified model parameters and their relations with the input current, the current signal loaded on the MR damper can be readily generated using the backpropagation (BP) neural network algorithm (Metered et al. 2010). Details of the current signal generation are illustrated in conjunction with the numerical example in Sect. 8.4.

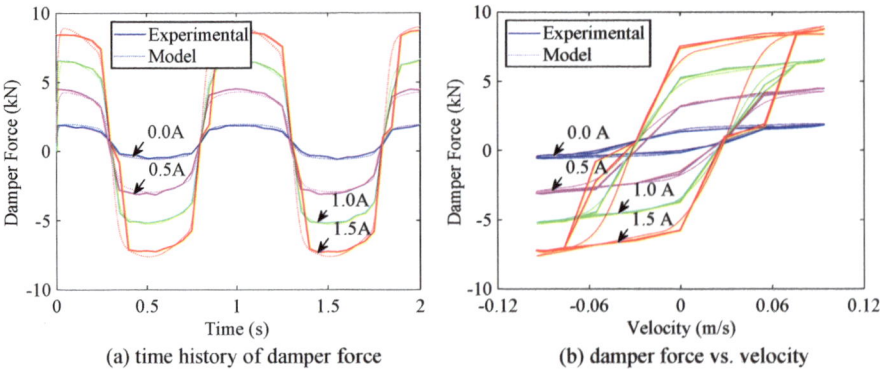

**Fig. 8.11** Comparison between modified Bouc–Wen hysteretic model and experimental data in the case of displacement amplitude 15 mm and excitation frequency 1.00 Hz (Case 3)

### 8.3.3   Microscale Mechanism of MR Dampers

The investigation of microstructured behaviors of MR suspensions exhibits a significance for revealing the physical essence of complicated dynamics of MR dampers and carrying out the optimization of control law of semiactive modality. The reference (Peng et al. 2012) addressed this issue for the first time.

The magnetorheological fluid is viewed as an elementary material assembling MR dampers. It consists of micrometer-sized magnetizable particles and nonmagnetic fluid, which shows a unique ability to experience the phase separation in a rapid and completely reversible manner. The physical origin of this behavior is that under the external magnetic field with specified intensity, the particles acquire a magnetic dipole moment, resulting in particle aggregation to form chain-like structures parallel to the external magnetic field and form cluster-like and sheet-like structures perpendicular to the external magnetic field. These properties of the phase separation in the microstructure, moreover, are always accompanied by significant changes in flow behavior and optical properties with the increase in the viscosity of the suspension and generation of optical anisotropy. The essence of the MR damper control is thus setting the identified input current as the control law to drive the magnetic field upon the magnetic fluids so that the microstructured behaviors of MR suspensions change and prevent the flow from transporting induced by external excitations. A schematic describing the generation of input current signal and its influence upon the microstructured behaviors of MR suspensions is shown in Fig. 8.12.

A large-scale atomic/molecular massive parallel simulator (LAMMPS) is employed, which provides an embedded routine for large-scale and three-dimensional Brownian dynamics simulation. LAMMPS facilitates the simulations of millions of particles, which may include gas, liquid, solid, and complex phases. Its library of potential functions and force fields is extensive, and it has been applied to the simulation of a wide spectrum of particles, including atomic polymers,

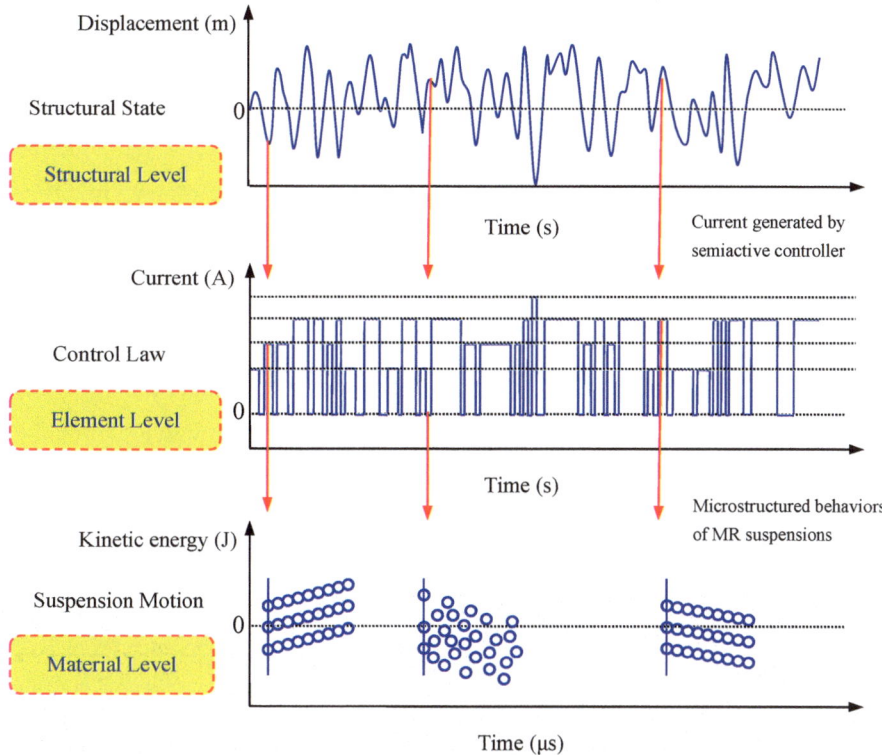

**Fig. 8.12** Schematic of input current generation and its influence upon microstructured behaviors of MR suspensions

bead-spring polymers, organic molecules, proteins, granular materials, and point dipolar particles. Moreover, LAMMPS can be readily parallelized for various computer architectures (Plimpton 1995). The initial position and initial velocity of suspensions are generated as statistically independent with a uniform distribution and a Maxwell–Boltzmann distribution, respectively (Liu et al. 2006). The neighbor list algorithm with the strategy of radius cutoff (truncated radius) is used to assess the interaction between particles (Verlet 1967). The motion of MR suspensions can be described by a Langevin equation, of which the numerical solution is derived using the velocity Verlet integral scheme (Swope et al. 1982). These numerical techniques can be readily implemented in conjunction with the LAMMPS.

A shear-valve mode MR damper with double rods in specification of MRD-9000 (Yang 2001) is used for the investigation. The involved MR fluid consists of the silicone oil and the emerged double suspensions with micrometer iron carbonyl particles. The relevant physical parameters of the simulated MR fluid are listed in Table 8.2.

**Table 8.2** Physical parameters of simulated MR fluid

| Physical quantities | | Parameter values |
|---|---|---|
| Volume ratio between suspensions and matrix | | 0.3 |
| Radius of particles | Large particle $a_L$ | $5 \times 10^{-6}$ m |
| | Small particle $a_S$ | $2.5 \times 10^{-6}$ m |
| Mass of particles | Large particle $M_L$ | $1 \times 10^{-13}$ kg |
| | Small particle $M_S$ | $1.25 \times 10^{-14}$ kg |
| Volume ratio of bidisperse, large/small particles | | 75:25 |
| Relative permeability | Matrix $\mu_c$ | 1.0 |
| | Particles $\mu_p$ | $1 \times 10^3$ |
| Saturation magnetization of particles | | 2 T |
| Viscosity of matrix | | 0.3 Pa s |
| Density of silicon oil | | $3.6 \times 10^3$ kg m$^{-3}$ |
| Temperature | | 298 K |
| Magnitude of steady magnetic field $H_0$ | | 100 k Am$^{-1}$ |

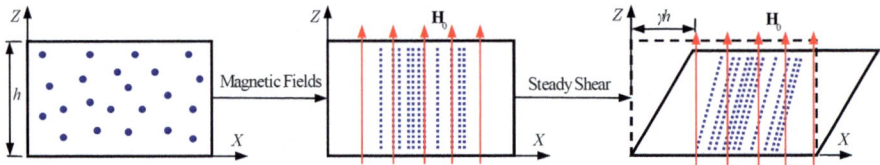

**Fig. 8.13** Schematic of magnetic field and steady shear loading on simulated cell

According to the volume ratio between suspensions and carrier fluid, these spheres are uniformly distributed in the space of a simulation cell with dimensions $(L_X^*, L_Y^*, L_Z^*) = (20, 10, 10)$, where the asterisk "*" represents dimensionless quantities. In this study, the length, timescales, and mass in the dimensionless units have specified relation with those in SI units; see dimensionless unit length $10^{-5}$ m, dimensionless unit time $1.7 \times 10^{-3}$ s, and dimensionless unit mass $2.43 \times 10^{-8}$ kg. The amount of particles is 3120 which includes 860 large particles and 2260 small particles. Sheared periodic boundaries are included at $X^* = \pm L_X^*$, $Y^* = \pm L_Y^*$ and $Z^* = \pm L_Z^*$. The shear flow is applied along the $X$-direction, and the magnetic field is applied along the $Z$-direction. The time step length of simulations is $\Delta t^* = 10^{-7}$. The schematic of simulation procedure is shown in Fig. 8.13.

Figure 8.14 shows the cluster–sheet phase of the MR suspensions at 10.0 μs, where "o" represents the large particles, and "." represents the small particles. It is seen that the direction of most cluster–sheet structures is parallel to the shorter axis $Y$, not the longer axis $X$. In view of the phenomenon of nematic-like ordering of the MR suspensions toward particular directions, one might realize that these magnetic dipoles exhibit some intelligent behaviors, and they always align to clusters along

**Fig. 8.14** Cluster–sheet phase of MR suspensions at 10.0 μs

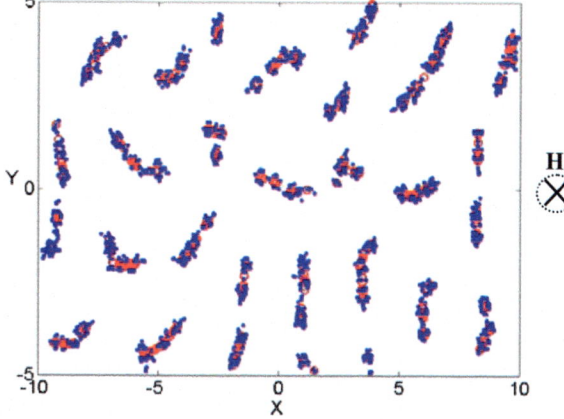

**Fig. 8.15** Structural anisotropy of aggregates of MR suspensions at shear strain of 1.0 under steady magnetic field and shear field with shear rate 1000 s$^{-1}$

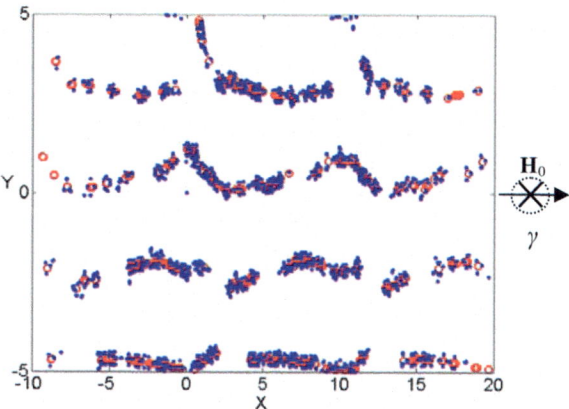

the direction whereby the sheet tends to be formed, though the initial configuration admits to the uniform distribution and the external magnetic fields are steady (Peng et al. 2012).

Figure 8.15 shows the structural anisotropy of aggregates of MR suspensions at shear strain of 1.0 under steady magnetic field and shear field with 1000 s$^{-1}$. It is seen that the suspensions move along the flow field and connect to long sheets along flow direction toward restraining the transportation of the shear flow, though the suspension structure suffers from yielding and the sheets tend to be ripped presenting as arch structures.

In order to reveal the dynamic performance of MR dampers from a microscale, the dynamic yielding stress of MR fluid is simulated. According to the previous studies, the strain energy of macroscale yielding stress of MR fluid in a unit volume equals the kinetic energy of microscale MR suspensions in the volume. On this principle, a multiscale constitutive relation of MR fluids can be established (Peng and Li 2011). Figure 8.16 shows the relation between yielding stress and shear rate of MR fluid

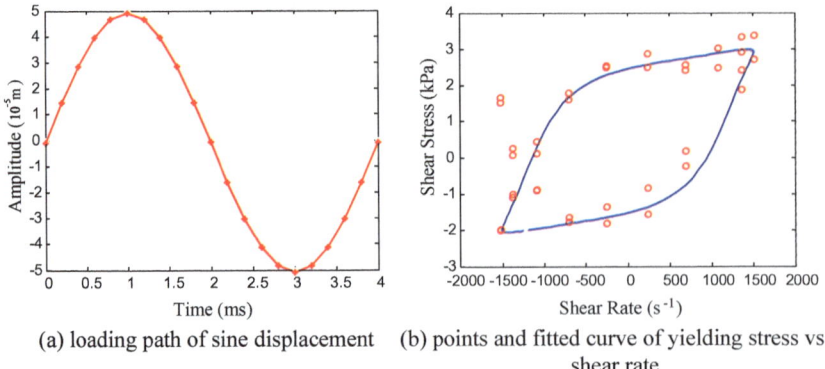

(a) loading path of sine displacement   (b) points and fitted curve of yielding stress vs. shear rate

**Fig. 8.16** Relation between yielding stress and shear rate of MR fluid under sine displacement loading on MR dampers

under sine displacement loading with period of 4 ms on MR dampers. It is seen that the fitted curve on the simulated data points shows a good consistency with the Bouc–Wen hysteretic model, which exhibits obvious similarities as the relation curves between damper force and damper velocity shown in Figs. 8.10, 8.11 and 8.12. These similarities, as a matter of fact, are just the representation of the macroscale performance of MR dampers under sine displacement loading and current driving on the microscale structured behaviors of MR fluids.

## 8.4   Numerical Example

The semiactive stochastic optimal control of single-story shear frame shown in Fig. 3.4 is carried out. The physically motivated random seismic ground motion model is used as the external excitation, of which the peak ground acceleration is set as $0.11g$. Design and optimization of a shear-valve mode MR damper are performed for implementing the semiactive control.

In order to attain the desired structural performance, the control algorithm in formulation of Eq. (8.2.6) is employed. The tunable times of the damper force are set as $s = 8$. The design parameters for the MR damper control are viscous damping coefficient and maximum Coulombic force. Since the multilinearity property of the semiactive control algorithm, the original state equation of the structural system shall be discretized into a discrete state equation, and the associated coefficients need to be identified so as to derive the control gain. In this study, the coefficient identification is performed using the precise integration method (Zhong 2004). For the purpose of a uniform numerical framework, the calculation of the reference active optimal control force and its relevant interstory drift refers to the methodology of physically based stochastic optimal control in kernel of discrete dynamic programming, i.e., the solving of a so-called matrix difference Riccati equation; see Appendix E.

The deterministic dynamic analysis with respect to the velocity quantities in the generalized probability density evolution equations employs a first-order forward difference scheme, as same as the discrete dynamic programming. The optimization of weighting matrices in the cost function is carried out as the criterion on system second-order statistics evaluation (SSSE); see Eq. (3.3.21): the interstory drift serves as the constraint, and the quantities of evaluation include the interstory drift, story acceleration, and interstory control force. The quantile function is defined as the sum of mean and three times of standard deviation. The threshold of interstory drift is set as 10 mm.

Since the structural system and seismic ground motion are consistent with the numerical example shown in Sect. 3.4.1, the relation between the statistical moments of equivalent extreme values of system quantities and the ratio of coefficients of weighting matrices, see Fig. 3.11, can be used straightforwardly in this study. In order to reduce the structural displacement in a more serious extent, the weighting matrices pertaining to the system state and control force are denoted by

$$\mathbf{Q_Z} = 80 \begin{bmatrix} 1 & 0 \\ 0 & 1 \end{bmatrix}, \mathbf{R_U} = 10^{-12} \tag{8.4.1}$$

By virtue of Eq. (3.3.18), the active stochastic optimal control of structure is proceeded. The probability density function of the extreme value of active optimal control force using the parameters of control law, say Eq. (8.4.1), is shown in Fig. 8.17.

It is seen that the reference active optimal control force exhibits a large range of distribution, of which the mean and standard deviation are 115.44 and 34.68 kN. It is revealed in Eq. (8.2.16) that the parameters of MR dampers are determined by the reference active optimal control force and its relevant interstory velocity. Due to the randomness inherent in the active optimal control force, the parameters of

**Fig. 8.17** Probability density function of extreme value of active optimal control force

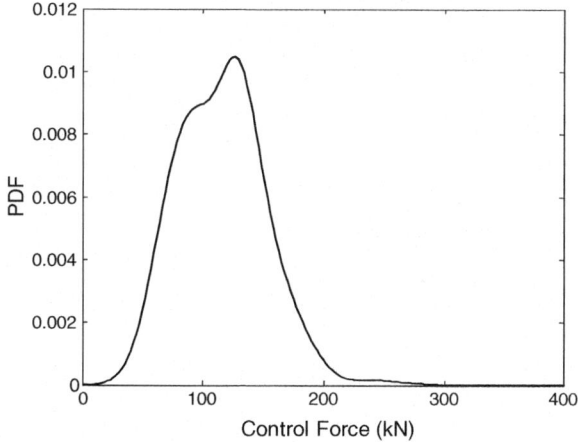

MR damper as the traditional deterministic control scheme give rise to uncertainty. According to the general principle of structural design, the mean or a certain quantile of the active optimal control force can be used as the design control force. One might recognize that this treatment lacks the accurate assessment of structural performance. A logical manner defining the reference active optimal control force is in conjunction with the physical causes.

Figure 8.18 shows the maximum active control force, the maximum interstory drift, and maximum story acceleration of semiactively controlled structure with respect to samples. It is seen that the interstory drift changes not obviously along with the active control force, which nearly distributes in the range of 2–5 mm; the story acceleration changes positively along with the active control force. The tendancy of third-order fitted curves in the figure shows that the application of a larger design active control force cannot attain a further reduction of the structural displacement. It is thus remarked that the semiactive controller and the passive controller lack ability for significantly reducing the structural displacement. However, the active controller exhibits a benefit of reducing the structural displacement. Besides, the fitted curve of relation between story acceleration and active control force has the similarity as that of actively controlled structures.

Figure 8.19 shows the relation between design parameters of MR damper and the active optimal control force with respect to samples. In comparison with Fig. 8.18, it is seen that the viscous damping coefficient is low and insensitive to the active control force when the control force magnitude is more than 100 kN, indicating that an accurate MR damper control system can be constructed using a series of low-cost components; the maximum Coulombic force exposes to be linearly relevant to the reference active control force owing to a linear regulation assumption with respect to the Coulombic force.

In fact, the definition of reference active control force needs to consider the physical mechanism and practical capacity of MR dampers. For example, the most eco-

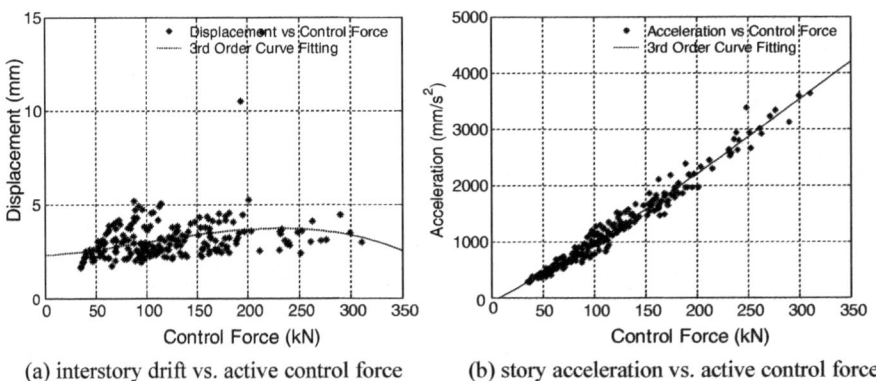

(a) interstory drift vs. active control force    (b) story acceleration vs. active control force

**Fig. 8.18** Relation between maximum active control force, maximum interstory drift, and maximum story acceleration of semiactively controlled structure with respect to samples

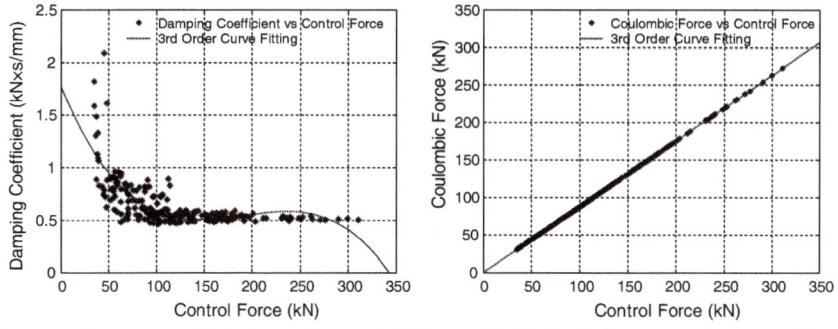

(a) viscous damping coefficient vs. active control force    (b) Coulomb force vs. active control force

**Fig. 8.19** Relation between maximum active control force, viscous damping coefficient, and maximum Coulombic force with respect to samples

**Fig. 8.20** All samples with relation between maximum active control forces and assigned probabilities

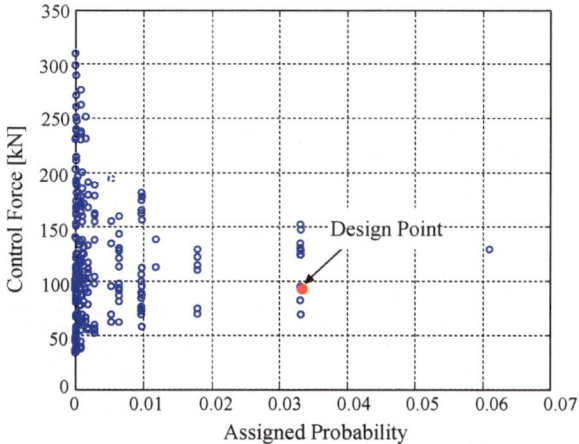

nomic output of the MR damper investigated in this study is 200 kN, and a high-viscous rheological liquid is inconvenient for maintenance. Therefore, the reference active control force is suggested to be defined on the sample with the largest assigned probability, i.e., with high occurrence rate, and in the neighborhood of 100 kN. Figure 8.20 shows all the samples with the relation between maximum active control forces and assigned probabilities. It is ready to recognize that the reference active control force is 94.03 kN, which is around the mean of the maximum active control force. The damping coefficient and the maximum Coulombic force are thus designed as 0.6119 kNs/mm, 82.28 kN, respectively.

Time histories of root-mean-square displacement of the structural system with and without controls are shown in Fig. 8.21. It is seen that the structural performance gains a significant improvement both using the semiactive and active stochastic optimal controls. In comparison with the active control, the semiactive control attains an almost same gain in the time domain with smaller response of uncontrolled structure.

**Fig. 8.21** Time histories of root-mean-square displacement of structural system with and without controls

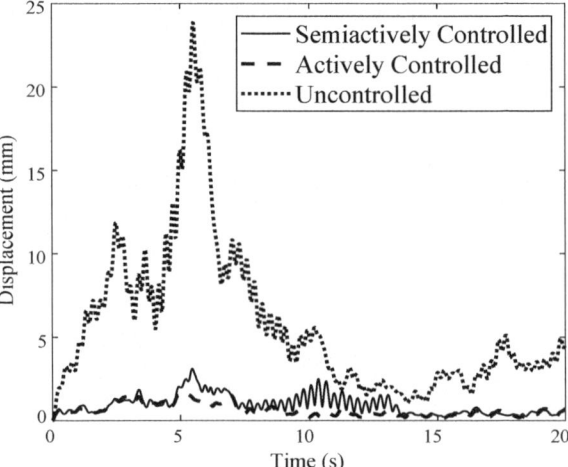

**Fig. 8.22** MR damping force tracing active optimal control force in sense of root mean square

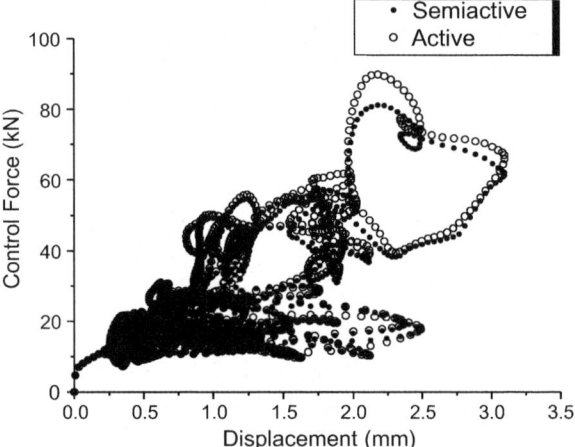

In the time domain with larger response of uncontrolled structure, the active control has a better control gain than the semiactive control. This is understood that the semiactive control algorithm employed in this study belongs to a family of bound amplitude control schemes, due to the fact that the MR damper exhibits magnetic saturation (Xu et al. 2012). In the case that the control requirement over the response domain exceeds the capacity of MR damper, the semiactive control will stick on the maximum output of the damper other than timely tracing active optimal control. The detail of MR damper force tracing active optimal control force in root-mean-square sense is shown in Fig. 8.22. It is seen that the bound Hrovat algorithm-based semiactive control has the capacity of tracing the active optimal control in real time.

More accurate probabilistic representation is the probability density function, as shown in Fig. 8.23. It is seen that the curves of PDFs of semiactive and active optimal

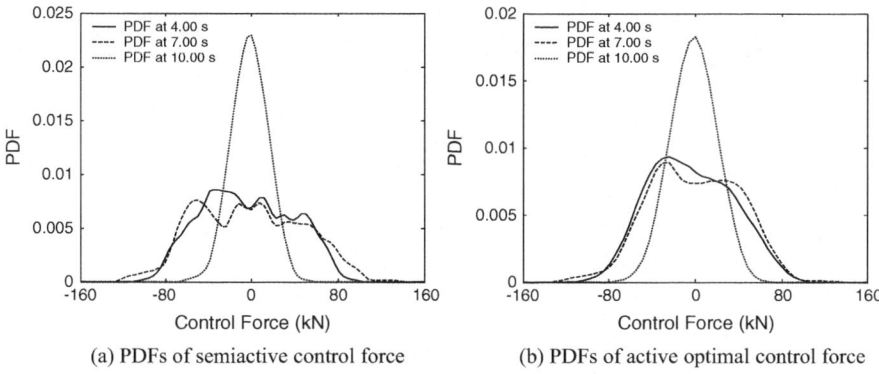

(a) PDFs of semiactive control force      (b) PDFs of active optimal control force

**Fig. 8.23** PDFs of semiactive and active optimal control forces at typical instants of time

control forces at typical instants of time are similar, where a slight difference lies in that the output of semiactive control merely relies upon the maximum active control force other than tracing the active control force in real time, such as the case that the active control force has a same direction with the interstory velocity or although the active control is opposed to the interstory velocity, the active control force is larger than the maximum output that the MR damper is able to actualize.

Figure 8.24 shows the probability density functions of interstory drift at typical instants of time with and without the MR damper controls. It is seen that the MR damper control can reduce the structural displacement significantly, where the distribution range of PDFs becomes narrower. Meanwhile, by comparison with Fig. 3.13b, the reduction of displacement amplitude of active control arises to more significant than that of semiactive control. One might see that the peak of PDFs of the former approaches to 0.9, while the peak of PDFs of the later approaches to 0.45. As mentioned previously, the semiactive and passive controllers still remain challenges in the displacement control of structures. Besides, the semiactive controller designed as the ratio of coefficients of weighting matrices $8 \times 10^{13}$ exhibits a worse instead of a better control effectiveness than the active controller design as a smaller ratio of coefficients of weighting matrices $8 \times 10^{12}$.

In order to reveal the influence of the semiactive control upon the dynamic performance of MR dampers, Fig. 8.25 shows the relation between the damper force and the damper displacement, damper velocity under a sample of seismic excitation. It is seen that the relation curves between damper velocity and damper force nearly all distribute in the first and third quadrants, indicating that the semiactive control force always remains an opposite direction to the damper velocity. The rationality is thus proposed that the MR damper can change the output timely so as to trace the active optimal control force in real time. The predictive and experimental data as an up-scaled profile of MR damper force curves are shown in Fig. 8.25 as well. In conjunction with the modified Bouc–Wen hysteretic model, the predictive and experimental data are derived from a 3-kN MR damper in type of VersaFlo MRX-135GD

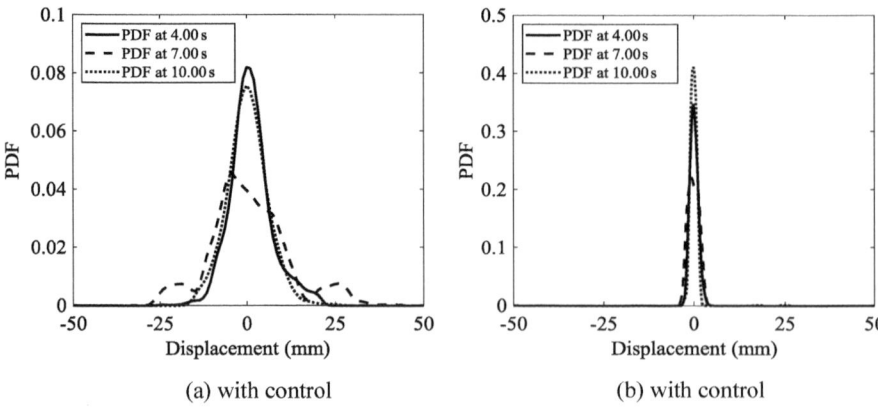

(a) with control                                    (b) with control

**Fig. 8.24**   PDFs of interstory drift at typical instants of time with and without MR damping controls

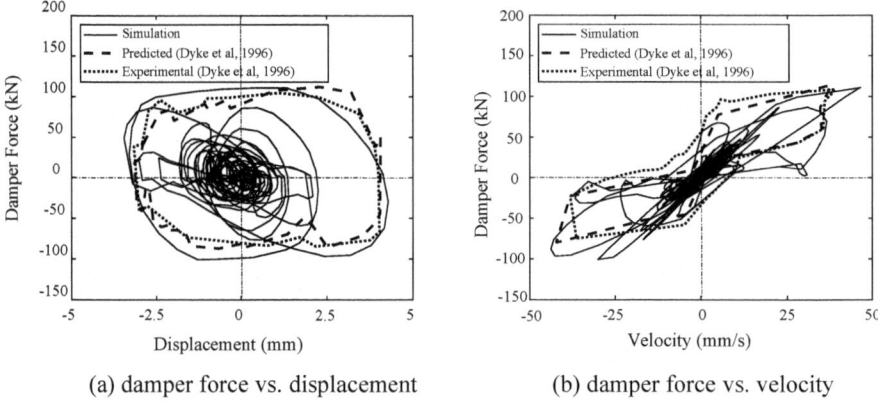

(a) damper force vs. displacement              (b) damper force vs. velocity

**Fig. 8.25**   Relation between damper force and damper displacement, damper velocity under the sample of seismic excitation

under a narrowband excitation of white Gaussian noise (Dyke et al. 1996; Spencer et al. 1997). It is revealed that the semiactive control can well accommodate the dynamic performance of MR dampers, which behaves a similarity as the Bouc–Wen hysteretic model with strength deterioration, stiffness degradation, and pinch effect (Wen 1976).

For illustrating the generation of input currents to the MR damper, the shear-valve mode MR damper with double rods in specification of MRD-9000 is employed which has a moderate performance as required by this numerical example. The simulated data of the MR damper under displacement amplitude of 25.4 mm, excitation frequency of 0.5 Hz, and input current of 0.0–2.0 A is utilized. A modified Bouc–Wen hysteretic model is applied to simulate the dynamic performance of the MR damper. Using the parameter identification procedure addressed in Sect. 8.3.2, the constant parameters of the modified Bouc–Wen hysteretic model for the MRD-9000 are denoted by $[x_0, k_1, c_0, k_0, g, n, b, A] = [0.0, 158.55, 469,042.70, 612.79, 98,540.12, 2.01, 73,224.85, 9,253.30]$, and the current variant parameters are denoted by $c_1 = -5.15 \times 10^6 I^2 + 1.73 \times 10^7 I + 3.63 \times 10^6$, $\alpha = -2.42 \times 10^5 I^2 + 8.71 \times 10^5 I + 7.34 \times 10^6$.

According to the parameterized model and the system state of structures, a numerical procedure by virtue of backpropagation (BP) neural network algorithm is employed to generate the current signal loaded on the MR damper. The BP algorithm can be readily implemented in conjunction with the MATLAB toolbox function *nftool*. The numbers of input nodes, output nodes, and implication nodes are 7, 1, and 13, respectively. For solving the differential equations of the Bouc–Wen hysteretic model in the numerical procedure, the four-order Runge–Kutta method is employed here and implemented by the solver *ode4* of MATLAB/Simulink as well.

Prior to the generation of input currents, a step of sample training needs to be proceeded to activate the BP neural network-based retrorse model of MR dampers. A 20-s time series with 1000 data points is used for training and validation, which consists of three segments: white Gaussian noise of current and displacement in the first 10 s, high-amplitude sine current and white Gaussian noise of displacement in the next 5 s, and low-amplitude sine current and white Gaussian noise of displacement in the last 5 s. The validation of sample data shows that the goodness of fit attains to more than 98%. Other three different cases are addressed to verify the effectiveness of the retrorse model of MR dampers, involving Case 1: white Gaussian noise of current and displacement, Case 2: sine current and white Gaussian noise of displacement, and Case 3: constant current and sine displacement. The current and displacement in each case are 20-s time series with 1000 data points. Figure 8.26 shows the expected and identified input currents and the associated outputs of MR dampers in the three cases. It is seen that the identified damper force matches well with the expected, of which the errors of three cases are 3.12%, 6.93%, and 2.54%, respectively.

Similarly, if the system state and the output of MR dampers are known, one can readily derive the input current. In this study, the identification of optimal current for MR damper control of structures under three different samples of seismic ground motions is carried out; see Fig. 8.27. It is seen that the optimal currents used for semiactive control of structures arise to irregularly fluctuate in the range of 0.0–0.6 A, which are significantly different from the samples of seismic ground motions. One might recognize that the real-time feedback control exhibits a practical significance for improving the structural performance; the development of highly efficient MR damper control ought to follow an accurate modality from the present simple Bang–Bang control or Passive-on and Passive-off step controls in practice. It is also noted that the methodology of physically based stochastic optimal control, the probabilistic optimization and design of controller parameters, and control device placement play an important role in this developing process.

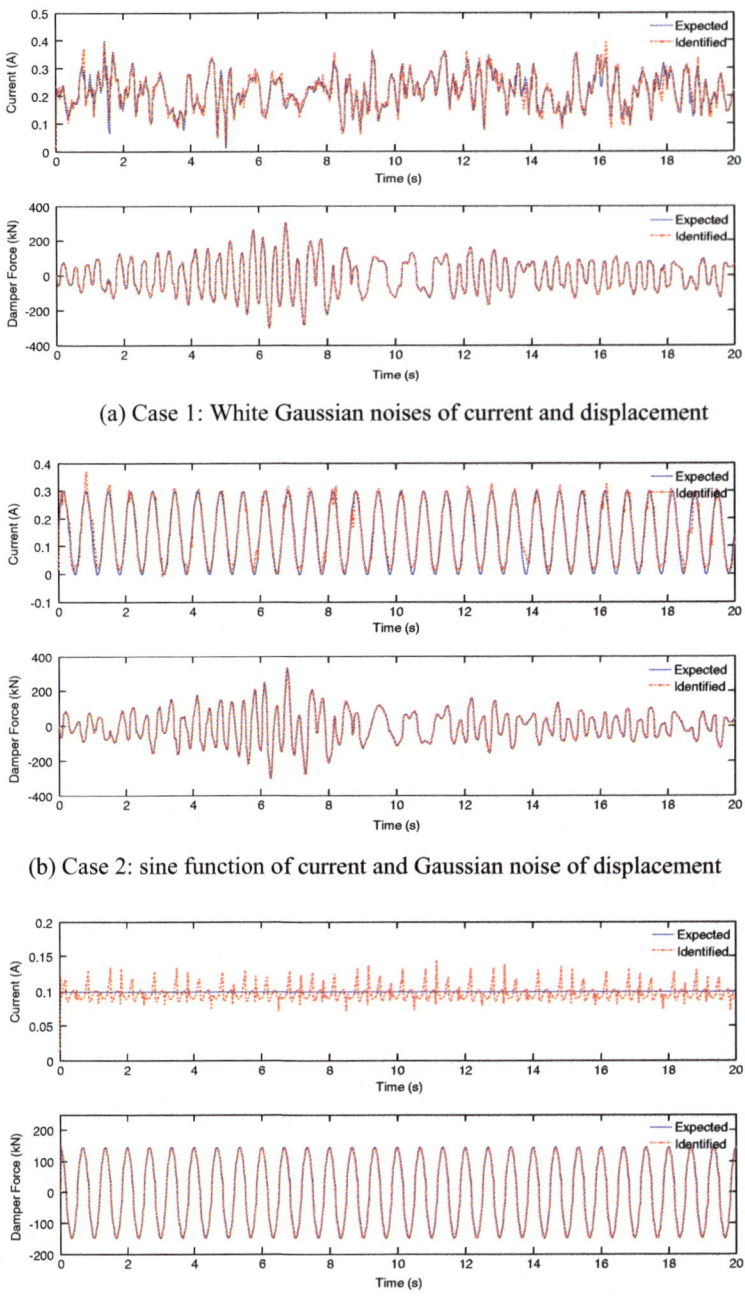

(a) Case 1: White Gaussian noises of current and displacement

(b) Case 2: sine function of current and Gaussian noise of displacement

(c) Case 3: Sine function of current and white Gaussian noise of displacement

**Fig. 8.26** Expected and identified input currents and associated outputs of MR dampers in three cases

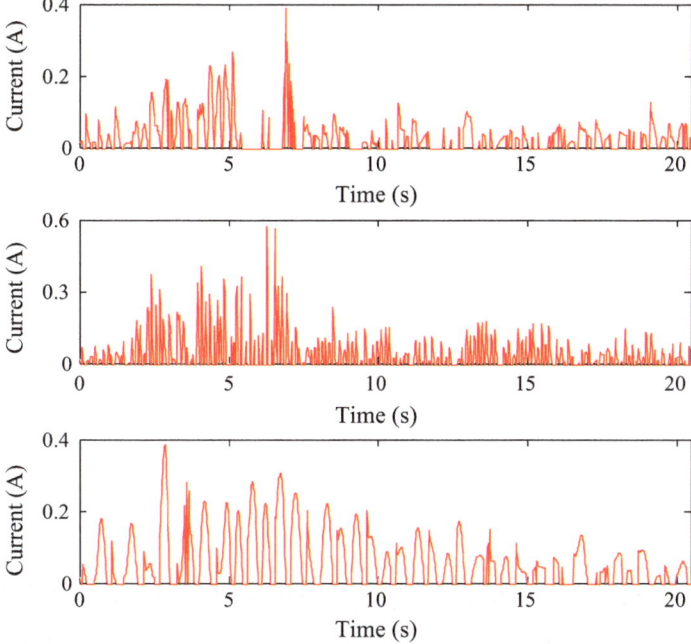

**Fig. 8.27** Optimal currents of MR damping control of structure under three different samples of seismic ground motions

## 8.5  Discussions and Summaries

The input current as the control law of regulating MR dampers is often derived from the inverse solution of damper models. How the dynamic performance of MR dampers can be fulfilled relies upon the effectiveness of the damper model. This chapter first addresses the stochastic optimal control of structures with MR dampers by virtue of bound Hrovat algorithm. In conjunction with the provided MR damper, the dynamic performance and parameterized models of MR dampers under input current, loading frequency, and amplitude of sine displacement are then illustrated. The parameter identification of MR damper model and the influence of input current on the microstructured behaviors of MR suspensions are investigated.

Molecular dynamics simulation reveals the mechanism of microstructured behaviors of MR suspensions in MR dampers, which provides a new perspective for the control law optimization and performance enhancement of MR dampers: using the numerical simulation and experimental analysis to explore the quantitative relationship between suspension structures of MR fluids, input current and material parameters of MR fluids such as dynamic viscosity and yield stress, and to reveal the physical essence of influence of eddy current effect of MR fluids and of nonlinear magnetization of MR suspensions upon the performance of MR dampers. This multiscale scheme allows for a high-efficient regulation of MR dampers so as to attain the desired performance of controlled structures.

For illustrative purposes, the stochastic optimal control of MR damping system subjected to random seismic ground motion is carried out. As the control criterion of tracing the reference active optimal control force, the parameter design of MR dampers and identification of input current are performed. Numerical results reveal that the appropriately designed semiactive controller can achieve almost the same effectiveness as the active controller; the dynamic performance of MR dampers, meanwhile, exhibits similarities to the Bouc–Wen hysteretic model with the strength deterioration, the stiffness degradation, and the pinch effect.

# References

Asai T, Chang CM, Spencer BF Jr (2015) Real-time hybrid simulation of a smart base-isolated building. J Eng Mech 141(3):04014128-1-10

Boada MJL, Calvo JA, Boada BL, Diaz V (2011) Modeling of a magnetorheological damper by recursive lazy learning. Int J Non-Linear Mech 46(3):479–485

Bouc R (1967) Forced vibration of mechanical system with hysteresis. In: Proceedings of 4th conference on nonlinear oscillations, Prague, Czechoslovakia

Carlson JD, Jolly MR (2000) MR fluid, foam and elastomer devices. Mechatronics 10(4–5):555–569

Carrion JE, Spencer BF Jr, Phill BM (2009) Real-time hybrid simulation for structural control performance assessment. Earthq Eng Eng Vibr 8(4):481–492

Casciati F, Magonette G, Marazzi F (2006) The technology of semiactive devices and applications in vibration mitigation. John Wiley & Sons, Ltd

Cha YJ, Zhang JQ, Agrawal AK, Dong BP, Friedman A, Dyke SJ, Ricles J (2013) Comparative studies of semiactive control strategies for MR dampers: pure simulation and real-time hybrid tests. J Struct Eng 139(7):1237–1248

Chae Y, Ricles JM, Sause R (2013) Modeling of a large-scale magneto-rheological damper for seismic hazard mitigation. Part II: semi-active mode. Earthq Eng Struct Dyn 42(5):687–703

Chang CC, Roschke P (1998) Neural network modeling of a magnetorheological damper. J Intell Mater Syst Struct 9(9):755–764

Charalampakis AE, Koumousis VK (2008) Identification of Bouc-Wen hysteretic systems by a hybrid evolutionary algorithm. J Sound Vib 314(3–5):571–585

Chu SY, Soong TT, Reinhorn AM (2005) Active, hybrid and semi-active structural control. Wiley, New York

Dan M, Ishizawa Y, Tanaka S, Nakahara S, Wakayama S, Kohiyama M (2015) Vibration characteristics change of a base-isolated building with semi-active dampers before, during, and after the 2011 Great East Japan earthquake. Earthq Struct 8(4):889–913

Dyke SJ, Spencer BF Jr, Sain MK, Carlson JD (1996) Modeling and control of magnetorheological dampers for seismic response reduction. Smart Mater Struct 5:565–575

Dyke SJ, Spencer BF Jr, Sain MK, Carlson JD (1998) An experimental study of MR dampers for seismic protection. Smart Mater Struct 7(5):693–703

Hogsberg J (2011) The role of negative stiffness in semi-active control of magneto-rheological dampers. Struct Control Health Monit 18(3):289–304

Hrovat D, Barak P, Rabins M (1983) Semi-active versus passive or active tuned mass dampers for structural control. ASCE J Eng Mech 109(3):691–705

Ikhouane F, Rodellar J (2005) On the hysteretic Bouc-Wen model. Nonlinear Dyn 42(1):63–78

Imaduddin F, Mazlan SA, Zamzuri H (2013) A design and modelling review of rotary magnetorheological damper. Mater Des 51:575–591

Jansen LM, Dyke SJ (2000) Semi-active control strategies for MR dampers comparative study. ASCE J Eng Mech 126(8):795–803

Jung HJ, Jang DD, Lee HJ, Lee IW, Cho SW (2010) Feasibility test of adaptive passive control system using MR fluid damper with electromagnetic induction part. J Eng Mech 136(2):254–259

Li J, Chen JB, Fan WL (2007) The equivalent extreme-value event and evaluation of the structural system reliability. Struct Saf 29(2):112–131

Liu WK, Karpov EG, Park HS (2006) Nano mechanics and materials: theory, multiscale methods and applications. Wiley, New York

Metered H, Bonello P, Oyadiji SO (2010) The experimental identification of magnetorheological dampers and evaluation of their controllers. Mech Syst Signal Process 24(4):976–994

Nagarajaiah S, Narasimhan S (2006) Smart base-isolated benchmark building. Part II: phase I sample controllers for linear isolation systems. Struct Control Health Monit 13(2–3):589–604

Ni YQ, Chen Y, Ko JM, Cao DQ (2002) Neuro-control of cable vibration using semi-active magneto-rheological dampers. Eng Struct 24:295–307

Patten WN, Mo C, Kuehn J, Lee J (1998) A primer on design of semi-active vibration absorbers (SAVA). ASCE J Eng Mech 124(1):61–68

Peng YB, Li J (2011) Multiscale analysis of stochastic fluctuation of dynamic yield of magnetorheological fluids. Int J Multiscale Comput Eng 9(2):175–191

Peng YB, Ghanem R, Li J (2012) Investigations of microstructured behaviors of magnetorheological suspensions. J Intell Mater Syst Struct 23(12):1349–1368

Peng YB, Yang JG, Li J (2017) Seismic risk based stochastic optimal control of structures using magnetorheological dampers. Nat Hazards Rev ASCE 18(1):UNSP B4016001

Peng YB, Yang JG, Li J (2018) Parameter identification of modified Bouc-Wen model and analysis of size effect of magnetorheological dampers. J Intell Mater Syst Struct 29(7):1464–1480

Plimpton SJ (1995) Fast parallel algorithms for short-range molecular dynamics. J Comput Phys 117:1–19. lammps.sandia.gov

Soong TT (1990) Active structural control: theory and practice. Longman Scientific & Technical, New York

Spencer BF Jr, Sain MK, Carlson JD (1997) Phenomenological model of magnetorheological damper. ASCE J Eng Mech 123(3):230–238

Swope WC, Andersen HC, Berens PH, Wilson KR (1982) A computer simulation method for the calculation of equilibrium constrains for the formation of physical cluster of molecules: application to small water clusters. J Chem Phys 76(1):637–649

Tsang HH, Su RKL, Chandler AM (2006) Simplified inverse dynamics models for MR fluid dampers. Eng Struct 28(3):327–341

Tse T, Chang CC (2004) Shear-mode rotary magnetorheological damper for small-scale structural control experiments. J Struct Eng 130(6):904–911

Verlet L (1967) Computer "experiments" on classical fluids I: thermodynamical properties of Lennard-Jones molecules. Phys Rev 159(1):98–103

Wang YM, Dyke S (2013) Modal-based LQG for smart base isolation system design in seismic response control. Struct Control Health Monit 20(5):753–768

Wen YK (1976) Method for random vibration of hysteretic systems. ASCE J Eng Mech Div 102(2):249–263

Xu ZD, Guo YQ (2008) Neuro-fuzzy control strategy for earthquake-excited nonlinear magnetorheological structures. Soil Dyn Earthq Eng 28(9):717–727

Xu ZD, Jia DH, Zhang XC (2012) Performance tests and mathematical model considering magnetic saturation for magnetorheological damper. J Intell Mater Syst Struct 23(12):1331–1349

Yang G (2001) Large-scale magnetorheological fluid damper for vibration mitigation: modeling, testing and control. PhD Thesis, University of Notre Dame, USA

Yang G, Spencer BF Jr, Carlson JD, Sain MK (2002) Large-scale MR fluid dampers: modeling and dynamic performance considerations. Eng Struct 24(3):309–323

Yang MG, Li CY, Chen ZQ (2013) A new simple non-linear hysteretic model for MR damper and verification of seismic response reduction experiment. Eng Struct 52:434–445

Ying ZG, Ni YQ, Ko JM (2009) A semi-active stochastic optimal control strategy for nonlinear structural systems with MR dampers. Smart Struct Syst 5(1):69–79

Yoshioka H, Ramallo JC, Spencer BF Jr (2002) "Smart" base isolation strategies employing magnetorheological dampers. J Eng Mech 128(5):540–551

Zhong WX (2004) On precise integration method. J Comput Appl Math 163:59–78

# Chapter 9
# Experimental Studies of Stochastic Optimal Control

## 9.1 Preliminary Remarks

Experimental studies provide a validating means for revealing the effectiveness of control systems upon structures. Utilizing the shaking table to carry out the simulation of seismic response aims at exploring the damage mechanism and seismic capacity of engineering structures. As the practical demand and the sustaining development of shaking-table test techniques, the motivation of shaking-table tests has been transferred to the validations of structural control and soil-structure interaction from the traditional investigations on seismic capacity of structures and facilities for nearly 30 years. For instance, Chung and his colleagues performed the shaking-table test on the Clough testing model attached with active control systems (Chung et al. 1989). Spencer and Dyke carried out the shaking-table tests of structures with active mass dampers (AMD) and active tendon systems (ATS), respectively, whereby two scale-reduced structural models used in the tests were introduced to the benchmark problem of structural control (Dyke et al. 1996; Spencer et al. 1998a, b). For more than 10 years, the hybrid simulation technique and the comprehensive simulation method have received extensive attention in the experimental studies of structural control (Wu et al. 2007; Carrion et al. 2009; Asai et al. 2015).

It is worth noting that the previous shaking-table tests of structural control almost employ one or several typical recorded ground motions with different peak ground accelerations as the base excitation (Dyke et al. 1996; Nagarajaiah et al. 2000; Kim et al. 2006; Lee et al. 2008; Jung et al. 2009). The effectiveness of vibration mitigation of seismic structures is mostly represented by the reduction of peak responses. The uncertainties, however, inherent in the occurrence and propagation of earthquakes are not taken into account. It has been proved that considering the randomness of dynamic actions upon engineering structures is one of critical challenges for structural control.

© Springer Nature Singapore Pte Ltd. and Shanghai Scientific and Technical Publishers 2019
Y. Peng and J. Li, *Stochastic Optimal Control of Structures*,
https://doi.org/10.1007/978-981-13-6764-9_9

For this reason, this chapter devotes to the shaking-table test of structural control considering the random seismic ground motions for the first time, aiming at verifying the applicability and effectiveness of the presented theory and methods of stochastic optimal control of structures.

## 9.2  Design of Test Model

### 9.2.1  Dynamics of Test Model

The structural model used for the shaking-table test is a six-story and single-bay steel structure with a geometrically similar constant 1:5 to the prototype structure. Overall dimensions of the model are 1.6 m × 1.6 m in plane, 1.0 m for the height of first story, and 0.8 m for the height of other stories. The total mass is 10.0 tons consisting of the self-weight of 2.8 tons and the artificial mass of 7.2 tons distributed uniformly on the six stories that consider the structural fitment and story live loads of the prototype structure. This treatment allows for a similar dynamic behavior to the prototype structure. The material of columns and beams both employs Q345 channel steel of which the nominal yield strength is 345 MPa. The channel steel for the columns and for the frame and non-frame beams is M8 with height 80 mm, width 43 mm, and height 5.0 mm and M6.3 with height 63 mm, width 40 mm, and height 4.8 mm, respectively. The material of stories employs Q235 steel plate with height 10 mm, of which the nominal yield strength is 235 MPa.

The model-to-prototype ratios on time, mass, and length are 0.4472, 0.04, and 0.2, respectively. The dynamic similarity of physical quantities between the model and the prototype are shown in Table 9.1.

Using the commercial software ANSYS, the modeling and analysis of test structure are carried out. As shown in Fig. 9.1, the finite element model includes 3000 elements, involving the column and beam elements BEAM188, the story elements SHELL63, and the additional mass elements MASS21. The former six-order natural frequencies in unit Hz are denoted by 1.46, 4.62, 8.38, 12.46, 16.79, and 20.25, respectively. It is shown in the modal analysis that the lateral deformation of the model under the one-dimensional horizontal excitation gives rise to a shear-type structure along the vibrational direction, since the in-plane stiffness of stories is far larger than the column stiffness.

### 9.2.2  Representative Seismic Ground Motions

Physically motivated random seismic ground motion model is employed to generate the representative seismic ground motions used for the shaking-table test (Peng et al. 2014). In this model, four basic random variables are involved, i.e., the amplitude of seismic ground motion at the bedrock, the fundamental frequency of local site,

**Table 9.1** Dynamic similarity of physical quantities between model and prototype

| Classes | Physical quantities | Physical relationships | Similarity ratios | Remarks |
|---|---|---|---|---|
| Material property | Stress $\sigma$ | $S_\sigma = S_E$ | 1 | |
| | Strain $\varepsilon$ | 1 | 1 | |
| | Elastic modulus $E$ | $S_E$ | 1 | Model control |
| | Poisson ratio $\upsilon$ | 1 | 1 | |
| | Density $\rho$ | $S_\rho = S_E/S_l$ | 5 | Model control |
| Geometrical property | Length $l$ | $S_l$ | 0.2 | Model control |
| | Linear displacement $x$ | $S_x = S_l$ | 0.2 | |
| | Angular displacement $\theta$ | 1 | 1 | |
| | Area $A$ | $S_A = S_l^2$ | 0.04 | |
| | Moment of inertia $I$ | $S_I = S_l^4$ | 0.0016 | |
| Loadings | Concentrated load $P$ | $S_P = S_E S_l^2$ | 0.04 | |
| | Linear load $\omega$ | $S_\omega = S_E S_l$ | 0.2 | |
| | Area load $q$ | $S_q = S_E$ | 1 | |
| | Moment of force $M$ | $S_M = S_E S_l^3$ | 0.008 | |
| Dynamic property | Mass $m$ | $S_m = S_\rho S_l^3$ | 0.04 | |
| | Stiffness $k$ | $S_k = S_E S_l$ | 0.2 | |
| | Damping $c$ | $S_c = S_m/S_t$ | 0.0894 | |
| | Time $t$, Period $T$ | $S_t = S_T = (S_m/S_k)^{1/2}$ | 0.4472 | Loading control |
| | Velocity $\dot{x}$ | $S_{\dot{x}} = S_x/S_t$ | 0.4472 | |
| | Acceleration $\ddot{x}$ | $S_{\ddot{x}} = S_x/S_t^2 = S_{\ddot{x}_g}$ | 1 | Loading control |

equivalent damping ratio of local site, and the initial phase angle. In view of the Chinese Code for Seismic Design of Building Structures (GB50011-2010) and the engineering background of the prototype structure, the local site belongs to site class II, the seismic fortification intensity is 7, and the Fourier amplitude of seismic ground motion at the bedrock is set as a deterministic value through introducing the conditional seismic ground motion, i.e., the earthquake occurrence period 475 years, the peak ground acceleration of seismic ground motions $0.1g$. The associated parameters with the random seismic ground motion model are shown in Table 9.2.

**Fig. 9.1** Finite element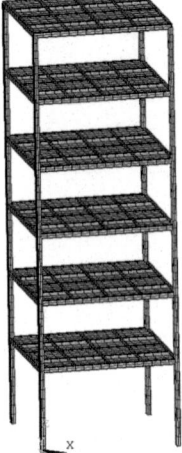
model of test structure

**Table 9.2** Parameters of physically motivated random seismic ground motion model

| Random variables | Fourier amplitude | Circular frequency | Damping ratio | Initial phase angle |
|---|---|---|---|---|
| Mean | 0.25 (m s$^{-2}$) | 20 (rad s$^{-1}$) | 0.7 | $\pi$ |
| Coefficient of variation | 0 | 0.4 | 0.3 | 1.2 |

According to the model parameters, 120 representative seismic ground motions and their assigned probabilities are derived by means of the tangent spheres method (Chen and Li 2008). For illustrative purposes, the seismic ground motion with mean parameters is denoted as the mean-parameterized seismic ground motion. All these seismic ground motions have the same sampling frequency of 50 Hz and the same duration of 20.48 s.

In order to validate that the test model always remains in a linear elastic state during the whole experiment so that similar initial conditions of structure can be guaranteed for each case of seismic ground motions, the peak ground accelerations of representative ground motions need to be regulated. According to the results of dynamic analysis of test model by ANSYS, it is found that when the mean-parameterized seismic ground motion with peak ground acceleration 2.15 m/s$^2$ (0.22$g$) serves as the input, the peak of the bottom interstory drift is 12.0 mm, and the von Mises stress at the bottom column attains 307 MPa which approaches to the design strength of the Q345 steel, i.e., 310 MPa. Meanwhile, the response of test model needs to be moderate so that the seismic mitigation of the test model with control is significant. Under these conditions, the peak ground accelerations of the 120 representative seismic ground motions are regulated, of which the minimum, the maximum, the mean, and the coefficient of variation are set as 0.39 m/s$^2$, 2.30 m/s$^2$, 1.09 m/s$^2$, and 0.256, respectively.

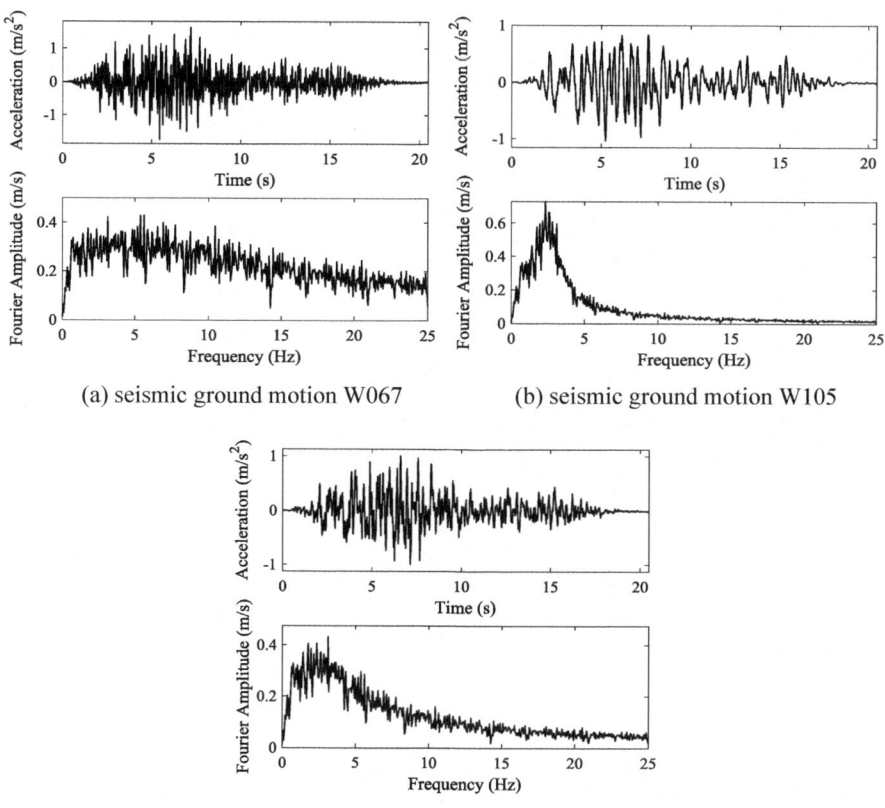

(a) seismic ground motion W067          (b) seismic ground motion W105

(c) mean-parameterized seismic ground motion W000

**Fig. 9.2** Time histories of representative and mean-parameterized seismic ground motions and their Fourier amplitude spectra

The peak ground acceleration of the mean-parameterized seismic ground motion is 0.98 m/s$^2$ (0.1$g$), which is in accordance with the basic design seismic acceleration of seismic fortification intensity 7 provided by the Chinese Code for Seismic Design of Building Structures (GB50011-2010).

Figure 9.2 shows the time histories of two representatives, the mean-parameterized seismic ground motions, and their Fourier amplitude spectra. The two representative seismic ground motions are labeled as W067 and W105, respectively. The mean-parameterized seismic ground motion is labeled as W000. One might recognize that the time histories and Fourier amplitude spectra are significantly different from each other, which is just the consideration of randomness inherent in seismic ground motions.

## 9.2.3  Design Parameters of Viscous Dampers

Owing to the effective energy dissipation and the excellent durability, the viscous dampers have been widely used in the vibration control of engineering structures (Symans and Constantinou 1998; McNamara and Taylor 2003). Three viscous dampers, labeled VD-A, VD-B, and VD-C, respectively, are employed as the control devices for the present shaking-table test; see Fig. 9.3. According to the results of finite element analysis, these dampers are designed as the same specification: a maximum output 10 kN, a stroke range ±50 mm, a length 670 mm in their equilibrium position, and the diameter of cylinder 85 mm. The hysteretic curves of the three viscous dampers under the axial loading of sine displacement are shown in Fig. 9.4. It is seen that these curves give rise to be pretty plump, indicating that the viscous dampers exhibit a good energy dissipation behavior. Table 9.3 shows the testing data of the dampers under the axial loading of sine displacement.

It is seen from the table that under the same loading condition, the outputs of the three viscous dampers are slightly different. With the concern of the specimen difference and velocity relevance, the design parameters of the viscous damper, in conjunction with its dynamic model, are derived through an optimization analysis (Yun et al. 2008): $F_D = c_D V^\alpha$, where $F_D$ denotes the damper force; $\alpha$ denotes the velocity exponent, which is set as 0.3; $c_D$ denotes the damping coefficient, which is set as 20 kN/(m/s)$^{0.3}$; and $V$ denotes the piston velocity. Figure 9.5 shows the testing data and design curve of the viscous dampers.

**Fig. 9.3**  Photo of viscous dampers VD-A, VD-B, and VD-C

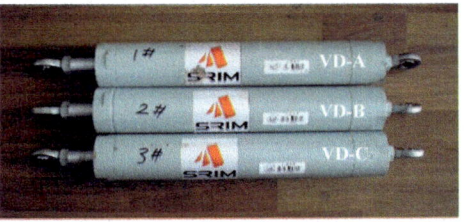

**Fig. 9.4**  Hysteretic curves of three viscous dampers under axial loading of sine displacement

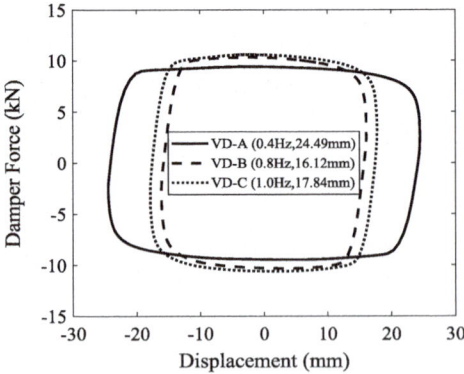

**Table 9.3** Testing data of viscous dampers under axial loading of sine displacement

| Dampers | Sine displacement | | $V_{max}$ (mm/s) | $F_{D,act}$ (kN) | $F_{D,des}$ (kN) | $(F_{D,act} - F_{D,des})/F_{D,des}$ (%) | On average (%) |
|---------|-----------|-----------|---|---|---|---|---|
| | Frequency (Hz) | Amplitude (mm) | | | | | |
| VD-A | 0.4 | 16.56 | 41.63 | 8.83 | 7.71 | 14.58 | 6.3 |
| | 0.4 | 24.49 | 61.56 | 9.44 | 8.67 | 8.93 | |
| | 0.8 | 16.61 | 83.50 | 9.36 | 9.50 | −1.43 | |
| | 1.0 | 17.86 | 112.23 | 10.71 | 10.38 | 3.21 | |
| VD-B | 0.4 | 16.60 | 41.73 | 9.05 | 7.71 | 17.35 | 10.6 |
| | 0.8 | 16.12 | 81.04 | 10.32 | 9.41 | 9.66 | |
| | 1.0 | 17.88 | 112.36 | 10.89 | 10.38 | 4.91 | |
| VD-C | 0.4 | 16.60 | 41.73 | 8.77 | 7.71 | 13.72 | 6.9 |
| | 0.8 | 18.09 | 90.94 | 10.21 | 9.74 | 4.80 | |
| | 1.0 | 17.84 | 112.11 | 10.59 | 10.37 | 2.09 | |

$V_{max}$ denotes the maximum velocity; $F_{D,act}$ denotes the actual damper force; $F_{D,des}$ denotes the design damper force

**Fig. 9.5** Testing data and design curve of viscous dampers

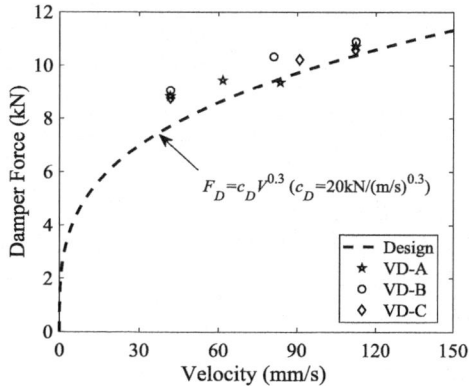

$$F_D = c_D V^{0.3} \ (c_D = 20 \text{kN}/(\text{m/s})^{0.3})$$

## 9.3 Experimental Layout and Case Verification

### 9.3.1 Experimental Layout

The shaking-table test is carried out on the platform provided by the State Key Laboratory of Disaster Reduction in Civil Engineering at Tongji University, China. The experimental facility includes a 4.0 m × 4.0 m MTS shaking table with a capacity of $2.5 \times 10^4$ kg. The motion of the shaking table involves $X$, $Y$, and $Z$ which are the three spatial dimensions and six degrees of freedoms. In the case of bearing $1.5 \times 10^4$ kg specimen, the maximum accelerations exerted on the horizontal direction of the table, $X$ and $Y$, are up to $1.2g$ and $0.8g$, respectively. The schematic and photo of the test structural model are shown in Figs. 9.6 and 9.7, respectively.

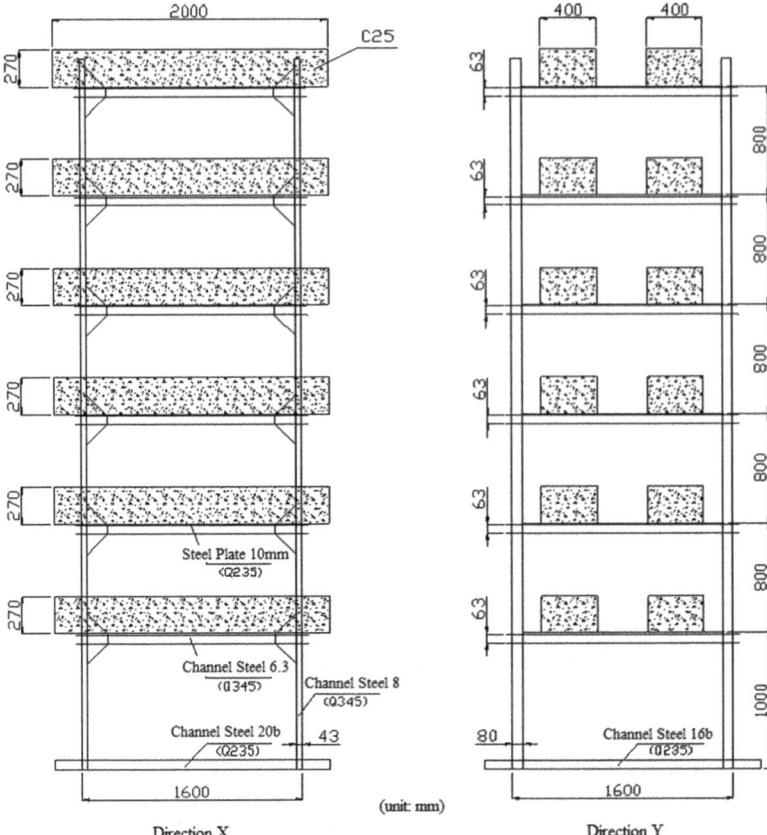

**Fig. 9.6** Schematic of test structural model

In the shaking-table test, the data are recorded by an automatic data acquisition system. The piezoelectric accelerometers, displacement transducers, and resistance strain gages are employed to measure the dynamic responses of the structural model during the shaking-table test. The displacement transducers and the piezoelectric accelerometers are deployed both along two horizontal directions, i.e., $X$ and $Y$. $X$ denotes the main direction of table motion, which is the input direction of the one-dimensional seismic ground motion. The distribution of monitoring points with displacement transducers and piezoelectric accelerometers is shown in Fig. 9.8. Displacement and acceleration measurements involve 10 monitoring points, respectively. Along the direction of table motion, a displacement transducer and a piezoelectric accelerometer are both deployed on the shaking table and each story. Perpendicular to direction of table motion, a displacement transducer and a piezoelectric accelerometer are both placed on the shaking table, the third and sixth stories. The monitoring points along the direction $X$ are used to assess the dynamic responses of structural model subjected to the one-dimensional seismic ground motion, while the monitor-

**Fig. 9.7** Photo of test structural model

ing points along the direction $Y$ are used to check whether the model exhibits an obvious vibration at the direction perpendicular to direction of table motion.

Besides, resistance strain gages are used to measure the dynamic strain at the positions with larger stress such as at the bottom column and the joint between beams and columns, validating whether the state of the structural model under seismic ground motions remains in the linear elastic stage. Totally eight resistance strain gages are used and their layout is shown in Fig. 9.9.

It has been proved in shaking-table tests that the D/A transfer of loading cell during the test loading tends to cause an error that might result in an inconsistency between the real input of seismic ground motion on the shaking table and the design of seismic ground motion. In order to validate the accuracy of loading cell of the shaking table, a representative seismic ground motion is tested. Figure 9.10 shows the data of the design and the recorded seismic ground motion at the shaking table subjected to the representative seismic ground acceleration W056. It is seen that the design matches well with the recorded on both the time series and the Fourier amplitude spectrum, revealing that the loading cell of shaking table exhibits a good accuracy.

**Fig. 9.8** Setup of
displacement transducers
and piezoelectric
accelerometers on test model

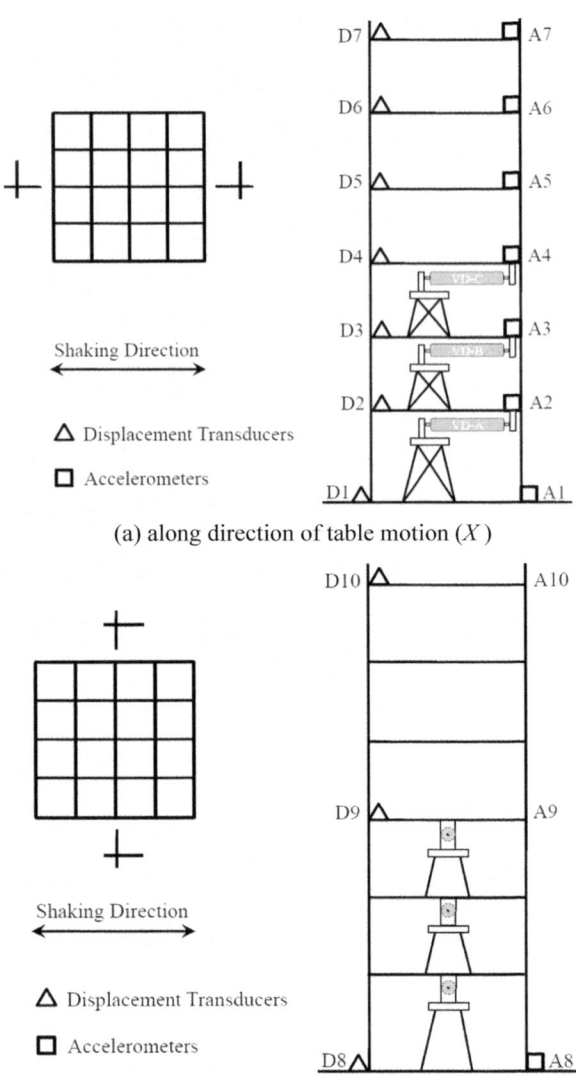

(a) along direction of table motion ($X$)

(b) perpendicular direction of table motion ($Y$)

## *9.3.2   Case Verification*

The cases of shaking-table test are shown in Table 9.4. Among these cases, Case
1 and Case 243 refer to the mean-parameterized seismic ground acceleration, i.e.,
W000, with the peak ground acceleration 0.1$g$. These two cases are considered for
identifying the parameters of the mass-lumped system so as to perform the numerical
analysis of test structural model and the optimization of damper placement. Also,
the measure on the two cases helps to validate the linear elastic state of the structural

**Fig. 9.9** Layout of resistance strain gages

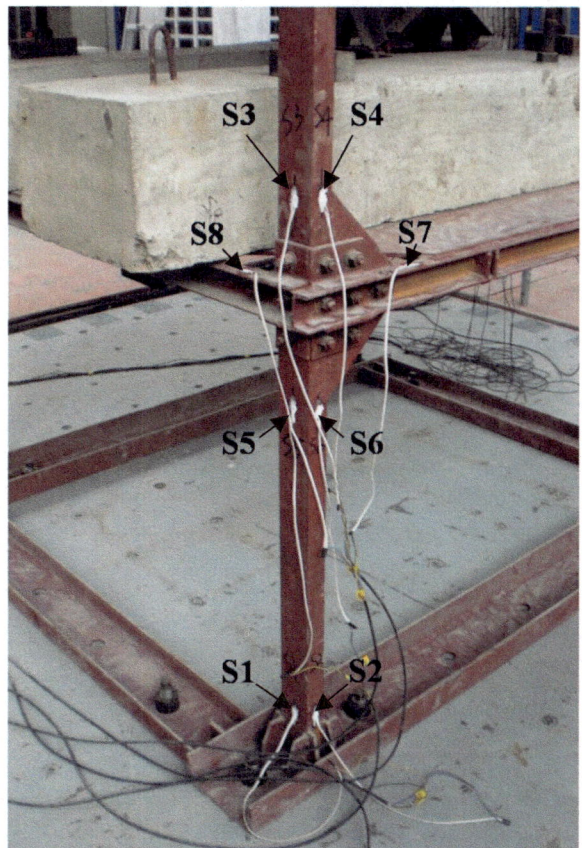

**Fig. 9.10** Design and recorded ground motions of shaking table subjected to W056

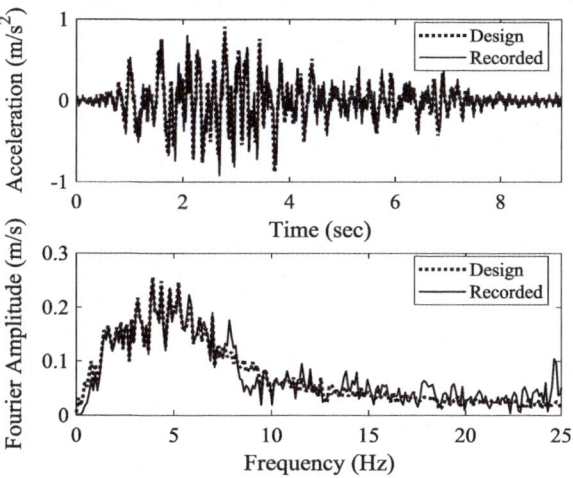

**Table 9.4**  Cases of shaking-table tests on structural control

| Cases | Seismic input | Peak ground acceleration (m/s²) | Remark |
|---|---|---|---|
| 1 | W000 | 1.00 | Without control |
| 2–121 | W001-W120 | Minimum 0.39, Maximum 2.30 Mean 1.09, Std. 0.28 | Passive control |
| 122 | W000 | 1.00 | Passive control |
| 123–242 | W001-W120 | Minimum 0.39, Maximum 2.30 Mean 1.09, Std. 0.28 | Without control |
| 243 | W000 | 1.00 | Without control |

**Fig. 9.11**  Amplitude–frequency curve of story accelerations in Case 1

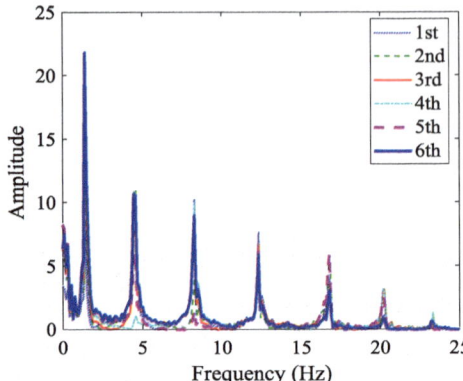

model during the entire shaking-table test. Cases 2–12 and Case 122 are defined as the tests of controlled structures, and the viscous dampers are all deployed at the low three interstories. Cases 123–242 are defined as uncontrolled structures which serve as a reference for the seismic mitigation of structures with viscous dampers.

Considering that the time similarity ratio between the test and prototype structures is 0.4472, the sampling interval of the seismic acceleration at the shaking table needs to change to 0.0089 s from the original 0.02 s. Accordingly, the time length of seismic acceleration process changes to 9.16 s from the original 20.48 s.

The modal frequencies of the mass-lumped system can be derived from the amplitude–frequency curve of story acceleration. For illustrative purposes, the acceleration data of the structural model measured at Cases 1 and 243 is thus utilized to carry out the parameter identification of test model. Figure 9.11 shows the amplitude–frequency curve of the story accelerations in Case 1. One might recognize that the frequencies corresponding to the peak of amplitude–frequency curve are related to the modal frequencies. The former six-order modal frequencies of the structural model are shown in Table 9.5. It is seen that the modal frequencies of the test model change slightly during the entire shaking-table test. For instance, the fundamental period of the structure is 1.460 Hz at Case 1 and 1.453 Hz at Case 243, revealing that the structural fundamental period reduces 0.48% after the shaking-table test.

**Table 9.5** Identified former six-order modal frequencies of structural model

| Cases | Modal frequencies (Hz) | | | | | |
|---|---|---|---|---|---|---|
|  | 1st-order | 2nd-order | 3rd-order | 4th-order | 5th-order | 6th-order |
| 1 | 1.460 | 4.624 | 8.365 | 12.452 | 16.803 | 20.277 |
| 243 | 1.453 | 4.605 | 8.338 | 12.385 | 16.745 | 20.184 |

**Table 9.6** Identified model parameters of structural model

| Parameters | Matrices | | | | | |
|---|---|---|---|---|---|---|
| Mass matrix **M** (kg) | 1650.0 | 0.0 | 0.0 | 0.0 | 0.0 | 0.0 |
|  | 0.0 | 1640.0 | 0.0 | 0.0 | 0.0 | 0.0 |
|  | 0.0 | 0.0 | 1570.0 | 0.0 | 0.0 | 0.0 |
|  | 0.0 | 0.0 | 0.0 | 1560.0 | 0.0 | 0.0 |
|  | 0.0 | 0.0 | 0.0 | 0.0 | 1560.0 | 0.0 |
|  | 0.0 | 0.0 | 0.0 | 0.0 | 0.0 | 1570.0 |
| Stiffness matrix **K** (N/m) | 8,015,844.2 | −6,684,376.6 | 1,453,296.1 | −158,645.8 | 91,651.8 | 40,338.2 |
|  | −6,684,376.6 | 12,082,761.6 | −7,307,937.6 | 1,349,673.9 | −248,284.3 | 56,120.5 |
|  | 1,453,296.1 | −7,307,937.6 | 11,821,305.5 | −6,988,277.6 | 1,390,059.2 | −224,321.8 |
|  | −158,645.8 | 1,349,673.9 | −6,988,277.6 | 11,630,474.0 | −7,012,874.1 | 1,155,521.9 |
|  | 91,651.8 | −248,284.3 | 1,390,059.2 | −7,012,874.1 | 11,286,009.9 | −5,454,943.7 |
|  | 40,338.2 | 56,120.5 | −224,321.8 | 1,155,521.9 | −5,454,943.7 | 4,393,435.1 |
| Damping matrix **C** (N/(m/s)) | 1018.1 | −350.2 | 76.1 | −8.3 | 4.8 | 2.1 |
|  | −350.2 | 1227.5 | −382.8 | 70.7 | −13.0 | 2.9 |
|  | 76.1 | −382.8 | 1188.4 | −366.1 | 72.8 | −11.8 |
|  | −8.3 | 70.7 | −366.1 | 1174.8 | −367.4 | 60.5 |
|  | 4.8 | −13.0 | 72.8 | −367.4 | 1156.8 | −285.8 |
|  | 2.1 | 2.9 | −11.8 | 60.5 | −285.8 | 799.3 |

Meanwhile, the first three modal frequencies almost remain unchanged. Therefore, the test model remains the linear elastic state during the shaking-table test.

According to the identified results of the modal frequencies and shapes, one can readily attain the model parameters of the mass-lumped system. The mass matrix is calculated straightforwardly from the sum of story self-weight of the test model and the additional mass. The stiffness matrix is then given by

$$\mathbf{K} = [\mathbf{\Phi}^T]^{-1}\mathbf{\Omega}[\mathbf{\Phi}]^{-1} \tag{9.3.1}$$

where **K** denotes the stiffness matrix; **Φ** denotes the modal matrix, i.e., the normalized modal shape with respect to mass matrix; **Ω** denotes the diagonal matrix with elements $\omega_i^2$; and $\omega_i$ denotes the circular frequency of the $i$th mode. Besides, the structural damping is considered as the Rayleigh damping. The identified model parameters are shown in Table 9.6.

**Fig. 9.12** Comparison of
structural responses between
shaking-table test and
dynamic analysis under Case
122

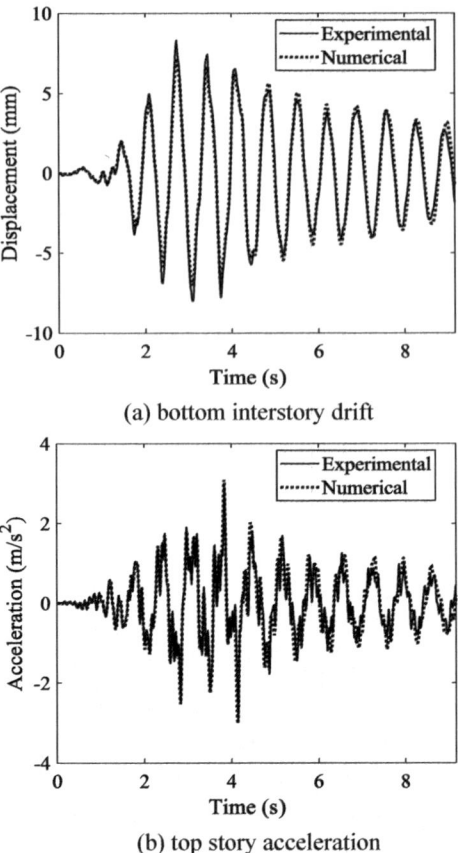

(a) bottom interstory drift

(b) top story acceleration

**Fig. 9.12** Comparison of structural responses between shaking-table test and dynamic analysis under Case 122

The model parameters are verified by virtue of the experimental data and the numerical simulation. Figures 9.12 and 9.13 show the structural responses between the shaking-table test and the dynamic analysis under Cases 122 and 241, respectively. It is ready to see that the mass-lumped system with model parameters identified from the experimental data is reliable to reveal the dynamics of the test model.

Using the mass-lumped system, the optimization of viscous damper deployment in the structural space under the random seismic ground motion is carried out. In the framework of physically based stochastic optimal control, the generalized optimal control policy is applied. The optimization of viscous damper deployments is carried out as the probabilistic criterion on minimizing the mean of equivalent extreme values of interstory drifts, which is given by

$$I_{D,i}^* = \arg\min_{I_{D,i}}\{J\} = \arg\min_{I_{D,i}}\left\{ E[\tilde{X}(\Theta)]\middle|\left\{\sum_i I_{D,i} = N_D\right\}\right\}, \quad i = 1, 2, \ldots, n$$

$$(9.3.2)$$

**Fig. 9.13** Comparison of structural responses between shaking-table test and dynamic analysis under Case 241

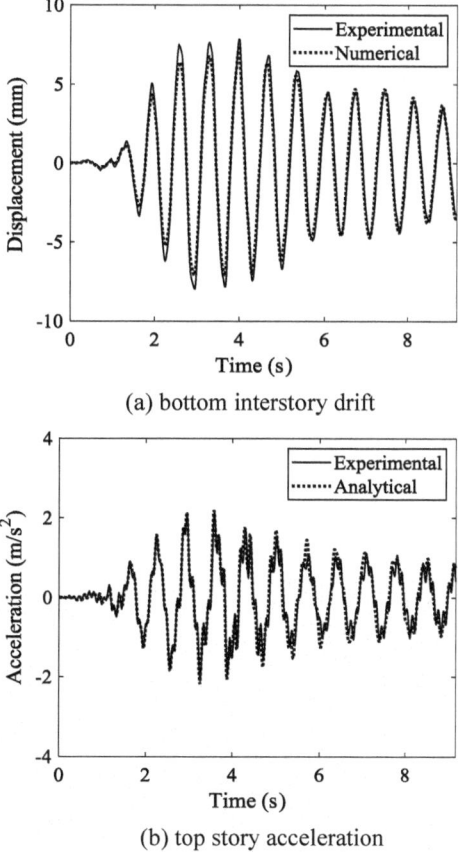

(a) bottom interstory drift

(b) top story acceleration

where $\tilde{X}(\boldsymbol{\Theta}) = \max_t[\max_i|X_i(\boldsymbol{\Theta}, t)|]$, $X_i(\boldsymbol{\Theta}, t)$ denotes the interstory drift of the $i$th story; $I_{D,i}$ denotes an indicative function representing that there is a damper, i.e., $I_{D,i} = 1$ or not, i.e., $I_{D,i} = 0$ in the interstory drift of the $i$th story; and $N_D$ denotes the total number of viscous dampers that needs to be placed.

Numerical results reveal that the preferable placement of the three dampers is the low three interstories which can gain the best seismic mitigation. The structural model with the viscous dampers is shown in Fig. 9.7.

## 9.4   Experimental Analysis

### 9.4.1   Samples and Ensemble

The dynamic responses of the structural model with and without controls under three representative seismic ground motions are shown in Figs. 9.14, 9.15 and 9.16. The seismic mitigation ratio of the root-mean-square responses of structures is shown in Table 9.7.

It is seen that under different representative seismic ground motions, the seismic mitigation ratio of structural responses changes significantly even in the case of the same damper deployment. When the representative seismic ground motion W067 is employed as the input, the bottom interstory drift gains a significant reduction while the top story acceleration has no such improvement. When the representative seismic ground motion W105 is employed as the input, the top story acceleration gives rise to be enlarged although the bottom interstory drift gains a reduction. When the mean-

**Fig. 9.14** Responses of structural model with and without controls under seismic ground motion W067

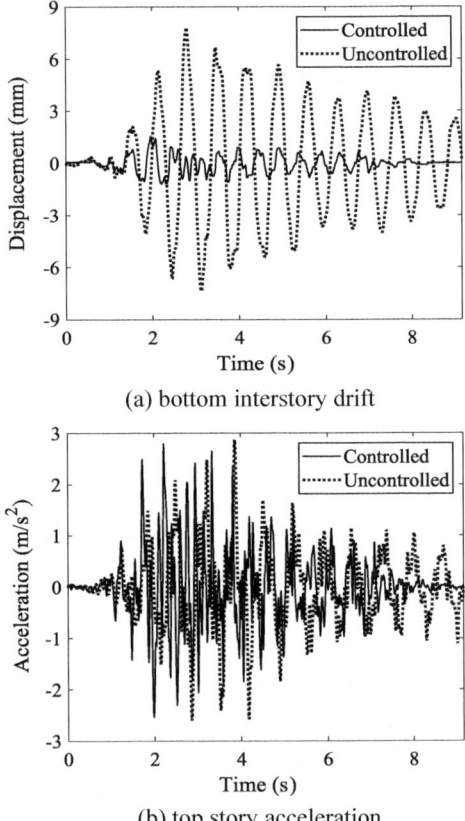

(a) bottom interstory drift

(b) top story acceleration

**Fig. 9.15** Responses of structural model with and without controls under seismic ground motion W105

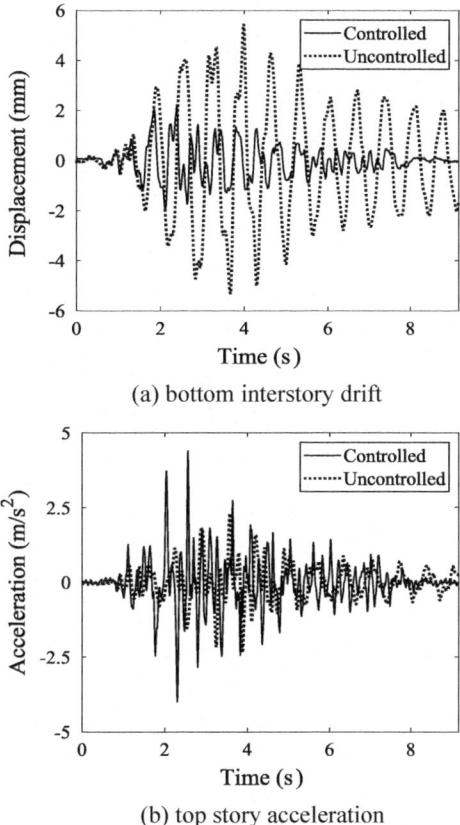

(a) bottom interstory drift

(b) top story acceleration

parameterized seismic ground motion W000 is employed as the input, both the top story acceleration and the bottom interstory drift gain a better improvement. It is thus remarked that the randomness inherent in the seismic ground motion indeed poses a significant influence upon the control effectiveness of viscous dampers. In fact, the probabilistic criterion used in the deployment optimization of viscous dampers exhibits a global trade-off; thereby the control effectiveness of structures is the most modest as to the ensemble of seismic ground motions.

Figures 9.17 and 9.18 show the mean and standard deviation of the root-mean-square interstory drift and the root-mean-square story acceleration along story level, respectively. The relevant seismic mitigation ratios are shown in Table 9.8. It is seen that the structural performance gains a significant improvement after the viscous dampers are deployed. Therefore, the optimally designed control system considering the randomness inherent in external excitation exhibits a robustness that prevents the structure from potential seismic hazards in the future. The stochastic optimal control of structures involves optimization and design with respect to control parameters, of which the objective can be readily implemented.

**Fig. 9.16** Response of
structural model with and
without controls under
mean-parameterized seismic
ground motion W000

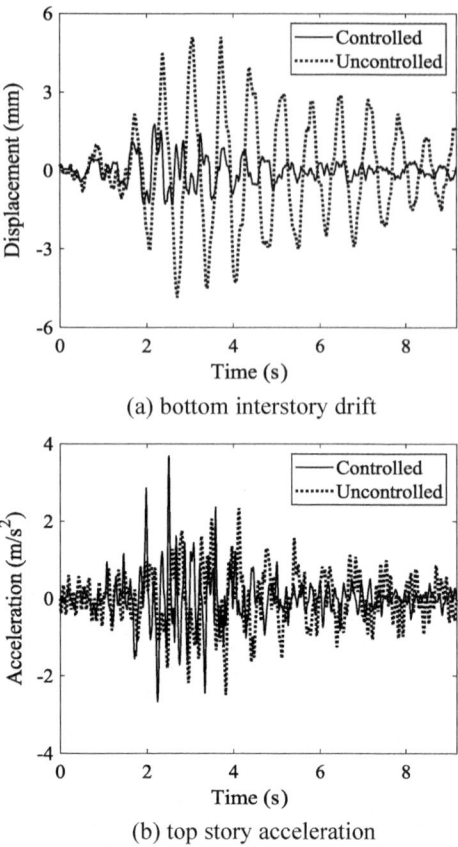

(a) bottom interstory drift

(b) top story acceleration

**Table 9.7**  Seismic mitigation ratio of root-mean-square responses of structural model

| Structural responses | Seismic ground motion | Root-mean-square responses | | |
|---|---|---|---|---|
| | | Uncontrolled | Controlled | Effectiveness[a] |
| Bottom interstory drift (mm) | W067 | 3.22 | 0.47 | −85.42% |
| | W105 | 2.21 | 0.63 | −71.48% |
| | W000 | 1.95 | 0.48 | −75.38% |
| Top story acceleration (m/s²) | W067 | 0.75 | 0.68 | −0.10% |
| | W105 | 0.57 | 0.80 | 40.79% |
| | W000 | 0.69 | 0.59 | −14.49% |

[a]Effectiveness is defined as (Con. − Unc.)/Unc.

**Fig. 9.17** Mean and standard deviation of root-mean-square interstory drift along story level

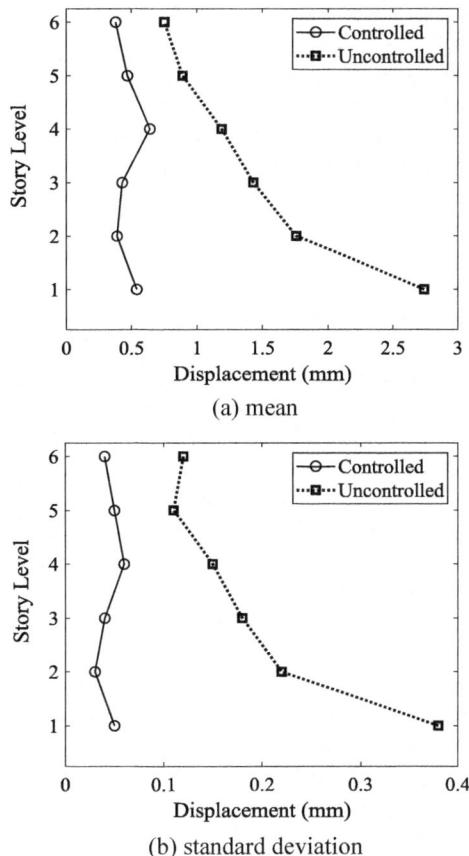

(a) mean

(b) standard deviation

One might recognize as well from Tables 9.7 and 9.8 that no matter on the samples or on the ensembles, the control effectiveness of interstory drift is pretty better than that of story acceleration. It owes to the fact that the optimization criterion of damper deployment used in this study is minimizing the mean of equivalent extreme values of interstory drifts. In fact, the effectiveness of structural control highly relies upon the physical meaning of the design criterion.

The statistical moments of root-mean-square interstory shear force varying along story level are shown in Fig. 9.19. It is seen that the interstory shear force has received extensive reduction, especially at the low stories, indicating a more smooth structural performance after the control. Table 9.9 shows the statistical moments of root-mean-square stress derived from the strain data at the measure points. It is seen that the seismic mitigation ratio of the stress attains to around 80%, which is in consistently with the control effectiveness of interstory drift of structures.

**Fig. 9.18** Mean and
standard deviation of
root-mean-square story
acceleration along story level

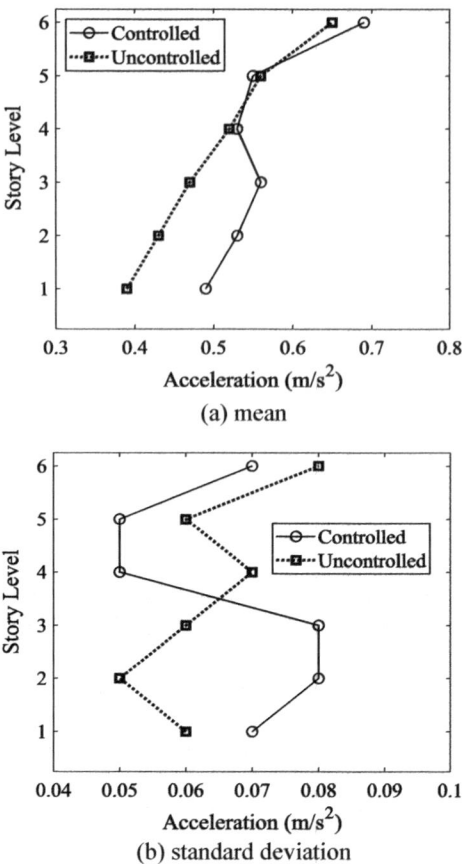

(a) mean

(b) standard deviation

According to the testing data of interstory drifts, the working performance of the viscous dampers for seismic mitigation of structures can be analyzed. For illustrative purposes, the testing data under the representative seismic ground motion W105 is used. Figure 9.20 shows the time history of damper force, relations between damper force, damper displacement, and velocity of the damper VD-A at the bottom interstory. It is seen that the nonlinear damper exhibits a good working performance, which arises an excellent energy dissipation capacity even in the case of small interstory velocities. Moreover, the other two viscous dampers deployed at the second and third interstories, i.e., VD-B and VD-C, exhibit a good working performance as well. Besides, the time histories of damper forces of the three dampers have similarities since the structural model behaves mostly in shear-type vibration as the first mode.

**Table 9.8** Mean and standard deviation of root-mean-square responses of structural model

| RMS responses | | | Story number | | | | | |
|---|---|---|---|---|---|---|---|---|
| | | | 1 | 2 | 3 | 4 | 5 | 6 |
| Interstory drift (mm) | Mn | Unc. | 2.74 | 1.76 | 1.43 | 1.19 | 0.89 | 0.75 |
| | | Con. | 0.54 | 0.39 | 0.43 | 0.64 | 0.47 | 0.38 |
| | | Eff.[a] | 80.3% | 77.8% | 69.9% | 46.2% | 47.2% | 49.3% |
| | Std. d | Unc. | 0.38 | 0.22 | 0.18 | 0.15 | 0.11 | 0.12 |
| | | Con. | 0.05 | 0.03 | 0.04 | 0.06 | 0.05 | 0.04 |
| | | Eff. | 86.8% | 86.4% | 77.8% | 60.0% | 54.5% | 66.7% |
| Story acceleration $(m/s^2)$ | Mn | Unc. | 0.39 | 0.43 | 0.47 | 0.52 | 0.56 | 0.65 |
| | | Con. | 0.49 | 0.53 | 0.56 | 0.53 | 0.55 | 0.69 |
| | | Eff. | −25.6% | −23.3% | −19.1% | −1.9% | 1.8% | −6.2% |
| | Std. d | Unc. | 0.06 | 0.05 | 0.06 | 0.07 | 0.06 | 0.08 |
| | | Con. | 0.07 | 0.08 | 0.08 | 0.05 | 0.05 | 0.07 |
| | | Eff. | −16.7% | −60.0% | −33.3% | 28.6% | 16.7% | 12.5% |

[a]Effectiveness is defined as (Unc. − Con.)/Unc.

## 9.4.2 Regulation of Probability Density

In order to reveal the control effectiveness in the sense of probability density, the testing data of structural dynamic responses at typical instants of time is first plotted into statistical histogram and the probability density is then estimated using Gaussian kernel density estimation (GKDE) (Bowman and Azzalini 1997). Figures 9.21 and 9.22 show the statistical histograms and the estimated probability density curves of the bottom interstory drift, at typical instants of time 3 s and 8 s, respectively, of the structural model with and without controls.

It is seen that the probability densities estimated by the GKDE have a sound consistency with the statistical histograms. Comparing the probability densities at same instants of time, the distribution range of the bottom interstory drift of the controlled structure is far less than that of the uncontrolled structure. It is thus revealed that the variation of structural displacement significantly decreases and the structural safety gains a remarkable enhancement. Therefore, the regulation of probability density of stochastic responses of structures can be implemented through the logical definition of probabilistic criterion.

## 9.5 Reliability Assessment

Interstory drift ratio is a critical index used for the reliability assessment of building structures. In conjunction with the equivalent extreme-value event criterion addressed in Sect. 2.4.2, the component and system reliabilities of the interstory drift ratio,

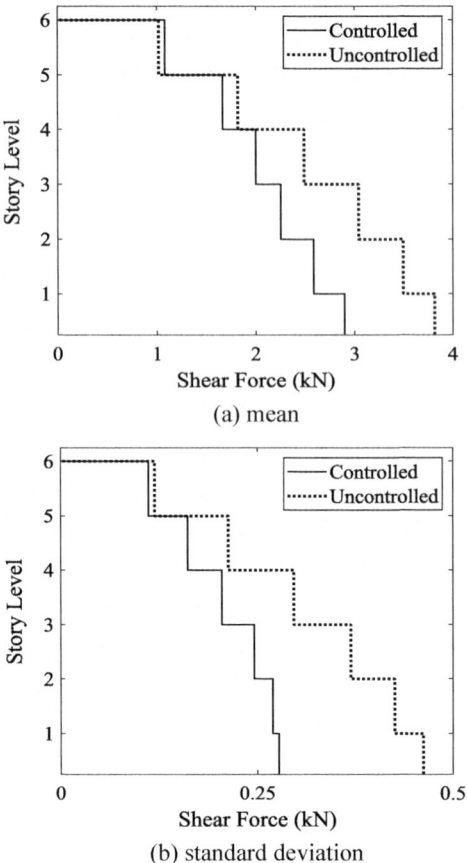

**Fig. 9.19** Mean and standard deviation of root-mean-square interstory shear force along story level

(a) mean

(b) standard deviation

under the condition of provided thresholds, of the linear elastic structure with and without controls can be readily assessed using the testing data of interstory drifts; see Table 9.10. Definition of the threshold of interstory drift ratio refers to the Chinese Code for Seismic Design of Building Structures (GB50011-2010). It is seen that the component and system reliabilities of the interstory drift ratio both gain a significant enhancement after the viscous dampers are deployed, revealing that the optimization and design of damper control attains the desired structural performance.

The probability density functions (PDFs) and cumulative density functions (CDFs) of the equivalent extreme values of interstory drift ratios of structure are shown in Figs. 9.23 and 9.24, respectively. It is seen that the equivalent extreme value of interstory drift ratio is much smaller after the viscous dampers are deployed. Moreover, the distribution range of the interstory drift ratio becomes far narrower, indicating that the system reliability of the structural model with control gains

**Table 9.9** Mean and standard deviation of root-mean-square stress at measure points

| Measure points | | S1 | S2 | S5 | S6 | S7 |
|---|---|---|---|---|---|---|
| Mean (MPa) | Unc. | 49.49 | 27.95 | 26.91 | 29.22 | 42.13 |
| | Con. | 11.54 | 6.10 | 5.56 | 7.38 | 9.16 |
| | Eff.[a] | 76.68% | 78.18% | 79.34% | 74.74% | 78.26% |
| Standard deviation (MPa) | Unc. | 6.04 | 3.39 | 3.23 | 3.60 | 5.12 |
| | Con. | 1.02 | 0.55 | 0.54 | 0.67 | 0.77 |
| | Eff. | 83.11% | 83.78% | 83.28% | 81.39% | 84.96% |

[a]Effectiveness is defined as (Unc. − Con.)/Unc.

a significant enhancement, especially in the case that the threshold of interstory drift ratio is valued in the range from 0.002 to 0.009. For instance, when the threshold is set as 0.004, the system reliability of the structure approaches to 1.0 after control but is almost 0.0 before control. It is thus revealed that the system reliability of the structure can be enhanced significantly with the application of viscous dampers. Besides, one might recognize that the control gain would be lost if the threshold of interstory drift ratio was less than 0.002 rad or was more than 0.009. This range, e.g., the thresholds 0.002–0.009 in this case, is the efficient domain for the reinforcement of viscous dampers. One might understand that a small threshold denotes a high requirement of the interstory drift ratio; while a large threshold denotes a low requirement of the interstory drift ratio. Therefore, a minimum design requirement needs to be satisfied allowing for an effective control gain; and the damper control would be improvident if a sufficient design was satisfied. According to the performance objective of structures, the decision-maker can determine whether an additional damper is needed or not through the trade-off analysis.

## 9.6  Discussions and Summaries

In this chapter, the experimental studies of stochastic optimal control are carried out. Complete shaking-table tests on a randomly base-excited framed structure with viscous dampers are involved, where the physically motivated random seismic ground motion model is employed. The control effectiveness has been verified through a variety of aspects, including the analysis of samples and ensembles, the regulation of probability density, and the reliability assessment. The findings are summarized as follows:

(i)   The physically based stochastic optimal control can accommodate the globally optimal performance of controlled structure: the interstory drift and interstory shear force become more uniform along story level than those without control.

(ii)  The critical step of the stochastic optimal control of structures is the optimization and design of parameters and placement of control devices, which just

**Fig. 9.20** Working
performance of viscous
damper VD-A for seismic
mitigation of structural
model

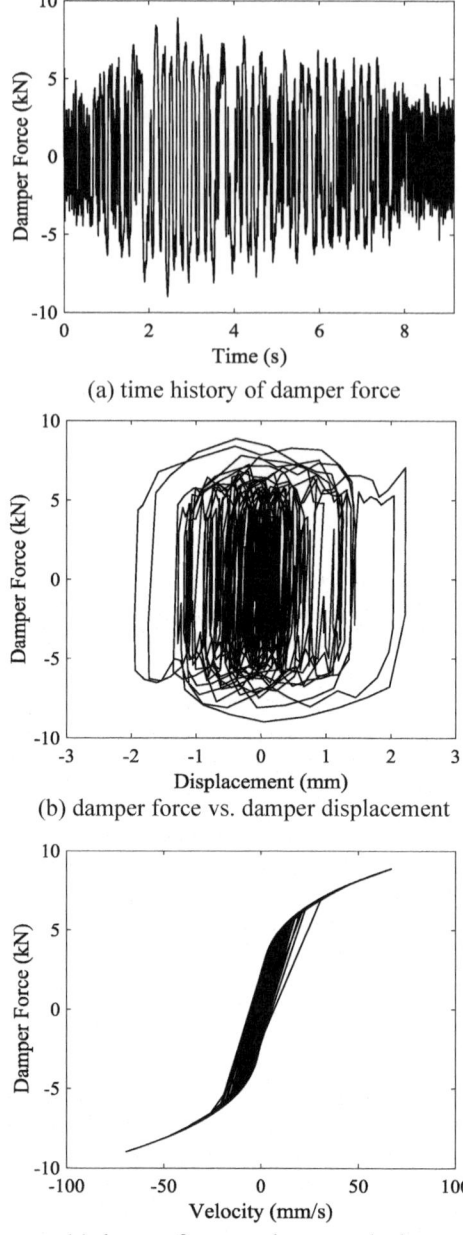

(a) time history of damper force

(b) damper force vs. damper displacement

(c) damper force vs. damper velocity

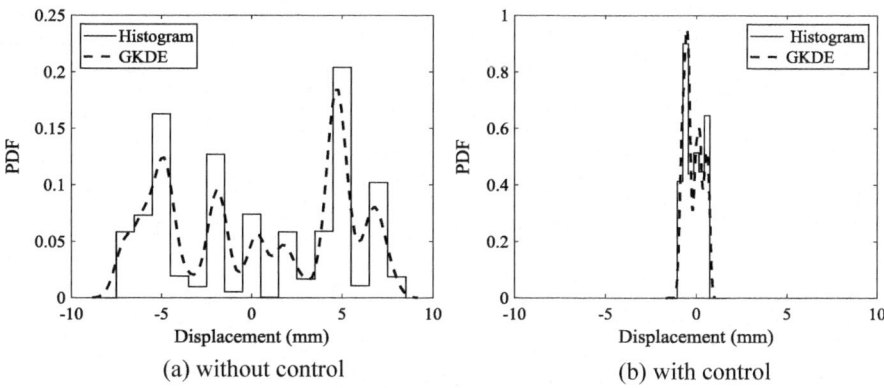

(a) without control        (b) with control

**Fig. 9.21** Statistical histograms and estimated probability density curves of bottom interstory drift at typical instant of time 3 s

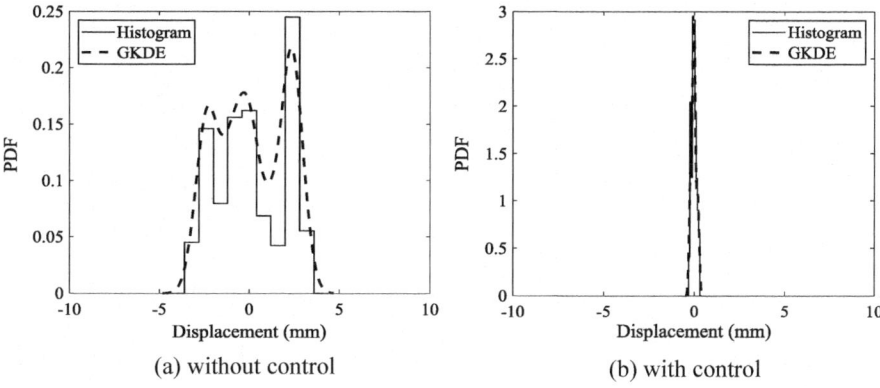

(a) without control        (b) with control

**Fig. 9.22** Statistical histograms and estimated probability density curves of bottom interstory drift at typical instant of time 8 s

**Table 9.10** Component and system reliabilities of interstory drift ratio of structure with and without controls

| Threshold | Cases | Component reliabilities | | | | | | System reliability |
|-----------|-------|-----------|-----------|-----------|-----------|-----------|-----------|-------------------|
| | | 1st story | 2nd story | 3rd story | 4th story | 5th story | 6th story | |
| 0.004 | Unc. | 0.0555 | 0.4029 | 0.7569 | 0.8736 | 1.0000 | 1.0000 | 0.0492 |
| | Con. | 1.0000 | 1.0000 | 1.0000 | 1.0000 | 1.0000 | 1.0000 | 1.0000 |

**Fig. 9.23** PDFs of
equivalent extreme values of
interstory drift ratios of
structure with and without
controls

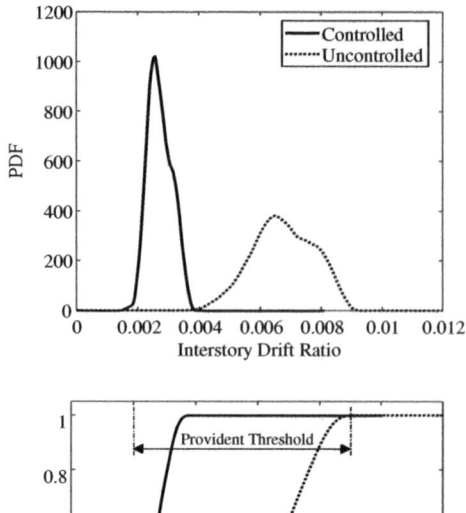

**Fig. 9.24** CDFs of
equivalent extreme values of
interstory drift ratios of
structure with and without
controls

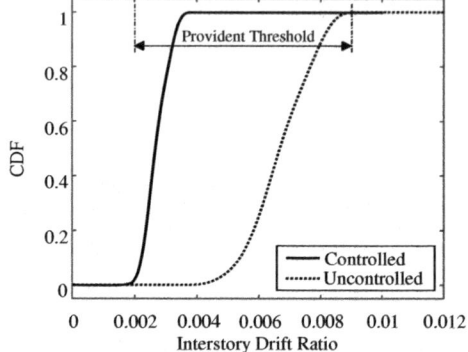

relies upon the probabilistic criterion linking to the performance objective of
structures. The probabilistic criterion of optimizing the deployment of viscous
dampers can attain a good control effectiveness, owing to its specified formulation that is in function of the mean of equivalent extreme values of interstory
drifts.

# References

Asai T, Chang CM, Spencer BF Jr (2015) Real-time hybrid simulation of a smart base-isolated
    building. J Eng Mech 141(3):04014128-1-10
Bowman AW, Azzalini A (1997) Applied Smoothing Techniques for Data Analysis. Oxford Uni-
    versity Press, Oxford
Carrion JE, Spencer BF Jr, Phill BM (2009) Real-time hybrid simulation for structural control
    performance assessment. Earthquake Engineering & Engineering Vibration 8(4):481–492
Chen JB, Li J (2008) Strategy for selecting representative points via tangent spheres in the probability
    density evolution method. Int J Numer Meth Eng 74(13):1988–2014
Chung LL, Lin RC, Soong TT, Reinhorn AM (1989) Experimental study of active control for MDOF
    seismic structures. J Eng Mech 115(8):1609–1627

Dyke SJ, Spencer BF Jr, Quast P et al (1996) Implementation of an active mass driver using acceleration feedback control. Microcomput Civil Eng 11(5):305–323

Jung HJ, Jang DD, Choi KM, Cho SW (2009) Vibration mitigation of highway isolated bridge using MR damper-based smart passive control system employing an electromagnetic induction part. Struct Control Health Monit 16(6):613–625

Kim HS, Roschke PN, Lin PY, Loh CH (2006) Neuro-fuzzy model of hybrid semi-active base isolation system with FPS bearings and an MR damper. Eng Struct 28(7):947–958

Lee HJ, Jung HJ, Cho SW, Lee IW (2008) An experimental study of semiactive modal neuro-control scheme using MR damper for building structure. J Intell Mater Syst Struct 19(9):1005–1015

McNamara RJ, Taylor DP (2003) Fluid viscous dampers for high-rise buildings. Struct Des Tall Spec Build 12:145–154

Nagarajaiah S, Sahasrabudhe S, Iyer R (2000) Seismic response of sliding isolated bridges with MR dampers. Proc Am Control Conf 6:4437–4441

Peng YB, Mei Z, Li J (2014) Stochastic seismic response analysis and reliability assessment of passively damped structures. J Vib Control 20(15):2352–2365

Spencer BF Jr, Dyke SJ, Deoskar HS (1998a) Benchmark problems in structural control: part I-active mass driver system. Earthq Eng Struct Dyn 27(11):1127–1139

Spencer BF Jr, Dyke SJ, Deoskar HS (1998b) Benchmark problems in structural control: part II-active tendon system. Earthq Eng Struct Dyn 27(11):1141–1147

Symans MD, Constatinou MC (1998) Passive fluid viscous passive systems for seismic energy dissipation. J Earthq Technol 35(4):185–206

Wu B, Wang QY, Shing B, Ou JP (2007) Equivalent force control method for generalized real-time substructure testing with implicit integration. Earthq Eng Struct Dyn 36:1127–1149

Yun HB, Tasbighoo F, Masri SF, Caffrey JP, Wolfe RW (2008) Comparison of modeling approaches for full-scale nonlinear viscous dampers. J Vib Control 14(1–2):51–76

# Appendix A
# Bouc–Wen Model for Hysteretic Component of Structures

In engineering practice, the structural system can be viewed as an assembling of a set of components, e.g., beams, columns, etc. These components usually behave in hysteretic states and suffer from serious damages when the structure experiences hazardous actions. The logical representation of restoring forces of the component and system is thus an important step for securing the dynamic behaviors of the structure. Among the proposed hysteretic models, the Bouc–Wen model is widely used in practice due to its ready application with equivalent linearization techniques.

The equation of motion of hysteretic structures subjected to random excitation is given by

$$\mathbf{M}\ddot{\mathbf{X}}(t) + \mathbf{C}_t\dot{\mathbf{X}}(t) + \mathbf{R}_t(\mathbf{X}, \mathbf{z}) = \mathbf{F}(t) \tag{A.1}$$

where $\mathbf{X}(t)$ is the $n$-dimensional column vector denoting structural displacement; $\mathbf{M}$ is the mass matrix; $\mathbf{C}_t$ is the instantaneous damping matrix; $\mathbf{F}(t)$ denotes the external excitation; $\mathbf{R}_t(\mathbf{X}, \mathbf{z})$ is the $n$-dimensional vector denoting the restoring force, which is a function of hysteretic displacement $\mathbf{z}$. The restoring force can thus be modeled as a combination of a linear elastic term and a hysteretic term:

$$\mathbf{R}_t(\mathbf{X}, \mathbf{z}) = \alpha\mathbf{K}_0\mathbf{X} + (1 - \alpha)\mathbf{K}_0\mathbf{z} \tag{A.2}$$

where $\alpha$ denotes the stiffness ratio between the post-yielding stiffness $\mathbf{K}_1$ and the pre-yielding stiffness $\mathbf{K}_0$. It is seen that the restoring force pertinently relies upon the definition of hysteretic displacement $\mathbf{z}$.

In view of the basic formulation of Osgood–Ramberg model, a model of restoring force in differential equation of smooth hysteretic displacement was proposed by Bouc and later modified by Wen (Bouc 1967; Wen 1976), where a component hysteretic displacement is given by

$$\dot{z} = A\dot{x} - \beta|\dot{x}||z|^{n-1}z + \gamma\dot{x}|z|^n \tag{A.3}$$

© Springer Nature Singapore Pte Ltd. and Shanghai Scientific and Technical Publishers 2019
Y. Peng and J. Li, *Stochastic Optimal Control of Structures*,
https://doi.org/10.1007/978-981-13-6764-9

where $A, \beta, \gamma, n$ denote the parameters of the basic formulation of component hysteretic displacement.

The basic Bouc–Wen model reveals the hysteretic behaviors of restoring force, which cannot, however, reveal the strength deterioration, stiffness degradation and pinching effect that are usually inherent in the restoring force curves of structural components (Ma et al. 2004). An extended Bouc–Wen model was thus developed and the updated formulation of component hysteretic displacement is written as (Foliente 1995)

$$\dot{z} = h(z)\left\{ \frac{A\dot{x} - v(\beta|\dot{x}||z|^{n-1}z + \gamma\dot{x}|z|^{n})}{\eta} \right\} \tag{A.4}$$

where $h(z), v, \eta$ are the indices describing the pinching effect, strength deterioration, and stiffness degradation, respectively. These indexes are all relevant to the non-linearities evolution of structural systems, and are related to energy dissipation. An index denoting energy dissipation is defined as

$$\varepsilon(t) = \int_0^t z\dot{x}\mathrm{d}t \tag{A.5}$$

then

$$v(\varepsilon) = 1 + \delta_v \varepsilon \tag{A.6a}$$

$$\eta(\varepsilon) = 1 + \delta_\eta \varepsilon \tag{A.6b}$$

where $\delta_v, \delta_\eta$ denote the parameters associated with the strength deterioration and the stiffness degradation.

Further, the index describing the pinching effect is defined as

$$h(z) = 1 - \zeta_1 e^{-[z\mathrm{sgn}(\dot{x}) - qz_u]^2/\zeta_2^2} \tag{A.7}$$

where $z_u$ denotes the extreme value of component:

$$z_u = \left( \frac{1}{v(\beta + \gamma)} \right)^{\frac{1}{n}} \tag{A.8}$$

and the two indices $\zeta_1(\varepsilon)$ and $\zeta_2(\varepsilon)$ take the forms, respectively, as follows:

$$\zeta_1(\varepsilon) = \zeta_s(1 - e^{-p\varepsilon}) \tag{A.9}$$

$$\zeta_2(\varepsilon) = (\psi + \delta_\psi \varepsilon)(\lambda + \zeta_1) \tag{A.10}$$

It is seen that the parameters associated with the pinching effect are $q, \zeta_s, p, \psi, \delta_\psi, \lambda$.

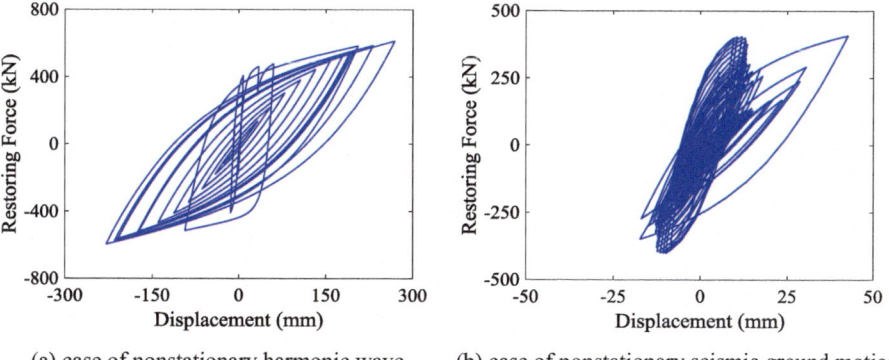

(a) case of nonstationary harmonic wave   (b) case of nonstationary seismic ground motion

**Fig. A.1** Typical hysteretic curves of component of structure subjected to nonstationary harmonic wave and seismic ground motion.

Therefore, a total of 13 parameters are included in the extended Bouc–Wen model for hysteretic components when the parameters associated with mass, damping, and stiffness of structures at initial state are provided, i.e., $\alpha$, $A$, $\beta$, $\gamma$, $n$, $\delta_v$, $\delta_\eta$, $q$, $\zeta_s$, $p$, $\psi$, $\delta_\psi$, $\lambda$.

For illustrative purposes, an eight-story shear frame with hysteretic components is investigated. The structural parameters are listed as follows: $m_1 = m_2 = 1.0 \times 10^5$ kg, $m_3 = m_4 = 0.9 \times 10^5$ kg, $m_5 = m_6 = 0.9 \times 10^5$ kg, $m_7 = m_8 = 0.8 \times 10^5$ kg; $k_1 = k_2 = 36$ kN/mm, $k_3 = k_4 = 32$ kN/mm, $k_5 = k_6 = 32$ kN/mm, $k_7 = k_8 = 28$ kN/mm; Rayleigh's damping $\mathbf{C} = a\mathbf{M} + b\mathbf{K}$ is employed, where $\mathbf{K}$ denotes stiffness matrix, $a = 0.01$, $b = 0.005$, whereby the damping ratio of the first vibrational mode is 1.05%. The extended Bouc–Wen model describing the behaviors of hysteretic components of the structure is employed. Parameters of the extended Bouc–Wen model are valued by $\alpha = 0.01$, $A = 1.0$, $\beta = 140.0$, $\gamma = 20.0$, $n = 1.0$, $\delta_v = 0.002$, $\delta_\eta = 0.001$, $q = 0.25$, $\zeta_s = 0.95$, $p = 2000$, $\psi = 0.2$, $\delta_\psi = 0.005$, $\lambda = 0.1$, respectively. Figure A.1 shows the typical hysteretic curves of a component of the structure subjected to nonstationary harmonic wave and seismic ground motion, respectively. It is seen that the extended Bouc–Wen model has the benefit of revealing the strength deterioration, stiffness degradation and pinching effect.

# References

Bouc R (1967) Forced vibration of mechanical system with hysteresis. In: Proceedings of 4th conference on nonlinear oscillations, Prague, Czechoslovakia

Foliente GC (1995) Hysteresis modeling of wood joints and structural systems. ASCE J Struct Eng 121(6):1013–1022

Ma F, Zhang H, Bochstedte A, Foliente GC, Paevere P (2004) Parameter analysis of the differential model of hysteresis. J Appl Mech 71:342–349

Wen YK (1976) Method for random vibration of hysteretic systems. ASCE J Eng Mech Div 102 (2):249–263

# Appendix B
# Relation Between Costate and Excitation Vectors

If $\lambda(t)$ is uncorrelated to $\mathbf{F}(\mathbf{\Theta}, t)$, the Hamiltonian function Eq. (3.3.7) under the minimum cost function $J_1$ satisfies

$$\frac{\partial H(\mathbf{Z}^*, \mathbf{U}^*, \lambda^*, \mathbf{F}, \mathbf{\Theta}, t)}{\partial \mathbf{F}} = \mathbf{D}^T \lambda(t) = \mathbf{0} \tag{B.1}$$

In view of the location matrix of external excitation $\mathbf{D}$, one has

$$\begin{bmatrix} \mathbf{0} \\ \mathbf{M}^{-1}\mathbf{D}_s \end{bmatrix}^T \begin{bmatrix} \lambda_1(t) \\ \lambda_2(t) \end{bmatrix} = \mathbf{0} \tag{B.2}$$

Then

$$\mathbf{M}^{-1}\mathbf{D}_s\lambda_2(t) = \mathbf{0} \tag{B.3}$$

where $\mathbf{M}^{-1}\mathbf{D}_s$ denotes a matrix with full rank. The unique condition satisfying with Eq. (B.3) is thus $\lambda_2(t) = \mathbf{0}$. However, the costate vector $\lambda(t)$ and the state vector $\mathbf{Z}(t)$ exhibit a one-to-one mapping relation; while the condition of zero components in state vector does not come into existence. Moreover, $\lambda_2(t) = \mathbf{0}$ would result in the collapse of the canonical equation set; see Eqs. (2.2.14) and (2.2.16). Therefore, the costate vector $\lambda(t)$ must have a pertinent relation with the excitation vector $\mathbf{F}(\mathbf{\Theta}, t)$ in the closed-open-loop control system.

Assuming that the costate vector $\lambda(t)$ and the excitation vector $\mathbf{F}(\mathbf{\Theta}, t)$ are linearly correlated for convenience of input feedback control, i.e.,

$$\lambda(t) = \mathbf{S}_\mathbf{F}(t)\mathbf{F}(\mathbf{\Theta}, t) \tag{B.4}$$

where $\mathbf{S}_\mathbf{F}(t)$ denotes a predefined matrix, which satisfies

© Springer Nature Singapore Pte Ltd. and Shanghai Scientific and Technical Publishers 2019
Y. Peng and J. Li, *Stochastic Optimal Control of Structures*,
https://doi.org/10.1007/978-981-13-6764-9

$$\frac{\partial H}{\partial \mathbf{F}} = \mathbf{S}_{\mathbf{F}}^{\mathrm{T}}(t)[\mathbf{A}\mathbf{Z}(t) + \mathbf{B}\mathbf{U}(t) + \mathbf{D}\mathbf{F}(\mathbf{\Theta}, t)] + \mathbf{D}^{\mathrm{T}}\mathbf{S}_{\mathbf{F}}(t)\mathbf{F}(\mathbf{\Theta}, t) = \mathbf{0} \qquad (B.5)$$

One might recognize that there has $\mathbf{S}_{\mathbf{F}}(t_f) = \mathbf{0}$ when the problem of infinite-time optimal control is of concern.

According to the Lagrange multiplier formula, the minimum increment of the cost function is zero, and a terminal condition exists:

$$\lambda(t_f) = \frac{1}{2}\frac{\partial[\mathbf{Z}^{\mathrm{T}}(t_f)\mathbf{P}(t_f)\mathbf{Z}(t_f)]}{\partial \mathbf{Z}(t_f)} = \mathbf{P}(t_f)\mathbf{Z}(t_f) \qquad (B.6)$$

Therefore, the Lagrange multiplier vector $\lambda(t)$ and the state vector $\mathbf{Z}(t)$ exhibit a linear relation as follows (Bryson and Ho 1975):

$$\lambda(t) = \mathbf{P}(t)\mathbf{Z}(t) \qquad (B.7)$$

where $\mathbf{P}(t)$ denotes a predefined matrix, and $\mathbf{P}(t_f) = \mathbf{0}$.

Consequently, as to the closed-open-loop control system, the control force $\mathbf{U}(t)$ simultaneously by the state and the input feedbacks can be implemented through building a linear transform between $\lambda(t)$, $\mathbf{Z}(t)$ and $\mathbf{F}(\mathbf{\Theta}, t)$:

$$\lambda(t) = \mathbf{P}(t)\mathbf{Z}(t) + \mathbf{S}_{\mathbf{F}}(t)\mathbf{F}(\mathbf{\Theta}, t) \qquad (B.8)$$

# Reference

Bryson AE Jr, Ho YC (1975) Applied optimal control. Hemisphere, New York

# Appendix C
# Statistical Linearization-Based LQG Control

As a substitute for Eq. (6.3.1), the equation of motion of the equivalent linear system is given by

$$\ddot{x}(t) + 2\zeta\omega_0\dot{x}(t) + \omega_{eq}^2 x(t) = u(t) + F(\mathbf{\Theta}, t), \; x(t_0) = \dot{x}(t_0) = 0 \qquad (C.1)$$

where $\omega_{eq}$ denotes the natural frequency of the equivalent linear system, obtained by minimizing the expected value of the difference between Eqs. (C.1) and (6.3.1) in a least square sense, i.e.,

$$\frac{\mathrm{d}}{\mathrm{d}\omega_{eq}^2} E[\{\omega_0^2[x(t) + \mu x^3(t)] - \omega_{eq}^2 x(t)\}^2] = 0 \qquad (C.2)$$

Then, it yields

$$\omega_{eq}^2 = \omega_0^2 \left(1 + \mu \frac{E[x^4(t)]}{E[x^2(t)]}\right) \qquad (C.3)$$

It is seen that $\omega_{eq}$ depends on the former fourth-order statistical moments of $x(t)$, the exact evaluation of $\omega_{eq}^2$ thus requires a knowledge of the probability density function of $x(t)$. As an approximation to the exact solution, the process might be assumed to be Gaussian (Roberts and Spanos 1990), and Eq. (C.3) can be simplified as

$$\omega_{eq}^2 = \omega_0^2 \left(1 + 3\mu E[x^2(t)]\right) \qquad (C.4)$$

Here, a decomposition formula has been used for a Gaussian vector $\mathbf{\eta}$ (Kazakov 1965)

$$E[f(\mathbf{\eta})\mathbf{\eta}] = E[\mathbf{\eta}\mathbf{\eta}^{\mathrm{T}}]E[\nabla f(\mathbf{\eta})] \qquad (C.5)$$

where $\nabla$ denotes the gradient operator defined by

© Springer Nature Singapore Pte Ltd. and Shanghai Scientific and Technical Publishers 2019
Y. Peng and J. Li, *Stochastic Optimal Control of Structures*,
https://doi.org/10.1007/978-981-13-6764-9

$$\nabla = \left[\frac{\partial}{\partial \eta_1}, \quad \frac{\partial}{\partial \eta_2}, \quad \cdots, \quad \frac{\partial}{\partial \eta_n}\right]^{\mathrm{T}} \tag{C.6}$$

In the state space, Eq. (C.1) can be written as

$$\dot{\mathbf{Z}}(t) = \mathbf{A}(t)\mathbf{Z}(t) + \mathbf{B}u(t) + \mathbf{D}F(\mathbf{\Theta}, t) \tag{C.7}$$

where

$$\mathbf{Z}(t) = \begin{bmatrix} x(t) \\ \dot{x}(t) \end{bmatrix}, \mathbf{A}(t) = \begin{bmatrix} 0 & 1 \\ -\omega_{eq}^2 & -2\zeta\omega_0 \end{bmatrix}, \mathbf{B} = \begin{bmatrix} 0 \\ 1 \end{bmatrix}, \mathbf{D} = \begin{bmatrix} 0 \\ 1 \end{bmatrix} \tag{C.8}$$

A cost function involved in the stochastic linear quadratic regulator problem is considered (Chen et al. 1998)

$$J(\mathbf{Z}, u) = E\left[\mathbf{S}(\mathbf{Z}(t_f), t_f) + \frac{1}{2}\int_{t_0}^{t_f} [\mathbf{Z}^{\mathrm{T}}(t)\mathbf{Q}_{\mathbf{Z}}\mathbf{Z}(t) + \mathbf{R}_U u^2(t)]dt\right] \tag{C.9}$$

subjected to

$$d\mathbf{Z}(t) = [\mathbf{A}(t)\mathbf{Z}(t) + \mathbf{B}u(t)]dt + \mathbf{L}dw(t), \mathbf{Z}(t_0) = \mathbf{0} \tag{C.10}$$

where $\mathbf{L}$ denotes the $(2 \times 1)$ force influence matrix; $w(t)$ denotes a one-dimensional Brownian motion process, which is modeled by a Gaussian white noise with

$$E[dw(t)] = 0, E[dw^2(t)] = 2\pi S_0 dt \tag{C.11}$$

where $S_0$ denotes the spectral intensity factor of random excitation $F(\mathbf{\Theta}, t)$. In the numerical example of Duffing oscillator systems, the spectral intensity factor pertains to seismic ground motion, and its value refers to Table 3.2.

The minimization of the cost function Eq. (C.9) results in the Hamilton–Jacobi–Bellman equation in stochastic scenario; see Eq. (3.5.8). Utilizing the dynamic programming method, one has

$$u(t) = -\mathbf{R}_U^{-1}\mathbf{B}^{\mathrm{T}}\mathbf{P}(t)\mathbf{Z}(t) \tag{C.12}$$

where the Riccati matrix $\mathbf{P}(t)$ is a function of time since it hinges upon $\mathbf{A}(t)$ of which $\omega_{eq}$ is related to $E[x^2(t)]$. However, the response of linearized system remains as a Gaussian process, and its mean square is a constant, indicating that the Riccati matrix $\mathbf{P}(t)$ is still time invariant.

Taking Eq. (C.12) into Eq. (C.1) and using Fourier transform, one has

$$\{[(\omega_{eq}^2 + \widehat{K}) - \omega^2] + (2\zeta\omega_0 + \widehat{C})(i\omega)\}x(\omega) = F(\mathbf{\Theta}, t) \tag{C.13}$$

where $\widehat{C}, \widehat{K}$ denote the numerical damping and numerical stiffness provided by the optimal control force $u(t)$, respectively.

$$\widehat{C} = \mathbf{R}_U^{-1}(B_1 P_{12} + B_2 P_{22}), \ \widehat{K} = \mathbf{R}_U^{-1}(B_1 P_{11} + B_2 P_{21}) \qquad (C.14)$$

where $B_i (i = 1, 2)$ denote the elements of control force location vector; $P_{ij}(i, j = 1, 2)$ denote the elements of the Riccati matrix.

According to the Wiener–Khintchine theorem, the mean square displacement can be deduced as

$$E[x^2(t)] = \int_{-\infty}^{\infty} \frac{S_0}{[(\omega_{eq}^2 + \widehat{K}) - \omega^2]^2 + (2\zeta\omega_0 + \widehat{C})^2\omega^2} \, d\omega \qquad (C.15)$$

The close solution of Eq. (C.15) can be attained as a specific rule (Roberts and Spanos 1990), i.e.,

$$E[x^2(t)] = \frac{\pi S_0}{(\omega_{eq}^2 + \widehat{K})(2\zeta\omega_0 + \widehat{C})} \qquad (C.16)$$

Taking into account Eq. (C.4), a single algebraic equation for $E[x^2(t)]$ appears

$$E[x^2(t)] = \frac{\sqrt{(\omega_0^2 + \widehat{K})^2 + \frac{12\pi\mu\omega_0^2 S_0}{(2\zeta\omega_0 + \widehat{C})}} - (\omega_0^2 + \widehat{K})}{6\mu\omega_0^2} \qquad (C.17)$$

The relation between the state and the control force in frequency domain is

$$u(\omega) = [-\widehat{C}(i\omega) - \widehat{K}]x(\omega) \qquad (C.18)$$

and the mean square control is deduced as

$$E[u^2(t)] = \int_{-\infty}^{\infty} \frac{(\widehat{K}^2 + \widehat{C}^2\omega^2)S_0}{[(\omega_0^2 + \widehat{K}) - \omega^2]^2 + (2\zeta\omega_0 + \widehat{C})^2\omega^2} \, d\omega \qquad (C.19)$$

Likewise, one can readily gain the closed solution of the mean square control force

$$E[u^2(t)] = \frac{\pi S_0 \widehat{C}^2}{2\zeta\omega_0 + \widehat{C}} + \frac{\left[\sqrt{(\omega_0^2 + \widehat{K})^2 + \frac{12\pi\mu\omega_0^2 S_0}{(2\zeta\omega_0 + \widehat{C})}} - (\omega_0^2 + \widehat{K})\right]\widehat{K}^2}{6\mu\omega_0^2} \qquad (C.20)$$

# References

Chen SP, Li XJ, Zhou XY (1998) Stochastic linear–quadratic regulators with indefinite control weight costs. SIAM J Control Optim 36:1685–1702

Kazakov IE (1965) Generalization of the method of statistical linearization to multidimensional systems. Automation and Remote Control 26:1201–1206

Roberts JB, Spanos PD (1990) Random vibration and statistical linearization. Wiley, West Sussex

# Appendix D
# Equivalent Damping Ratio of Viscously Damped Structures

The equation of motion of structure attached with viscous dampers as shown in Fig. D.1 can be written as

$$\mathbf{M}\ddot{\mathbf{X}}(t) + \mathbf{C}\dot{\mathbf{X}}(t) + \mathbf{K}\mathbf{X}(t) + \mathbf{f}(\dot{\mathbf{X}}(t)) = \mathbf{F}(t) \tag{D.1}$$

where $\mathbf{M}$, $\mathbf{C}$ and $\mathbf{K}$ are $n \times n$ mass, damping, and stiffness matrices, respectively; $\mathbf{f}(\dot{\mathbf{X}}(t))$ denotes the damping force provided by viscous dampers; $\mathbf{F}(t)$ denotes external excitation.

The component expression of the damping force can be figured out as

$$f_j(\dot{X}) = c_{D,j}\left|\dot{X}_j - \dot{X}_{j-1}\right|^{\alpha_j}\mathrm{sgn}(\dot{X}_j - \dot{X}_{j-1}) - c_{D,j}\left|\dot{X}_{j+1} - \dot{X}_j\right|^{\alpha_{j+1}}\mathrm{sgn}(\dot{X}_{j+1} - \dot{X}_j) \tag{D.2}$$

where $\mathrm{sgn}(\cdot)$ denotes the sign of arguments; $c_{D,j}$ denotes damping coefficient; and $\alpha_j$ denotes velocity exponent; see Fig. D.1.

According to the statistical linearization technique, the linearized system of Eq. (D.1) can be expressed as (Di Paola and Navarra 2009)

$$\mathbf{M}\ddot{\mathbf{X}}(t) + \mathbf{C}\dot{\mathbf{X}}(t) + \mathbf{C}^{(\text{S-E})}\dot{\mathbf{X}}(t) + \mathbf{K}\mathbf{X}(t) = \mathbf{F}(t) \tag{D.3}$$

where $\mathbf{C}^{(\text{S-E})}$ denotes the additional equivalent damping matrix, and is assumed to be Rayleigh's damping matrix.

The difference between Eqs. (D.1) and (D.3) can be measured by the following formula:

$$\mathbf{e} = \mathbf{f}(\dot{\mathbf{X}}) - \mathbf{C}^{(\text{S-E})}\dot{\mathbf{X}} \tag{D.4}$$

where $\mathbf{e}$ denotes the error vector between the original and equivalent damping forces.

© Springer Nature Singapore Pte Ltd. and Shanghai Scientific and Technical Publishers 2019
Y. Peng and J. Li, *Stochastic Optimal Control of Structures*,
https://doi.org/10.1007/978-981-13-6764-9

**Fig. D.1** MDOF structural system attached with viscous dampers.

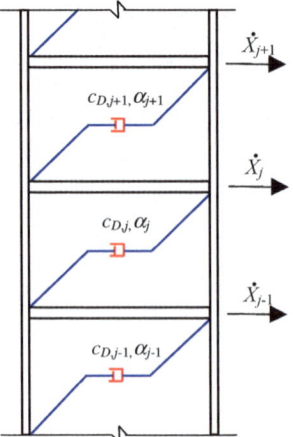

Minimization of the covariance matrix of the error vector, i.e.,

$$\frac{\partial E[\mathbf{ee}^{\mathrm{T}}]}{\partial \mathbf{C}^{(\mathrm{S\text{-}E})}} = 0 \tag{D.5}$$

The detailed expansion of $\mathbf{e}$ is

$$
\mathbf{e} = \left\{
\begin{array}{l}
c_{D,1}\left|\dot{X}_1\right|^{\alpha_1}\mathrm{sgn}(\dot{X}_1) - c_{D,2}\left|\dot{X}_2 - \dot{X}_1\right|^{\alpha_2}\mathrm{sgn}(\dot{X}_2 - \dot{X}_1) \\[2mm]
c_{D,2}\left|\dot{X}_2 - \dot{X}_1\right|^{\alpha_2}\mathrm{sgn}(\dot{X}_2 - \dot{X}_1) - c_{D,3}\left|\dot{X}_3 - \dot{X}_2\right|^{\alpha_3}\mathrm{sgn}(\dot{X}_3 - \dot{X}_2) \\[2mm]
\vdots \\[2mm]
c_{D,n-1}\left|\dot{X}_{n-1} - \dot{X}_{n-2}\right|^{\alpha_{n-1}}\mathrm{sgn}(\dot{X}_{n-1} - \dot{X}_{n-2}) - c_{D,n}\left|\dot{X}_n - \dot{X}_{n-1}\right|^{\alpha_n}\mathrm{sgn}(\dot{X}_n - \dot{X}_{n-1}) \\[2mm]
c_{D,n}\left|\dot{X}_n - \dot{X}_{n-1}\right|^{\alpha_n}\mathrm{sgn}(\dot{X}_n - \dot{X}_{n-1})
\end{array}
\right\}
$$

$$
- \begin{bmatrix}
C_{1,1}^{(\mathrm{S\text{-}E})} & C_{1,2}^{(\mathrm{S\text{-}E})} & \cdots & 0 & 0 \\
C_{2,1}^{(\mathrm{S\text{-}E})} & C_{2,2}^{(\mathrm{S\text{-}E})} & \cdots & 0 & 0 \\
\vdots & \vdots & & \vdots & \vdots \\
0 & 0 & \cdots & C_{n-1,n-1}^{(\mathrm{S\text{-}E})} & C_{n-1,n}^{(\mathrm{S\text{-}E})} \\
0 & 0 & \cdots & C_{n,n-1}^{(\mathrm{S\text{-}E})} & C_{n,n}^{(\mathrm{S\text{-}E})}
\end{bmatrix}
\begin{Bmatrix}
\dot{X}_1 \\
\dot{X}_2 \\
\vdots \\
\dot{X}_{n-1} \\
\dot{X}_n
\end{Bmatrix}
\tag{D.6}
$$

Substituting Eq. (D.6) into Eq. (D.5) leads to $\mathbf{C}^{(\mathrm{S\text{-}E})}$. It is observed that deducing from the $n$th component backward to the first component is convenient. The $n$th component is extracted to be

$$\frac{\partial E\left[\left(c_{D,n}\left|\dot{X}_n - \dot{X}_{n-1}\right|^{\alpha_n}\mathrm{sgn}(\dot{X}_n - \dot{X}_{n-1}) + C_{n,n}^{\mathrm{(S\text{-}E)}}\dot{X}_{n-1} - C_{n,n}^{\mathrm{(S\text{-}E)}}\dot{X}_n\right)^2\right]}{\partial C_{n,n}^{\mathrm{(S\text{-}E)}}} = 0$$

$$\Rightarrow C_{n,n}^{\mathrm{(S\text{-}E)}} = c_{D,n} \cdot \frac{E\left[\left|\dot{X}_n - \dot{X}_{n-1}\right|^{\alpha_n+1}\right]}{E\left[\left(\dot{X}_n - \dot{X}_{n-1}\right)^2\right]}, \quad C_{n-1,n}^{\mathrm{(S\text{-}E)}} = C_{n,n-1}^{\mathrm{(S\text{-}E)}} = -c_{D,n} \cdot \frac{E\left[\left|\dot{X}_n - \dot{X}_{n-1}\right|^{\alpha_n+1}\right]}{E\left[\left(\dot{X}_n - \dot{X}_{n-1}\right)^2\right]}$$

(D.7)

Similarly, the $(n-1)$th component is extracted to be

$$C_{n-1,n-2}^{\mathrm{(S\text{-}E)}} = -c_{D,n-1} \cdot \frac{E\left[\left|\dot{X}_{n-1} - \dot{X}_{n-2}\right|^{\alpha_{n-1}+1}\right]}{E\left[\left(\dot{X}_{n-1} - \dot{X}_{n-2}\right)^2\right]}$$

(D.8)

and the expression of $C_{n-1,n-1}^{\mathrm{(S\text{-}E)}}$ can be easily obtained

$$C_{n-1,n-1}^{\mathrm{(S\text{-}E)}} = c_{D,n} \cdot \frac{E\left[\left|\dot{X}_n - \dot{X}_{n-1}\right|^{\alpha_n+1}\right]}{E\left[\left(\dot{X}_n - \dot{X}_{n-1}\right)^2\right]} + c_{D,n-1} \cdot \frac{E\left[\left|\dot{X}_{n-1} - \dot{X}_{n-2}\right|^{\alpha_{n-1}+1}\right]}{E\left[\left(\dot{X}_{n-1} - \dot{X}_{n-2}\right)^2\right]}$$

(D.9)

Likewise, the component of $\mathbf{C}^{\mathrm{(S\text{-}E)}}$ can be expressed in turn as below

$$C_{j,j}^{\mathrm{(S\text{-}E)}} = c_{D,j} \cdot \frac{E\left[\left|\dot{X}_j - \dot{X}_{j-1}\right|^{\alpha_j+1}\right]}{E\left[\left(\dot{X}_j - \dot{X}_{j-1}\right)^2\right]} + c_{D,j+1} \cdot \frac{E\left[\left|\dot{X}_{j+1} - \dot{X}_j\right|^{\alpha_{j+1}+1}\right]}{E\left[\left(\dot{X}_{j+1} - \dot{X}_j\right)^2\right]}$$

(D.10)

$$C_{j,j+1}^{\mathrm{(S\text{-}E)}} = -c_{D,j+1} \cdot \frac{E\left[\left|\dot{X}_{j+1} - \dot{X}_j\right|^{\alpha_{j+1}+1}\right]}{E\left[\left(\dot{X}_{j+1} - \dot{X}_j\right)^2\right]}$$

(D.11)

Assuming the modal vector of the $k$th mode is $\boldsymbol{\phi}_k$, we have

$$u_k = \boldsymbol{\phi}_k^{\mathrm{T}}\mathbf{X}$$

(D.12)

where $u_k$ denotes the modal displacement of the $k$th mode. If all the viscous dampers share a same velocity exponent $\alpha$, Eqs. (D.10) and (D.11) can be written as

$$C_{j,j}^{\mathrm{(S\text{-}E)}} = \left(c_{D,j}\Delta_j^{(k)} + c_{D,j+1}\Delta_{j+1}^{(k)}\right) \cdot \frac{E\left[\left|u_k\right|^{\alpha+1}\right]}{E\left[\left(u_k\right)^2\right]}$$

(D.13)

$$C_{j,j+1}^{\mathrm{(S\text{-}E)}} = -c_{D,j+1}\Delta_{j+1}^{(k)} \cdot \frac{E\left[\left|u_k\right|^{\alpha+1}\right]}{E\left[\left(u_k\right)^2\right]}$$

(D.14)

where $\Delta_j^{(k)} = \left| u_j^{(k)} - u_{j-1}^{(k)} \right|^{\alpha-1}$.

The modal equation of motion of the viscously damped structure is

$$\ddot{u}_k + 2\zeta_k \omega_k \dot{u}_k + 2\zeta_k^{(S\text{-}E)} \omega_k \dot{u}_k + \omega_k^2 u_k = \tilde{F}_k \tag{D.15}$$

where $\tilde{F}_k = \boldsymbol{\phi}_k^{\mathrm{T}} \mathbf{F}(t) / (\boldsymbol{\phi}_k^{\mathrm{T}} \mathbf{M} \boldsymbol{\phi}_k)$ denotes the generalized excitation. Substituting Eqs. (D.13) and (D.14) into Eq. (D.15), the additional equivalent modal damping ratio can be derived as

$$\zeta_k^{(S\text{-}E)} = \eta_k \cdot \frac{E\left[ |u_k|^{\alpha+1} \right]}{E\left[ (u_k)^2 \right]} \tag{D.16}$$

where $\eta_k = q_k / (2\bar{m}_k \omega_k)$, $\bar{m}_k$ denotes the modal mass of the $k$th mode, and $q_k$ has the formulation as follows:

$$q_k = \sum_{j=1}^{n} \left( c_{D,j} \Delta_j^{(k)} + C_{D,j+1} \Delta_{j+1}^{(k)} \right) u_j^{(k)2} - 2 \sum_{j=2}^{n} c_{D,j} \Delta_j^{(k)} u_j^{(k)} u_{j-1}^{(k)}, \quad \Delta_{n+1}^{(k)} = 0 \tag{D.17}$$

Applying that the structural response is assumed to be a Gaussian process, there is

$$\sigma_{u_k}^2 \approx \frac{\pi G_{\tilde{F}_k}(\omega_k)}{4\left( \zeta_k^{(S\text{-}E)} + \zeta_k \right) \omega_k} \tag{D.18}$$

The additional equivalent modal damping ratio of the MDOF system can be expressed as

$$\zeta_k^{(S\text{-}E)} = \eta_k \rho(\alpha) \left( \frac{G_{\tilde{F}_k}(\omega_k)}{\left( \zeta_k^{(S\text{-}E)} + \zeta_k \right) \omega_k} \right)^{(\alpha-1)/2} \tag{D.19}$$

where $\rho(\alpha) = \Gamma(1 + \alpha/2) \sqrt{2^{3-\alpha} \pi^{\alpha-2}}$

# Reference

Di Paola M, Navarra G (2009) Stochastic seismic analysis of MDOF structures with nonlinear viscous dampers. Struct Control Health Monit 16(3):303–318

# Appendix E
# Optimal Control of Discrete Structural Systems

Without loss of generality, the system equation of a linear structure is given by

$$\dot{\mathbf{Z}}(t) = \mathbf{A}\mathbf{Z}(t) + \mathbf{B}\mathbf{U}(t) + \mathbf{D}\mathbf{F}(t) \tag{E.1}$$

The equation in discrete state can be written as follows:

$$\mathbf{Z}(k+1) = \mathbf{A}_d\mathbf{Z}(k) + \mathbf{B}_d\mathbf{U}(k) + \mathbf{D}_d\mathbf{F}(k) \tag{E.2}$$

where

$$\mathbf{A}_d = e^{\mathbf{A}\Delta t}, \mathbf{B}_d = e^{\mathbf{A}\Delta t}\mathbf{A}^{-1}\left(\mathbf{I} - e^{-\mathbf{A}\Delta t}\right)\mathbf{B}, \mathbf{D}_d = e^{\mathbf{A}\Delta t}\mathbf{A}^{-1}\left(\mathbf{I} - e^{-\mathbf{A}\Delta t}\right)\mathbf{D} \tag{E.3}$$

The quadratic cost function in discrete state is given by

$$J(\mathbf{Z}, \mathbf{U}, N) = \frac{1}{2}\mathbf{Z}^{\mathrm{T}}(N)\mathbf{P}(N)\mathbf{Z}(N) + \frac{1}{2}\sum_{k=0}^{N-1}\left(\mathbf{Z}^{\mathrm{T}}(k)\mathbf{Q}(k)\mathbf{Z}(k) + \mathbf{U}^{\mathrm{T}}(k)\mathbf{R}(k)\mathbf{U}(k)\right) \tag{E.4}$$

where $\mathbf{Q}(\cdot), \mathbf{S}(\cdot)$ denote the symmetric and semi-positive state weighting matrices; $\mathbf{R}(\cdot)$ denotes the symmetric and positive control force weighting matrix. Theoretically, these weighting matrices are all time variant. However, they are usually defined as time-invariant parameters for the control law design in practice.

Therefore, the objective of optimal control is seeking for the optimal control sequence $\mathbf{U}^*(0), \mathbf{U}^*(1), \ldots, \mathbf{U}^*(N-1)$ so as to gain a minimum cost function $J$.

© Springer Nature Singapore Pte Ltd. and Shanghai Scientific and Technical Publishers 2019
Y. Peng and J. Li, *Stochastic Optimal Control of Structures*,
https://doi.org/10.1007/978-981-13-6764-9

## E.1 Matrix Difference Riccati Equation

In view of Pontryagin's maximum principle, the Riccati equation is deduced from the discrete system; see Eq. (E.2). Introducing the Hamiltonian function

$$H(k) = \frac{1}{2}\left(\mathbf{Z}^{\mathrm{T}}(k)\mathbf{Q}\mathbf{Z}(k) + \mathbf{U}^{\mathrm{T}}(k)\mathbf{R}\mathbf{U}(k)\right) + \lambda^{\mathrm{T}}(k+1)(\mathbf{A}_d\mathbf{Z}(k) + \mathbf{B}_d\mathbf{U}(k)) \quad (\text{E.5})$$

the costate equation

$$\lambda(k) = -\left(\frac{\partial H(k)}{\partial \mathbf{Z}(k)}\right)^{\mathrm{T}} = -\mathbf{Q}\mathbf{Z}(k) - \mathbf{A}_d^{\mathrm{T}}\lambda(k+1) \quad (\text{E.6})$$

and the control equation

$$\frac{\partial H(k)}{\partial \mathbf{U}(k)} = \mathbf{U}^{\mathrm{T}}(k)\mathbf{R} + \lambda^{\mathrm{T}}(k+1)\mathbf{B} = \mathbf{0} \quad (\text{E.7})$$

it yields

$$\mathbf{U}(k) = -\mathbf{R}^{-1}\mathbf{B}_d^{\mathrm{T}}\lambda(k+1) \quad (\text{E.8})$$

As to the open-closed-loop control, the state feedback and input feedback are usually considered simultaneously, i.e.,

$$\lambda(k) = \mathbf{P}(k)\mathbf{Z}(k) + \mathbf{S}(k-1)\mathbf{F}(k-1) \quad (\text{E.9})$$

Substituting Eqs. (E.8) and (E.9) into Eq. (E.2), one has

$$\mathbf{Z}(k+1) = \left(\mathbf{I} + \mathbf{B}_d\mathbf{R}^{-1}\mathbf{B}_d^{\mathrm{T}}\mathbf{P}(k+1)\right)^{-1}\left[\mathbf{A}_d\mathbf{Z}(k) - \mathbf{B}_d\mathbf{R}^{-1}\mathbf{B}_d^{\mathrm{T}}\mathbf{S}(k)\mathbf{F}(k) + \mathbf{D}_d\mathbf{F}(k)\right] \quad (\text{E.10})$$

It is seen that Eq. (E.10) is a backward recursive difference equation with respect to the system state.

Substituting Eq. (E.9) into Eq. (E.6), one has

$$\mathbf{P}(k)\mathbf{Z}(k) + \mathbf{S}(k-1)\mathbf{F}(k-1) = \mathbf{Q}\mathbf{Z}(k) + \mathbf{A}_d^{\mathrm{T}}[\mathbf{P}(k+1)\mathbf{Z}(k+1) + \mathbf{S}(k)\mathbf{F}(k)] \quad (\text{E.11})$$

then

$$\mathbf{P}(k)\mathbf{Z}(k) = -\mathbf{S}(k)\mathbf{F}(k) + \mathbf{Q} + \mathbf{A}_d^{\mathrm{T}}\mathbf{P}(k+1)\mathbf{Z}(k+1) + \mathbf{A}_d^{\mathrm{T}}\mathbf{S}(k)\mathbf{F}(k) \quad (\text{E.12})$$

It is seen that different from Eqs. (E.10) and (E.12) is a forward recursive difference equation with respect to the Riccati matrix $\mathbf{P}(k)$. Eq. (E.12) is the so-called matrix difference Riccati equation.

Substituting Eq. (E.9) into Eq. (E.8) and considering Eq. (E.2), one has

$$
\begin{aligned}
\mathbf{U}(k) = & -\left[\mathbf{B}_d^{\mathrm{T}}\mathbf{P}(k+1)\mathbf{B}_d + \mathbf{R}\right]^{-1}\mathbf{B}_d^{\mathrm{T}}\mathbf{P}(k+1)\mathbf{A}_d\mathbf{Z}(k) \\
& -\left[\mathbf{R}^{-1}\mathbf{B}_d^{\mathrm{T}} + \mathbf{B}_d^{-1}\mathbf{P}^{-1}(k+1)\right]\left[\mathbf{P}(k+1)\mathbf{D}_d + \mathbf{S}(k)\right]\mathbf{F}(k)
\end{aligned}
\tag{E.13}
$$

which in formulation of feedback gain can be written as follows:

$$
\mathbf{U}(k) = -\mathbf{G}_{\mathbf{Z}}(k)\mathbf{Z}(k) - \mathbf{G}_{\mathbf{F}}(k)\mathbf{F}(k)
\tag{E.14}
$$

where $\mathbf{G}_{\mathbf{Z}}(k)$ denotes the gain matrix of state feedback:

$$
\mathbf{G}_{\mathbf{Z}}(k) = \left[\mathbf{B}_d^{\mathrm{T}}\mathbf{P}(k+1)\mathbf{B}_d + \mathbf{R}\right]^{-1}\mathbf{B}_d^{\mathrm{T}}\mathbf{P}(k+1)\mathbf{A}_d
\tag{E.15}
$$

and $\mathbf{G}_{\mathbf{F}}(k)$ denotes the gain matrix of input feedback:

$$
\mathbf{G}_{\mathbf{F}}(k) = \left[\mathbf{R}^{-1}\mathbf{B}_d^{\mathrm{T}} + \mathbf{B}_d^{-1}\mathbf{P}^{-1}(k+1)\right]\left[\mathbf{P}(k+1)\mathbf{D}_d + \mathbf{S}(k)\right]
\tag{E.16}
$$

Besides, substituting Eq. (E.9) into control equation $\partial H(k)/\partial \mathbf{F}(k) = \mathbf{0}$, there is

$$
\frac{\partial H(k)}{\partial \mathbf{F}(k)} = \left[\mathbf{A}_d\mathbf{Z}(k) + \mathbf{B}_d\mathbf{U}(k)\right]^{\mathrm{T}}\mathbf{S}(k) + \left[\mathbf{S}^{\mathrm{T}}(k)\mathbf{D}_d + \mathbf{D}_d^{\mathrm{T}}\mathbf{S}(k)\right]\mathbf{F}(k) = \mathbf{0}
\tag{E.17}
$$

it yields

$$
\begin{aligned}
\mathbf{S}(k+1) = & \left(\mathbf{B}_d^{\mathrm{T}}\right)^{-1}\mathbf{R}\mathbf{B}_d^{-1}\left[\mathbf{A}_d\mathbf{Z}(k)\mathbf{F}^{-1}(k) + \mathbf{S}^{\mathrm{T}}(k)\mathbf{D}_d\mathbf{F}(k)\mathbf{S}^{-1}(k)\mathbf{F}^{-1}(k) \right. \\
& \left. + \mathbf{D}_d^{\mathrm{T}}\mathbf{S}(k)\mathbf{F}(k)\mathbf{S}^{-1}(k)\mathbf{F}^{-1}(k)\right]
\end{aligned}
\tag{E.18}
$$

One might recognize from Eqs. (E.12) and (E.18) that as to the open-closed-loop control with both considering the state feedback and input feedback, the state weighting matrix $\mathbf{P}(k)$ and input matrix $\mathbf{S}(k)$ are related to the system state $\mathbf{Z}(k)$ and system input $\mathbf{F}(k)$ at each instant of time. It is thus indicated that the control law needs to be calculated in real time, which is not really convenient for the applications in practice. Therefore, the closed-loop control with consideration of the state feedback and without consideration of the input feedback is usually concerned. The matrix difference Riccati equation is thus given by

$$
\mathbf{P}(k) = \mathbf{Q} + \mathbf{A}_d^{\mathrm{T}}\mathbf{P}(k+1)\left(\mathbf{I} + \mathbf{B}_d\mathbf{R}^{-1}\mathbf{B}_d^{\mathrm{T}}\mathbf{P}(k+1)\right)^{-1}\mathbf{A}_d
\tag{E.19}
$$

and the control force is simplified as

$$
\mathbf{U}(k) = -\mathbf{G}_{\mathbf{Z}}(k)\mathbf{Z}(k)
\tag{E.20}
$$

## E.2 Discrete Dynamic Programming

The discrete dynamic programming aims at deducing the recursive relation between $J(N)$ and $J(N-1)$ by virtue of the Bellman's optimality principle so as to decompose an $N$-step optimal problem into $N$ one-step optimal problem. The optimization starts from the $N$th step and moves backward to $(N-1)$th step.

First considering the $N$th step, the terminal condition is denoted by $\mathbf{Z}(N)$, and the control force $\mathbf{U}^*(N)$ is optimized so that the cost function

$$J(\mathbf{Z}, \mathbf{U}, N) = \frac{1}{2}\mathbf{Z}^{\mathrm{T}}(N)\mathbf{P}(N)\mathbf{Z}(N) \tag{E.21}$$

is minimum. This is ready to be recognized since the terminal condition of matrix difference Riccati equation exists

$$\mathbf{P}(N) = 0 \tag{E.22}$$

Further considering the $(N-1)$th step, the terminal condition is replaced by $\mathbf{Z}(N-1)$, and the control force $\mathbf{U}^*(N-1)$ is optimized so that the cost function

$$J(\mathbf{Z}, \mathbf{U}, N-1) = \frac{1}{2}\mathbf{Z}^{\mathrm{T}}(N-1)\mathbf{Q}\mathbf{Z}(N-1) + \frac{1}{2}\mathbf{U}^{\mathrm{T}}(N-1)\mathbf{R}\mathbf{U}(N-1) + \frac{1}{2}\mathbf{Z}^{\mathrm{T}}(N)\mathbf{P}(N)\mathbf{Z}(N) \tag{E.23}$$

is minimum.

Substituting Eq. (E.2) into Eq. (E.23), one has

$$\begin{aligned} J(\mathbf{Z}, \mathbf{U}, N-1) = &\frac{1}{2}\mathbf{Z}^{\mathrm{T}}(N-1)\mathbf{Q}\mathbf{Z}(N-1) + \frac{1}{2}\mathbf{U}^{\mathrm{T}}(N-1)\mathbf{R}\mathbf{U}(N-1) \\ &+ \frac{1}{2}[\mathbf{A}_d\mathbf{Z}(N-1) + \mathbf{B}_d\mathbf{U}(N-1) + \mathbf{D}_d\mathbf{F}(N-1)]^{\mathrm{T}} \\ &\mathbf{P}(N)[\mathbf{A}_d\mathbf{Z}(N-1) + \mathbf{B}_d\mathbf{U}(N-1) + \mathbf{D}_d\mathbf{F}(N-1)] \end{aligned} \tag{E.24}$$

Considering $\partial J\{\mathbf{Z}, \mathbf{U}, N-1\}/\partial \mathbf{U}(N-1) = \mathbf{0}$, one has

$$\mathbf{R}\mathbf{U}(N-1) + \mathbf{B}_d^{\mathrm{T}}\mathbf{P}(N)[\mathbf{A}_d\mathbf{Z}(N-1) + \mathbf{B}_d\mathbf{U}(N-1) + \mathbf{D}_d\mathbf{F}(N-1)] = 0 \tag{E.25}$$

Since $\mathbf{Q}$ is semi-positive and $\mathbf{R}$ is positive, $\mathbf{B}^{\mathrm{T}}\mathbf{Q}\mathbf{B}\Delta t^2 + \mathbf{R}$ is thus positive, then

$$\mathbf{U}^*(N-1) = -\left[\mathbf{B}_d^{\mathrm{T}}\mathbf{P}(N)\mathbf{B}_d + \mathbf{R}\right]^{-1}\mathbf{B}_d^{\mathrm{T}}\mathbf{P}(N)[\mathbf{A}_d\mathbf{Z}(N-1) + \mathbf{D}_d\mathbf{F}(N-1)] \tag{E.26}$$

Defining the gain matrix of state feedback

$$\mathbf{G_Z}(N-1) = \left[\mathbf{B}_d^{\mathrm{T}}\mathbf{P}(N)\mathbf{B}_d + \mathbf{R}\right]^{-1}\mathbf{B}_d^{\mathrm{T}}\mathbf{P}(N)\mathbf{A}_d \tag{E.27}$$

and the gain matrix of input feedback

$$\mathbf{G_F}(N-1) = \left[\mathbf{B}_d^{\mathrm{T}}\mathbf{P}(N)\mathbf{B}_d + \mathbf{R}\right]^{-1}\mathbf{B}_d^{\mathrm{T}}\mathbf{P}(N)\mathbf{D}_d \tag{E.28}$$

the control force of open-closed-loop control system is written as

$$\mathbf{U}^*(N-1) = -[\mathbf{G_Z}(N-1)\mathbf{Z}(N-1) + \mathbf{G_F}(N-1)\mathbf{F}(N-1)] \tag{E.29}$$

Substituting Eq. (E.29) into Eq. (E.24), one has

$$\begin{aligned}
J(\mathbf{Z},\mathbf{U},N-1) = {} & \frac{1}{2}\mathbf{Z}^{\mathrm{T}}(N-1)\Big[\mathbf{Q} + \mathbf{G}_\mathbf{Z}^{\mathrm{T}}(N-1)\mathbf{R}\mathbf{G_Z}(N-1) + (\mathbf{A}_d - \mathbf{B}_d\mathbf{G_Z}(N-1))^{\mathrm{T}} \\
& \mathbf{P}(N)(\mathbf{A}_d - \mathbf{B}_d\mathbf{G_Z}(N-1))]\mathbf{Z}(N-1) \\
& + \frac{1}{2}\mathbf{Z}^{\mathrm{T}}(N-1)\Big[\mathbf{G}_\mathbf{Z}^{\mathrm{T}}(N-1)\mathbf{R}\mathbf{G_F}(N-1) - (\mathbf{A}_d - \mathbf{B}_d\mathbf{G_Z}(N-1))^{\mathrm{T}} \\
& \mathbf{P}(N)(\mathbf{B}_d\mathbf{G_F}(N-1) - \mathbf{D}_d)]\mathbf{F}(N-1) \\
& + \frac{1}{2}\mathbf{F}^{\mathrm{T}}(N-1)\Big[\mathbf{G}_\mathbf{F}^{\mathrm{T}}(N-1)\mathbf{R}\mathbf{G_Z}(N-1) - (\mathbf{B}_d\mathbf{G_F}(N-1) - \mathbf{D}_d)^{\mathrm{T}} \\
& \mathbf{P}(N)(\mathbf{A}_d - \mathbf{B}_d\mathbf{G_Z}(N-1))]\mathbf{Z}(N-1) \\
& + \frac{1}{2}\mathbf{F}^{\mathrm{T}}(N-1)\Big[\mathbf{G}_\mathbf{F}^{\mathrm{T}}(N-1)\mathbf{R}\mathbf{G_F}(N-1) - (\mathbf{B}_d\mathbf{G_F}(N-1) - \mathbf{D}_d)^{\mathrm{T}} \\
& \mathbf{P}(N)(\mathbf{B}_d\mathbf{G_F}(N-1) - \mathbf{D}_d)]\mathbf{F}(N-1)
\end{aligned} \tag{E.30}$$

Ignoring the external excitation and defining

$$\mathbf{P}(N-1) = \mathbf{Q} + \mathbf{G}_\mathbf{Z}^{\mathrm{T}}(N-1)\mathbf{R}\mathbf{G_Z}(N-1) + (\mathbf{A}_d - \mathbf{B}_d\mathbf{G_Z}(N-1))^{\mathrm{T}}\mathbf{P}(N)(\mathbf{A}_d - \mathbf{B}_d\mathbf{G_Z}(N-1)) \tag{E.31}$$

one can attain a similar cost function as Eq. (E.21) as follows:

$$J(\mathbf{Z},\mathbf{U},N-1) = \frac{1}{2}\mathbf{Z}^{\mathrm{T}}(N-1)\mathbf{P}(N-1)\mathbf{Z}(N-1) \tag{E.32}$$

Therefore, the parameters of control law of closed-loop system are the solutions of the recursive equations:

$$\mathbf{G_Z}(k) = \left[\mathbf{B}_d^{\mathrm{T}}\mathbf{P}(k+1)\mathbf{B}_d + \mathbf{R}\right]^{-1}\mathbf{B}_d^{\mathrm{T}}\mathbf{P}(k+1)\mathbf{A}_d \tag{E.33}$$

$$\mathbf{U}^*(k) = -\mathbf{G_Z}(k)\mathbf{Z}(k) \tag{E.34}$$

$$\mathbf{P}(k) = \mathbf{Q} + \mathbf{G}_{\mathbf{Z}}^{\mathrm{T}}(k)\mathbf{R}\mathbf{G}_{\mathbf{Z}}(k) + [\mathbf{A}_d - \mathbf{B}_d\mathbf{G}_{\mathbf{Z}}(k)]^{\mathrm{T}}\mathbf{P}(k+1)[\mathbf{A}_d - \mathbf{B}_d\mathbf{G}_{\mathbf{Z}}(k)] \quad \text{(E.35)}$$

Equation (E.35) is the so-called matrix difference Riccati equation in Joseph stability (Lewis and Syrmos 1995), which can be written as

$$\mathbf{P}(k) = \mathbf{Q} + \mathbf{A}_d^{\mathrm{T}}\mathbf{P}(k+1)\left[\mathbf{I} - \mathbf{B}_d\left(\mathbf{B}_d^{\mathrm{T}}\mathbf{P}(k+1)\mathbf{B}_d + \mathbf{R}\right)^{-1}\mathbf{B}_d^{\mathrm{T}}\mathbf{P}(k+1)\right]\mathbf{A}_d \quad \text{(E.36)}$$

According to the matrix transpose rule (Lewis and Syrmos 1995)

$$\left(A_{11}^{-1} + A_{12}A_{22}A_{21}\right)^{-1} = A_{11} - A_{11}A_{12}\left(A_{21}A_{11}A_{12} + A_{22}^{-1}\right)^{-1}A_{21}A_{11} \quad \text{(E.37)}$$

Equation (E.36) can be readily changed to Eq. (E.19).

It is thus revealed that the Riccati controls derived from Pontryagin's maximum principle and from Bellman's optimality principle are exactly the same.

# Reference

Lewis FL, Syrmos VL (1995) Optimal control. Wiley

# Bibliography

Ang AH-S, Tang W (1984) Probability concepts in engineering planning and design (Volume II: decision, risk and reliability). Wiley, New York

Au SK, Beck JL (2001) First excursion probabilities for linear systems by very efficient importance sampling. Probabilistic Eng Mech 16:193–207

Crandall SH (1970) First-crossing probabilities of the linear oscillator. J Sound Vib 12:285–299

Davenport AG (1967) Gust loading factors. J Struct Div ASCE 93(3):11–34

Doi M, Edwards SF (1986) The theory of polymer dynamics. Oxford University Press, Oxford

Gardiner CW (1983) Handbook of stochastic methods for physics, chemistry and the natural sciences, 2nd edn. Springer, Berlin

Kim J, Choi H (2004) Behavior and design of structures with buckling-restrained braces. Eng Struct 26(6):693–706

Kittipoomwong D, Klingenberg DJ, Ulicny JC (2005) Dynamic yield stress enhancement in bidisperse magnetorheological fluids. J Rheol 49(6):1521–1538

Li J (2005) Several basic ideas on physical stochastic systems. Tongji University, Shanghai (in Chinese)

Li J, Chen JB (2003) The probability density evolution method for analysis of dynamic nonlinear response of stochastic structures. Acta Mech Sin 35(6):716–722 (in Chinese)

Li J, Chen JB (2010) Advances in the research on probability density evolution equations of stochastic dynamical systems. Adv Mech 40(2):170–188 (in Chinese)

Li J Yan Q. Research on stochastic Fourier wave-number spectrum of fluctuating wind speed. J Tongji Univ (Nat Sci) 39(12):1725–1731 (In Chinese)

Li J, Chen JB, Peng YB, Ai XQ, Fan WL (2005) Stochastic response analysis and seismic reliability evaluation of seismically-isolated hospital building of Su-Qian City. Tongji University, Shanghai (in Chinese)

Li J, Peng YB, Chen JB (2010) A physical approach to structural stochastic optimal controls. Probabilistic Eng Mech 25(1):127–141

Li J, Chen JB, Sun WL, Peng YB (2012) Advances of the probability density evolution method for nonlinear stochastic systems. Probabilistic Eng Mech 28:132–142

Liberzon D. Calculus of variations and optimal control theory: a concise introduction. Princeton University Press, Princeton

Lin WH, Chopra AK (2002) Earthquake response of elastic SDF systems with non-linear fluid viscous dampers. Earthq Eng Struct Dyn 31(9):1623–1642

Lin WH, Chopra AK (2003) Asymmetric one-storey elastic systems with non-linear viscous and viscoelastic dampers: earthquake response. Earthq Eng Struct Dyn 32(4):555–577

Loève M (1977) Probability theory I, 4th edn. Springer, New York

© Springer Nature Singapore Pte Ltd. and Shanghai Scientific and Technical Publishers 2019
Y. Peng and J. Li, *Stochastic Optimal Control of Structures*,
https://doi.org/10.1007/978-981-13-6764-9

Loève M (1978) Probability theoryII, 4th edn. Springer, New York

Peng YB, Chen JB, Li J (2007) Probability density evolution method and pseudo excitation method for random seismic response analysis. In: Proceedings of international symposium on innovation and sustainability of structures in civil engineering, Shanghai, China, pp 462–471

Reitz JR, Milford FJ (1960) Foundations of electromagnetic theory. Addison-Wesley Publishing Company Inc, Massachusetts

Scruggs JT, Taflanidi AA, Iwan WD (2007) Semi-active control strategies for MR dampers comparative study. Struct Control Health Monit 14:1101–1120

Shkel YM, Klingenberg DJ (2001) Magnetorheology and magnetostriction of isolated chains of nonlinear magnetizable spheres. J Rheol 45(2):351–368

Soong TT (1973) Random differential equations in science and engineering. Academic Press, New York

Spencer BF Jr, Elishakoff I (1988) Reliability of uncertain linear and nonlinear systems. J Eng Mech 114(1):135–148

Wang B, Xiao Z (2002) A general constitutive equation of an ER suspension based on the internal variable theory. Acta Mechanica 163:99–120

Wongprasert N, Symans MD (2000) Optimal distribution of fluid dampers for control of the wind benchmark problem. Citeseer

Yan Q, Peng YB, Li J (2013) Scheme and application of phase delay spectrum towards spatial stochastic wind fields. Wind Struct 16(5):433–455

Ying ZG, Zhu WQ, Soong TT (2003) A stochastic optimal semi-active control strategy for ER/MR dampers. J Sound Vib 259(1):45–62

Zabihollah A (2010) Effects of structural configuration on vibration control of smart laminated beams under random excitations. J Mech Sci Technol 24(5):1119–1125

Zhang J, Roschke PN (2008) Neural network simulation of magnetorheological damper behavior. In: Proceedings of international conference on vibration engineering, Dalian, China, 25–30.

# Index

© Springer Nature Singapore Pte Ltd. and Shanghai Scientific and Technical Publishers 2019
Y. Peng and J. Li, *Stochastic Optimal Control of Structures*,
https://doi.org/10.1007/978-981-13-6764-9

Printed by Printforce, the Netherlands